Scientific Standards
FOR STUDIES ON
Modified Risk Tobacco Products

Committee on Scientific Standards for Studies on
Modified Risk Tobacco Products

Board on Population Health and Public Health Practice

INSTITUTE OF MEDICINE
OF THE NATIONAL ACADEMIES

THE NATIONAL ACADEMIES PRESS
Washington, D.C.
www.nap.edu

THE NATIONAL ACADEMIES PRESS 500 Fifth Street, NW Washington, DC 20001

NOTICE: The project that is the subject of this report was approved by the Governing Board of the National Research Council, whose members are drawn from the councils of the National Academy of Sciences, the National Academy of Engineering, and the Institute of Medicine. The members of the committee responsible for the report were chosen for their special competences and with regard for appropriate balance.

This study was supported by Contract No. HHSF22301011T, Task Order #17 between the National Academy of Sciences and the Food and Drug Administration of the U.S. Department of Health and Human Services. Any opinions, findings, conclusions, or recommendations expressed in this publication are those of the author(s) and do not necessarily reflect the view of the organizations or agencies that provided support for this project.

International Standard Book Number-13: 978-0-309-22398-0
International Standard Book Number-10: 0-309-22398-9

Additional copies of this report are available from the National Academies Press, 500 Fifth Street, NW, Keck 360, Washington, DC 20001; (800) 624-6242 or (202) 334-3313; http://www.nap.edu.

For more information about the Institute of Medicine, visit the IOM home page at: **www.iom.edu.**

Printed in the United States of America

The serpent has been a symbol of long life, healing, and knowledge among almost all cultures and religions since the beginning of recorded history. The serpent adopted as a logotype by the Institute of Medicine is a relief carving from ancient Greece, now held by the Staatliche Museen in Berlin.

Suggested citation: IOM (Institute of Medicine). 2012. *Scientific Standards for Studies on Modified Risk Tobacco Products.* Washington, DC: The National Academies Press.

"Knowing is not enough; we must apply.
Willing is not enough; we must do."
—Goethe

INSTITUTE OF MEDICINE
OF THE NATIONAL ACADEMIES

Advising the Nation. Improving Health.

THE NATIONAL ACADEMIES

Advisers to the Nation on Science, Engineering, and Medicine

The **National Academy of Sciences** is a private, nonprofit, self-perpetuating society of distinguished scholars engaged in scientific and engineering research, dedicated to the furtherance of science and technology and to their use for the general welfare. Upon the authority of the charter granted to it by the Congress in 1863, the Academy has a mandate that requires it to advise the federal government on scientific and technical matters. Dr. Ralph J. Cicerone is president of the National Academy of Sciences.

The **National Academy of Engineering** was established in 1964, under the charter of the National Academy of Sciences, as a parallel organization of outstanding engineers. It is autonomous in its administration and in the selection of its members, sharing with the National Academy of Sciences the responsibility for advising the federal government. The National Academy of Engineering also sponsors engineering programs aimed at meeting national needs, encourages education and research, and recognizes the superior achievements of engineers. Dr. Charles M. Vest is president of the National Academy of Engineering.

The **Institute of Medicine** was established in 1970 by the National Academy of Sciences to secure the services of eminent members of appropriate professions in the examination of policy matters pertaining to the health of the public. The Institute acts under the responsibility given to the National Academy of Sciences by its congressional charter to be an adviser to the federal government and, upon its own initiative, to identify issues of medical care, research, and education. Dr. Harvey V. Fineberg is president of the Institute of Medicine.

The **National Research Council** was organized by the National Academy of Sciences in 1916 to associate the broad community of science and technology with the Academy's purposes of furthering knowledge and advising the federal government. Functioning in accordance with general policies determined by the Academy, the Council has become the principal operating agency of both the National Academy of Sciences and the National Academy of Engineering in providing services to the government, the public, and the scientific and engineering communities. The Council is administered jointly by both Academies and the Institute of Medicine. Dr. Ralph J. Cicerone and Dr. Charles M. Vest are chair and vice chair, respectively, of the National Research Council.

www.national-academies.org

COMMITTEE ON SCIENTIFIC STANDARDS FOR STUDIES ON MODIFIED RISK TOBACCO PRODUCTS

Jane E. Henney (*Chair*), Professor of Medicine and Public Health Sciences, University of Cincinnati, Ohio

Timothy B. Baker, Professor of Medicine and Associate Director, Center for Tobacco Research and Intervention, University of Wisconsin, Madison

Rebecca Bascom, Professor of Medicine, Milton S. Hershey Medical Center, The Pennsylvania State University, Hershey

Shyam Biswal, Professor, Department of Environmental Health Sciences, Johns Hopkins Bloomberg School of Public Health, Baltimore, Maryland

Daniel Carpenter, Professor of Government, Center for Government and International Studies, Harvard University, Boston, Massachusetts

Constantine Gatsonis, Henry Ledyard Goddard University Professor and Chair, Department of Biostatistics, Brown University, Providence, Rhode Island

Gary H. Gibbons,[1] Professor of Medicine and Director, Cardiovascular Research Institute, Morehouse School of Medicine, Atlanta, Georgia

Bonnie L. Halpern-Felsher, Professor, Department of Pediatrics, Division of Adolescent Medicine, University of California, San Francisco

Stephen S. Hecht, Wallin Professor of Cancer Prevention, Department of Laboratory Medicine and Pathology, Masonic Cancer Center, University of Minnesota, Minneapolis

Peter K. Honig, Head, Global Regulatory Affairs, AstraZeneca, Wilmington, Delaware

Richard J. O'Connor, Associate Member, Department of Health Behavior, Division of Cancer Prevention and Population Sciences, Roswell Park Cancer Institute, Buffalo, New York

Joel L. Schwartz, Professor, Oral Medicine and Pathology, Colleges of Dentistry and Medicine, University of Illinois at Chicago

Donna-Bea Tillman, Director of Regulations and Policy, Microsoft Corporation, Health Solutions Group, Chevy Chase, Maryland

Alastair J. J. Wood, Managing Director, Symphony Capitol LLC, New York, New York

Anna H. Wu,[2] Professor, Preventive Medicine, Division of Epidemiology, University of Southern California, Los Angeles

---

[1] Committee member resigned October 2011.

[2] Committee member resigned June 2011.

Study Staff

Kathleen Stratton, Study Director (through September 2011)
Joel Wu, Study Director (from September 2011)
Michelle C. Catlin, Senior Program Officer (from September 2011)
Erin Rusch, Research Associate (from September 2011)
Hannan Braun, Research Assistant
Malcolm Biles, Senior Program Assistant
Rose Marie Martinez, Director, Board on Population Health and Public Health Practice

Consultants

Suchitra Krishnan-Sarin, Associate Professor of Psychiatry and Vice Chair, Human Investigations Committee, Yale University School of Medicine
Holly E. R. Morrell, Assistant Professor, Department of Psychology, Loma Linda University
Gary Stoner, Professor of Medicine, Department of Medicine, Medical College of Wisconsin
Wendy Theobald, Researcher, Center for Tobacco Research and Intervention, University of Wisconsin Medical School
Robert B. Wallace, Irene Ensminger Stecher Professor of Epidemiology and Internal Medicine, Department of Epidemiology, College of Public Health, The University of Iowa

Reviewers

This report has been reviewed in draft form by individuals chosen for their diverse perspectives and technical expertise, in accordance with procedures approved by the National Research Council's Report Review Committee. The purpose of this independent review is to provide candid and critical comments that will assist the institution in making its published report as sound as possible and to ensure that the report meets institutional standards for objectivity, evidence, and responsiveness to the study charge. The review comments and draft manuscript remain confidential to protect the integrity of the deliberative process. We wish to thank the following individuals for their review of this report:

Linda S. Birnbaum, National Institute of Environmental Health Sciences
Richard J. Bonnie, University of Virginia
David B. Coultas, The University of Texas Health Science Center at Tyler
Louis Anthony Cox, Jr., Cox Associates
Sean P. David, Stanford University School of Medicine
Jonathan Foulds, Penn State College of Medicine
Mitchell H. Gail, National Cancer Institute
John R. Hughes, University of Vermont College of Medicine
Donald S. Kenkel, Cornell University
Caryn Lerman, University of Pennsylvania
Dean Lillard, Cornell University
Ana Navas-Acien, Johns Hopkins Bloomberg School of Public Health

Dan Romer, The Annenberg Public Policy Center of the University of Pennsylvania
Barry Sickels, AstraZeneca
Brian L. Strom, Perelman School of Medicine at the University of Pennsylvania
Clifford H. Watson, Centers for Disease Control and Prevention

Although the reviewers listed above have provided many constructive comments and suggestions, they were not asked to endorse the conclusions or recommendations, nor did they see the final draft of the report before its release. The review of this report was overseen by **May R. Berenbaum,** University of Illinois, and **Robert S. Lawrence,** Johns Hopkins Bloomberg School of Public Health. Appointed by the National Research Council and Institute of Medicine, they were responsible for making certain that an independent examination of this report was carried out in accordance with institutional procedures and that all review comments were carefully considered. Responsibility for the final content of this report rests entirely with the authoring committee and the institution.

Preface

Tobacco use remains the leading cause of preventable morbidity and mortality in the United States. While the adverse health effects of tobacco use are well established in the scientific literature, an understanding of the science is not required to appreciate the human cost; every day, people see close friends and family suffer with the consequences of tobacco use. Every day, cigarette smokers try to quit, and yet, the vast majority of them will fail. An estimated 70 percent of smokers want to quit completely, and while 45 percent attempt to quit each year, only 6 percent of smokers are able to successfully quit.

Instead of quitting, many cigarette smokers have sought a product with less risk, and for decades, the tobacco industry has purposefully misled the public into believing that there have been safer alternatives. The most prominent example is the "light" cigarette—a product implied to be safer, which in fact, when used, was as hazardous as "regular" cigarettes. The prospect of a less hazardous tobacco product is not in and of itself problematic. The fundamental issue is that if a product is going to be marketed as being "safer," then the claim must be true.

Section 911 of the Family Smoking Prevention and Tobacco Control Act of 2009 directly addresses the problem of false and unfounded claims for modified risk tobacco products (MRTPs). The law remains open to the possibility that less hazardous products that reduce harm to public health may enter the market, but it gives the government the authority and the power to assure that they are actually reducing risk and harm. The law also directed the U.S. Food and Drug Administration (FDA) to develop, in consultation with the Institute of Medicine (IOM), regulations and guidance on the *design* and *conduct* of scientific studies of MRTPs, which was the task of the committee.

Regulating tobacco products creates unique challenges. Unlike most products regulated by the FDA, tobacco is inherently hazardous and offers primarily risks rather than any significant physiological benefit to the user's health. Recognizing this, the law provides a public health standard and additional requirements of these products that must be considered as the FDA regulates these products. First, the law creates a public health standard that requires the FDA to evaluate the effect of the MRTP on not only users of the product, but also nonusers and the entire population as a whole. Second, the law requires postmarket observational studies of the MRTPs as a condition of approval, and also requires the annual submission of data about the MRTPs to the FDA. Finally, the law sets expiration dates on the orders to market the MRTPs. In addition, the FDA can revoke an order for any failure to comply with regulatory requirements or if there is evidence that the product is in fact harmful to public health.

The evaluation of the effect of MRTPs on public health will require a wide range of evidence and therefore will require many different types of study designs, including studies of the composition of MRTPs and studies of human exposure, human health effects, the likelihood of addiction and abuse, and the perception and understanding of the product by the public. Furthermore, the evidence must be able to reliably support predictions about the effect of marketing the product on public health, and therefore these studies must be properly designed and rigorously conducted. Study designs will need to include all relevant populations including populations at a high risk for tobacco use. Study designs must be able to support not only inferences about the mechanisms of the products effects, but also predictions about the products' effects in the real world.

Also, relevant to the committee's deliberations as it considered the conduct of studies is the history of the tobacco industry's past behavior. The tobacco industry has a long and well-documented history of illegal and improper conduct, and its practices have only recently been regulated. Because of the health impact of its products and the opaque practices that have been engaged by the tobacco industry, many academic institutions and their faculty that would normally be involved in a product's evaluation have been separated from conducting research related to tobacco products for many years. Thus, the committee concluded that the tobacco industry currently lacks not only the trustworthiness, but also the expertise, infrastructure, and other resources needed to independently produce the scientific evidence necessary to meet the public health standards set by the law. In the report, the committee explores the possibility of new governance mechanisms to address this problem, including the potential creation of a third-party governance entity. The committee does recognize that there are MRTPs that may not be developed by the tobacco industry and thus believes the need for third-party governance may not be applicable in all cases.

Overall, the committee's goal was to develop enduring guidelines and considerations for the production of credible and comprehensive evidence of the effects of MRTPs. The committee emphasized that the principle of public disclosure, which adds the sunshine of openness and transparency, must be applied to the entire process of product development. It is hoped that this report will provide guidance not only to the FDA but also to all stakeholders (the tobacco industry, academic researchers, and journal editors, etc.) on how the important work of evaluating these products can move ahead.

This committee has volunteered a great deal of time and energy into completing a remarkably complex task, and for that I am very appreciative. I thank its members for their collective and individual efforts. I would also like to extend my own and the committee's gratitude to Suchitra Krishnan-Sarin, Holly E. Morrell, Gary Stoner, Wendy Theobald, and Robert B. Wallace for their assistance and expertise as external consultants. On behalf of the committee, appreciation is also extended to each who provided information, data, or even an informed opinion during open sessions or by mail. Finally, the committee and I would like to thank the IOM staff for their hard work and diligence: Kathleen Stratton, Joel Wu, Michelle C. Catlin, Erin Rusch, Hannan Braun, Malcolm Biles, and Rose Marie Martinez.

Jane E. Henney, *Chair*
Committee on Scientific Standards for
Studies on Modified Risk Tobacco Products

Contents

BOXES

FIGURE

TABLES

Acronyms and Abbreviations

1-HOP	1-hydroxypyrene
^{3}HdT	tritiated thymidine
4-NQO	4-Nitroquinoline 1-oxide
ARISE	Associates for Research into the Science of Enjoyment
BaP	benzo[a]pyrene
bp	base pair
BrdU	5′-bromodeoxy-uridine
CDC	Centers for Disease Control and Prevention
CER	comparative effectiveness research
CFR	Code of Federal Regulations
CIAR	Center for Indoor Air Research
CISNET	Cancer Intervention and Surveillance Modeling Network
CO	carbon monoxide
CONSORT	Consolidated Standards of Reporting Trials
COPD	chronic obstructive pulmonary disease
CRP	C-reactive protein
CS	cigarette smoke
CSC	cigarette smoke condensate
CSE	cigarette smoke extract
CTP	Center for Tobacco Products
CTR	Council for Tobacco Research

DAPI	4′,6-diamidino-2-phenylindole
DMBA	dimethylbenz[a] anthracene
DMC	data monitoring committee
DMEM	Dulbecco's Modified Eagle Medium
DMSO	dimethyl sulfoxide
DSMB	data and safety monitoring board
DSM-IV	Diagnostic and Statistical Manual of Mental Disorders, 4th Edition
ELISA	enzyme-linked immunosorbent assay
EPA	Environmental Protection Agency
FD&C Act	Food, Drug, and Cosmetic Act
FDA	Food and Drug Administration
FDAAA	Food and Drug Administration Amendments Act
FSPTCA	Family Smoking Prevention and Tobacco Control Act of 2009
FTC	Federal Trade Commission
GC-MS	gas chromatography-mass spectrometry
GC-MS/MS	gas chromatography-tandem mass spectrometry
H1N1	Influenza A
HBMA	4-hydroxybut-2-yl mercapturic acid
HBSS	Hanks buffered salt solution
HEI	Health Effects Institute
HEMA	2-hydroxyethyl mercapturic acid
HHS	U.S. Department of Health and Human Services
HONC	Hooked on Nicotine Checklist
HPMA	3-hydroxypropyl mercapturic acid
hr	hour
HSV-1	herpes simplex virus 1
IAPS	International Affective Picture System
ICAM-1	inter-cellular adhesion molecule-1
IL-8	interleukin 8
INS-GAS	insulin-gastrin
IOM	Institute of Medicine
IRB	institutional review board
IVR	interactive voice response
LC-MS/MS	liquid chromatography-tandem mass spectrometry

MAPK	mitogen-activated protein kinase
MCA	methylcoanthrene
MCP-1	monocyte chemotactic protein-1
MHBMA	1-hydroxy-2-(*N*-acetylcysteinyl)-3-butene and 1-(*N*-acetylcysteinyl)-2-hydroxy-3-butene
mRNA	messenger RNA
MRTP	modified risk tobacco product
MSA	Master Settlement Agreement
MTS	3-(4,5-dimethylthiazol-2-yl)-5-(3-carboxymethoxyphenyl)-2-(4-sulfophenyl)-2H-tetrazolium
MTT	3-(4,5-Dimethylthiazol-2-yl)-2,5-Diphenyltetrazolium Bromide
NACDA	National Advisory Council on Drug Abuse
NCI	National Cancer Institute
NF-kB	nuclear factor kappaB
NG	not given
NIDA	National Institute on Drug Abuse
NNAL	4-(methylnitrosamino)-1-(3-pyridyl)-1-butanol
NNK	nicotine-derived nitrosamine ketone
NNN	*N*-nitrosonornicotine
NRC	National Research Council
NRT	nicotine replacement therapy
OSMB	observational study monitoring board
OTC	over-the-counter
PAMP	pathogen-associated molecular patterns
PBS	phosphate buffered saline
PCR	polymerase chain reaction
poly(I:C)	Polyinosinic:polycytidylic acid
ppm	parts per million
PREP	potential reduced-exposure product
RCT	randomized controlled trial
RFA	requests for application
RICO	Racketeer Influenced and Corrupt Organizations
RNS	reactive nitrogen species
ROS	reactive oxygen species
RUF	Reagan-Udall Foundation
SPMA	*S*-phenylmercapturic acid
ST	smokeless tobacco

STE	smokeless tobacco extract
STROBE	Strengthening the Reporting of Observational Studies in Epidemiology
TI	Tobacco Institute
TRADD	tumor necrosis factor receptor type 1-associated death domain protein
TRGE	tobacco research governance entity
TSNA	tobacco-specific *N*-nitrosamines
TUNEL	terminal deoxynucleotidyl transferase dUTP nick end labeling
VCAM-1	vascular cell adhesion molecule 1
VEGF-A	vascular endothelial growth factor A
wk	week
yrs	years

Summary

Smoking is the leading cause of preventable morbidity and mortality in the United States, contributing to approximately 443,000 premature deaths each year nationally (CDC, 2008). Smoking-related disease causes more deaths than alcohol, illicit drug use, homicide, and suicide combined (Mokdad et al., 2004). Another 8.6 million smokers in the United States live with a smoking-attributable illness (CDC, 2009a). In total, tobacco-related mortality amounts to approximately 5.1 million years of potential life lost per year (CDC, 2008). Smoking also imposes enormous costs on the U.S. health care system and economy, with an estimated $193 billion in losses due to health care costs and productivity losses per year (CDC, 2008).

The current prevalence of cigarette use is 20.6 percent among adults and 19.5 percent in youth (CDC, 2010, 2011). After substantial declines in adult smoking rates through the 1980s and 1990s, the rate of U.S. adult smokers has remained relatively static from 20.9 percent in 2004 to 20.6 percent in 2009 (CDC, 2010). Between 1997 and 2003, smoking prevalence among high school students declined substantially from 36.4 percent to 21.9 percent; this decline slowed from a 21.9 percent youth smoking rate in 2003 to 19.5 percent in 2009 (CDC, 2011). Of the 46 million adult smokers in the United States, an estimated 70 percent of smokers wish to quit completely, and approximately 45 percent of smokers attempt to quit each year (CDC, 2002, 2009b). However, only approximately 6 percent of the smokers who attempt to quit are successful for 1 month or more (HHS, 2000).

THE FAMILY SMOKING PREVENTION AND TOBACCO CONTROL ACT

The Family Smoking Prevention and Tobacco Control Act of 2009 (FSPTCA)[1] grants the Food and Drug Administration (FDA) broad authority to regulate the manufacturing, distribution, and marketing of tobacco products, including "modified risk tobacco products" (MRTPs). Generally, an MRTP is defined by the law as any tobacco product that is sold or distributed for use to reduce harm or the risk of tobacco-related disease.

Under the FSPTCA, no MRTP may be marketed without an order for sale from the U.S. Department of Health and Human Services (HHS). To be marketed, the product must meet one of two public health standards: either (1) an empirically demonstrated Modified Risk claim or (2) a Special Rule for Certain Products claim, specifying a reduced-exposure product.

To meet the Modified Risk standard, the applicant must prove with scientific evidence that the product, as actually used by consumers, will (1) significantly reduce harm and the risk of tobacco-related disease to individual users and (2) benefit the health of the population as a whole, taking into account both users and nonusers of the product.

Under the Special Rule for Certain Products, the Secretary of HHS may issue an order for the sale of a reduced-exposure product for which there is inadequate long-term epidemiologic data to support a finding under the Modified Risk standard but where the available evidence demonstrates that a substantial reduction in morbidity and mortality is "reasonably likely."

In regard to both standards, the law further specifies that the Secretary should also take into account how the marketing of the MRTP affects the likelihood of current users continuing tobacco product use with an MRTP who otherwise would have quit, nonusers initiating tobacco use with an MRTP, and the risks and benefits compared to other smoking-cessation products.

The concept of harm reduction informs the public health rationale for permitting the development and potential marketing of modified risk tobacco products. The basic premise of harm reduction is the continuation of a potentially hazardous or dangerous behavior, with the aim of decreasing the potentially adverse consequences of these behaviors (Marlatt, 2002). In the context of tobacco harm reduction, "a product is harm reducing if it lowers total tobacco-related morbidity and mortality, even though use of that product may involve continued exposure to tobacco-related toxicants" (IOM, 2001).

---

[1] *Family Smoking Prevention and Tobacco Control Act of 2009*, Public Law 111-31, 123 Stat. 1776 (June 22, 2009).

Modification of the risk profiles of tobacco products is only one component of a comprehensive, multifaceted strategy to minimize the negative health effects of tobacco use. Tobacco harm reduction efforts specifically target users that are unwilling or unable to quit. In conjunction with tactics to prevent initiation of tobacco use and to promote immediate cessation, MRTPs with reduced risk profiles may potentially lessen the harm of tobacco for the substantial portion of U.S. smokers who are unable or unwilling to abstain.

REQUIREMENT FOR REGULATIONS, GUIDANCE, AND CONSULTATION WITH THE INSTITUTE OF MEDICINE

The FSPTCA requires the FDA to issue guidance or regulation on the scientific evidence required for the assessment and ongoing review of an MRTP applicant. The law also specifically requires the FDA to consult with the Institute of Medicine (IOM) in developing guidance and regulation on the "design and conduct of such studies and surveillance." Box S-1 provides the statement of task.

COMMITTEE APPROACH

The IOM convened a multidisciplinary committee of 15 experts with backgrounds in addiction, cardiology, pulmonology, oncology, epidemiology, study design methodology, biostatistics, risk perception, adolescent behavior, drug or device regulation and law, population health, tobacco initiation and cessation, and toxicology. Over the course of 10 months, the committee held five meetings; extensively reviewed literature; heard representatives from the tobacco industry, public health advocacy groups, and regulatory agencies; and consulted with external subject area experts.

To fully grasp the nature of the task, the committee sought guidance not only from the statement of task, but also from the enabling statutory language of the FSPTCA Section 911(l)(2). While it is essential to address

BOX S-1
Statement of Task

The Institute of Medicine will establish a committee of 15 public health and medical experts to advise the Food and Drug Administration on the minimum standards for scientific studies to support the marketing of modified risk tobacco products and for postmarket studies of approved products.

minimum standards for scientific studies, the committee interpreted the task more broadly than the simple rearticulation of basic scientific principles or the review of current scientific methods. The committee was particularly wary of making "perishable" recommendations that may lose relevance as time passes and scientific methods and technologies evolve. Rather, the committee sought to provide enduring insights into what constitutes credible and meaningful evidence of the effect of tobacco products on public health. The committee's insights can be generally organized into three categories:

1. Types of studies and evidence on MRTPs
2. Design of studies on MRTPs and decision making
3. Governance

EVIDENCE AND STUDIES

Generally, the evidence to support the marketing approval of an MRTP will come from three categories: health effects of the MRTP, addictive potential of the MRTP, and perceptions about the MRTP.

Evidence and Studies of Health Effects

Laboratory analysis of the performance and of the constituents of tobacco products will be the first step in the evaluation of any new product. These analyses involve standard methods of extraction, sample cleanup, analyte identification, and quantitation. There are important limitations to laboratory analysis of product performance and composition. First, laboratory analysis of constituents may not reflect constituent uptake under conditions of use. In particular, smoking machines do not replicate human smoking conditions. There is currently no proven way to replicate the many ways humans use tobacco. As such, it is crucial to describe the smoking regimen or other extraction methods employed. Second, there may be other unidentified compounds in tobacco that contribute in important ways to adverse health effects. Also, seemingly innocuous compounds can exacerbate the effects of toxicants.

The second step in the evaluation of an MRTP will be preclinical studies of toxicity. These assays are essential in identifying particularly risky or toxic products that should not be tested in humans, and to identify products that have reasonable potential to reduce risk and harm and therefore should proceed to clinical evaluation. *In vitro* assays for cytotoxicity, genotoxicity, apoptosis and cell proliferation, oxidative stress, inflammation, mucus production, and endothelial activation are a standard step in evaluations of all combusted and noncombusted products.

Evaluation of products *in vitro* should precede *in vivo* assays. Furthermore, assays in animal models should precede human assays. Although it is not possible to make laboratory animals use tobacco products the way humans do, and there are inherent interspecies differences that prevent meaningful extrapolation of human effects, it is still informative to observe the effect of tobacco products in live animal models. Assays of toxicity in humans will also be essential, in particular assays of urinary mutagenicity and sister chromatid exchange in peripheral lymphocytes.

Biomarkers of exposure measure human exposure to constituents of tobacco. Biomarkers of human exposure to specific constituents of tobacco include the constituents themselves, their metabolites, or protein- or DNA-binding products of the constituents or their metabolites. These biomarkers have the potential to bypass many of the uncertainties in product composition analysis and provide a realistic and direct assessment of carcinogen and toxicant dose in an individual. The first step in employing biomarkers of exposure is analytical validation. The second step is validation with respect to product use. Finally, biomarkers can be validated with respect to disease risk; however, there is no proof that any individual constituent or group of constituents is responsible for a given disease. For a biomarker of exposure to be accepted as a biomarker of risk or a surrogate endpoint of disease, there should be a strong biological rational as well as compelling data from clinical and epidemiologic studies.

Experimental designs, in particular randomized controlled trials (RCTs), provide data that can support the strong inferences about the effect of an MRTP on human health relative to conventional tobacco products. The use of appropriately designed clinical trials will be important to establish whether the use of the MRTP reduces exposure to toxicants or induces positive changes in surrogate markers as claimed by the manufacturer. An RCT is an effective means of examining acceptability and use of the MRTP, the ability of the MRTP to increase cessation in users of conventional tobacco products, and the likelihood that availability of the MRTP will lead to dual use. RCT methods can also produce evidence on whether and how much individuals use an MRTP after they have used it to help them quit conventional products, changes in perception of the MRTP with its continued use, and the MRTP's ability to suppress tobacco withdrawal symptoms. It is important to recognize that no single RCT can address all of these areas, and each study should have a focused objective with a primary endpoint.

Observational epidemiologic studies play a critical and central role in the evaluation of MRTPs. Although they will rarely, if ever, have the compelling scientific credibility of experimental designs, these methods form the basis for most evaluation studies of regulated products in the community. Long, intensive, and robust observational studies of actual health outcomes may be required to fully evaluate the net effects of MRTPs relative to conventional tobacco products.

Prospective cohort studies are obvious candidates for the evaluation of MRTPs, and will also be an essential tool to validating anticipated or claimed effects of marketed MRTPs on both individuals and on the public's health. Cohort studies allow assessment of overall health status and outcomes, as well as offering the following important strengths:

- Biochemical tobacco and MRTP exposure assessment can be made at baseline, offering unbiased exposure assessment before health outcomes occur.
- There is less of a problem with retrospective recall of product use, because this information can be summarized at the start of the study, and followed prospectively.
- Changing product use habits can be monitored as the study progresses.
- Outcomes can be documented as they occur, and verification becomes more efficient.

A wide variety of outcomes can be evaluated in the same study, including both intermediate and clinical outcomes. In addition, other epidemiological study designs will be necessary to evaluate MRTPs and provide evidence on the public health effects of marketed MRTPs; these include retrospective cohort studies, case-control studies, crossover or case-crossover designs, and comparative effectiveness research methods. Case-control studies are commonly used because of their efficiency in assembling study participants, including when the disease outcomes are not common in general populations (e.g., varying levels of biomarkers). When the outcomes are short term and/or recurrent (particularly when using intermediate endpoints), an observational crossover or case-crossover design becomes feasible and informative. Comparative effectiveness research methods more critically inform health care and policy decisions, but these methods can also sharpen or extend observational studies comparing health outcomes associated with use of conventional tobacco products and use of certain MRTPs. Overall, different study designs will be necessary depending on the circumstances and the research question.

Evidence and Studies of Addictive Potential

Evaluation of the likelihood of initiation, maintenance, and persistence of use in both conventional tobacco users and nonusers is critical to estimating the public health effect of marketing an MRTP. Specifically, evaluation of the MRTP's ability to promote initiation and continuation of its regular use, switching to its use and cessation of the consumption of more harmful products, dual use, and to promote relapse back to more harmful tobacco use are all essential. All of these outcomes are logically related to the reinforcing value of the MRTP (that is, how rewarding it is).

There is a continuum of reinforcement value. In theory, the MRTP should be somewhat more reinforcing than nicotine replacement therapies but perhaps less reinforcing than conventional cigarettes. Ideally, an MRTP would be sufficiently reinforcing so as to attract smokers away from conventional cigarettes but not enough to encourage the widespread dependent use of the product by individuals who were previously nonusers, or who would have quit smoking.

Evaluation of the abuse and addiction potential of a product can be accomplished with a variety of experimental designs and in a variety of contexts, including subjective evaluations in laboratory contexts, acute self-administration studies in laboratory contexts, use in extended residence facilities, and natural environment contexts where long-term use can be studied in real-world circumstances via RCTs, cross-sectional survey studies, and longitudinal cohort studies.

Evaluation of reinforcement value in a laboratory setting is particularly important because the results of these studies reliably correspond to an agent's addictive potential in real-world use. A standard with regard to human abuse liability drug testing is acute dose-effect comparison studies, because of the correspondence between subjective ratings of drug effects and real-world abuse potential. Behavioral economic self-administration studies will also be important in evaluating the reinforcement potency of a product. The usefulness of all studies in forecasting the risk for initiation and abuse of a product depends on study design factors. Important design considerations include the size of the sample, the nature of the sample (whether the sample includes heavy smokers or light smokers, smokers who want to quit, and nonsmokers), the characterization of the sample (age, sex, gender, ethnicity, educational attainment, socioeconomic status, etc.), and the nature of the comparison product.

Evidence and Studies of Risk Perception and Communication

Judgments about risk, otherwise known as risk perceptions, are a fundamental element to most theoretical models of health behavior and behavioral decision making. In general, these models argue that individuals' perceptions about the value and likelihood of behavior-related positive and negative consequences and their vulnerability to those consequences play a key role in behavioral choices. As such, understanding individuals' perceptions of tobacco-related products, including MRTPs, whether such perceptions change over time, and whether such perceptions play a role in tobacco behavior, is critical. It will be important to identify consumers' perceptions of disease risk, likelihood of addiction, likelihood of reducing or increasing others' exposure to potentially hazardous compounds, and perceptions of risk compared to other products already on the market. It is also important to assess intentions of using the product. It is essential that the industry carefully crafts messages about risks and benefits of any MRTP and demonstrates through rigorous testing that people correctly understand and interpret the risks.

Studies evaluating risk perceptions and risk communication should be performed both before the marketing of an MRTP and after the MRTP has been marketed. Premarket research will play an essential role in developing the messages the tobacco industry can use to communicate information about MRTPs to consumers. This research will determine consumers' ability to accurately understand messages that communicate information about the risks, benefits, and conditions of using an MRTP compared to existing tobacco products. Studies should also test how these messages influence consumers' perceptions of the risks, benefits, and likelihood of addiction related to an MRTP. The first stage of premarket research will involve formative work using focus groups. The second stage should include discussions with groups of similar individuals to assess how the messages that were developed in the first stage are received by consumers. Finally, the effects of these messages on consumer perceptions should be tested. It will be important to evaluate consumer understanding and to compare consumer perceptions of an MRTP to conventional products. After the product is released on the market, it is vital to continue monitoring consumer perceptions and behavior related to that product. Conducting nationally representative cohort-sequential longitudinal surveys will be essential.

Table S-1 presents the evidence domains and example considerations for using evidence from the different domains.

STUDY DESIGN AND DECISION MAKING

Study Design

Studies should be designed appropriately to create an evidence base that can support a finding of public health benefit. The ultimate goal of studying the effect of an MRTP on human health and behavior is to be able to accurately predict the public health effects of allowing an MRTP to be marketed. In other words, the ultimate goal of scientific studies is to produce generalizable data. The "generalizability" of data, or the reliability of predictions that can be made about the real world based on scientific observations, will depend on the design of the studies.

TABLE S-1 Evidence Domains Relevant to an MRTP Application and Examples of Types of Findings

Class of Evidence	Examples of Types of Finding That May Be Required
Preclinical	• Assurance of manufacturing quality control • Significant and substantial reduction in toxicant and carcinogen content in product • Significant reduction in exposure to toxicants and carcinogens in limited human study • No significant evidence for offsetting increases in content of or exposure to other toxicants
Clinical trial	• Significant reduction in exposure to toxicants and carcinogens in relation to continued use of traditional product, preferably approaching nonsmoker levels • Significant rates of cessation of conventional tobacco product use, or significant decrease in the rates of conventional tobacco product use • Significant reduction in biomarkers or surrogates of disease
Abuse potential	• No more liable for abuse than currently marketed products • No significant evidence of attractiveness to nonusers of tobacco
Epidemiology	• Clear and consistent evidence of reduction in disease risk (e.g., cancer, cardiovascular disease, chronic obstructive pulmonary disease) or intermediate endpoint thereof • No significant evidence of offsetting increased risk for other diseases • No significant evidence of uptake among nonusers or relapse among former users (postmarketing)
Consumer and nonconsumer perceptions	• Evidence for accurate understanding of product claim • No significant evidence that consumers equate reduced exposure with reduced risk • No significant evidence of intention to use product among nonusers (especially adolescents) • No significant evidence of switching from MRTP to other tobacco product usage
Populations at high risk for tobacco use	• No significant evidence of risk of initiation among nonusers (especially adolescents) • Consistency of findings across relevant subpopulations of interest (e.g., low socioeconomic status, racial/ethnic minorities)
Modeling and synthesis	• Population predictions show reduction in smoking-related morbidity and mortality following the introduction of an MRTP with no significant evidence of uptake by nonusers (especially adolescents)

NOTE: This table is not comprehensive and is not intended to be a guideline or framework for the evaluation of MRTP applications.

Elements of study design that should be carefully evaluated include the size of the sample and the nature of the sample. Sample sizes should be carefully determined and tailored to the study design and the effects that are being studied. Statistically underpowered studies cannot support inferences or projections about the effects of a product. The nature of the study sample is critical to the usefulness of study results. Results from studies conducted in one population may not be applicable to other populations because the characteristics that define the study population either are related to or cause the responses to the product. As such, it is important to study a wide range of populations. It is particularly important to include populations that have a high risk of using tobacco and populations that will be affected by the marketing of the product.

Study designs should also carefully consider the degree of control imposed on experimental designs. Internal and external validity should be balanced not only within studies, but also across studies of the same product. Highly controlled experimental designs can eliminate many variables and confounders and support strong inferences, but simultaneously they can lose relevance to the real world as the conditions of product use do not reflect real-world circumstances and behaviors. Experimental designs that are less controlled can emulate circumstances that reflect real-world conditions and behaviors, and therefore they may be more relevant in predicting real-world effects, but uncontrolled variables may confound meaningful associations or inferences. Multiple complementary study designs will be necessary to provide the evidence necessary to meet the statutory public health standards.

Decision Making

It is clear that no single class of evidence (e.g., preclinical, RCTs, consumer perception, epidemiologic) in itself will be sufficient to support an MRTP application. The portfolio of evidence brought to the FDA to justify a modified risk or modified exposure claim will be substantial. To inform regulatory decision making, the FDA will need to process the evidence at a higher level, beyond merely amassing the evidence in support of the MRTP claims.

A key challenge facing the FDA will be integrating the various domains and levels of evidence provided by sponsors in support of an MRTP application. It would be helpful to have a systematic, explicit approach that weighs outcomes in terms of their public health importance, identifies the measures and data most relevant to those outcomes, and combines the available evidence in a manner that is psychometrically sound, objective, and reproducible. The approach to data integration that the FDA takes will be highly influential in determining whether an

MRTP is marketed, and the approach should be transparent, objective, and reproducible.

It is anticipated that modeling and simulation methods will play a role in integrating evidence to inform regulatory decisions. For modeling and simulation to be transparent, detailed information on all aspects of model structure, sources of evidence used, computational approach, construction of summaries, and reporting of the results should be available. Such information is essential not only for a proper scientific understanding of the modeling, but also for allowing researchers and other stakeholders in the regulatory process to critique and validate the model. It is important to ensure that the methods for data integration that inform decision making are neither arbitrary nor flawed.

Another critical factor in deciding whether to issue an order for the marketing of an MRTP is the amount of harm reduction claimed by an MRTP sponsor in an application. Harm reduction is inherently relative; a reduction claim is, by definition, relative to a comparison product. Selection of an appropriate comparison product is essential for informed and accurate decision making. The FSPTCA recognizes this, giving the Secretary of HHS authority to require product sponsors to compare their product to a commercially marketed representative product. The choice of appropriate comparison products will be driven by the type of MRTP being tested, the anticipated claim, and the study design. The comparison products may even differ between different classes of evidence. Two reference products come to the forefront in terms of integration and synthesis of evidence: leading brands and smoking-cessation products.

"Leading brands" represent a set of products that accounts for a significant portion of the market and could capture subgroups of interest (e.g., those of low socioeconomic status, who tend to use discount brands, and certain racial/ethnic minorities, who have higher rates of menthol cigarette use). Using leading brands increases the likelihood that the findings will have broader applicability to the population, which is crucial given the public health standard against which MRTPs are evaluated. Using leading brands as a comparator also avoids potential mischief in allowing comparisons between an MRTP and a product that is little used but inflates the apparent benefit of the MRTP.

Smoking-cessation products represent a standard (or tobacco-cessation products in the case of smokeless tobacco users) as a comparison product, because these products pose very few, if any risks to health. These products provide an aspirational goal for risk and exposure from MRTPs. In principle, the closer the risks and exposures from the MRTP are to cessation products, the more confident a regulator can be in the chances for net public health benefit. Note that the use of this comparison product is not the same as studying whether the MRTP acts as an aid to smoking cessation. Rather, the goal is to compare how the risk or exposure reduction attained with use of the MRTP compares to smoking-cessation product use of similar duration.

GOVERNANCE

The role of governance is to ensure the proper conduct of research. In addition to the essential role of protecting the interests of human research participants, governance of research is critical to the production of credible and reliable evidence. Governance and oversight of research conduct can prevent unethical behavior such as the falsification and manipulation of research data. Over time, the proper conduct of research can also build credibility and public trust.

There is profound distrust of the tobacco industry and of research supported by the tobacco industry. This distrust is the direct result of the tobacco industry's history of improperly influencing or manipulating scientific findings and messaging about the health effects of tobacco. This history and the lack of trust may prevent independent experts from participating in research on tobacco products and therefore may impede the production of data on MRTPs necessary to assess public health impact. Particularly important illustrations of the lack of public trust include the fact that many major universities have bans on the acceptance of tobacco industry funding, and that many journals will not accept or publish research supported by the tobacco industry. Establishing the tobacco industry as a legitimate participant in tobacco research is an important consideration in the overall goal of producing evidence on the effects of MRTPs.

The need for academic institutions or other experts to conduct research on tobacco products is particularly important in the current regulatory environment. First, the tobacco industry currently lacks the infrastructure and expertise to independently produce the necessary evidence to support an application to market an MRTP. The FSPTCA now requires tobacco products to undergo a premarket approval process similar to drugs and devices. Prior to the passage of the FSPTCA, the tobacco industry lacked robust regulation, and as a consequence, the industry may lack the institutional and organizational capacity to assemble a complete application to meet the requirements of the law. In addition, there are significant domains of evidence that should be addressed in an application to market an MRTP wherein the tobacco industry either lacks the expertise or the willingness to independently conduct the research. In particular, research involving populations with a high risk for tobacco use such as behavioral research, studies of adolescents, research on abuse liability, and observational studies of health effects will be very challenging for the tobacco industry.

The committee recommends several strategies to create an environment conducive to the production of reliable and credible evidence, in spite of the tobacco industry's reputation and currently limited infrastructure and expertise. The first strategy is to create a mechanism to distance the reputation and influence of the tobacco industry from experts, researchers, and institutions that will be critical to the production of evidence on MRTPs. Fear, either real or perceived, of being influenced by or aiding the tobacco industry prevents many institutions, researchers, and journals from having any association with the tobacco industry. Providing independence, autonomy, and separation from the industry addresses these fears. An independent third party that conducts research, provides oversight of research, distributes funding for research or manages research contracts, or otherwise provides governance of research may be a useful mechanism for reengaging the experts and institutions necessary to producing high-quality evidence on the effects of MRTPs. Relevant examples of third-party partnerships between industry and government include the Health Effects Institute and the Reagan-Udall Foundation. Currently, there are no independent entities that fulfill these roles for the tobacco industry.

The second strategy is to require that the conduct of research in support of MRTPs conform to ethical standards and that study information be made publicly available. Transparency and the proper conduct of research not only protect the interests of research participants, but also improve data quality. Requiring transparency and ethical conduct of research may also help change public perceptions of the tobacco industry, and subsequently engagement and support from key stakeholders may be more likely. Over time, requiring adherence to codes of ethics, and requiring the publication of study information and results, will improve the quality and availability of evidence about the effects of MRTPs on health.

FINDINGS AND RECOMMENDATIONS

In the committee's view, the fundamental problem that confronts the FDA is a shortage of credible and reliable evidence about the effects of MRTPs on both individual and public health. The history of deceptive behavior by the tobacco industry undermined the trust of the public as well as the public's confidence in the industry's ability to rigorously conduct studies that will generate the data needed to evaluate these products. Therefore, the committee's recommendations are designed to articulate the minimum standards for producing credible and reliable evidence to demonstrate that the marketing of an MRTP is consistent with the protection of public health. The committee articulates a strategy for the production of scientific evidence by making recommendations in three areas:

1. Types of evidence and studies
2. Design and integration
3. Governance of studies

Types of Evidence and Studies

Finding 1: Types of Evidence. The public health standard articulated by the FSPTCA requires collection of scientific evidence from a wide range of disciplines and research domains. Although the committee respects the FDA's independence and discretion in regulating MRTPs, the committee maintains there is a minimum range of research domains required to evaluate the effect of MRTPs on individuals and public health. Individual methods may change as the technology or state of the science may evolve, but the minimum standards for the domains of evidence will be relevant regardless of the state of the science in the future.

Recommendation 1: The FDA should require that studies submitted in support of an MRTP application address all key research domains needed to forecast and monitor the product's public health impact, including

- **product composition and performance;**
- **addiction potential and likelihood for initiation or persistence of use;**
- **human exposure to harmful and potentially harmful constituents;**
- **perceptions about the product's effects and likelihood of addiction; and**
- **effects of the product on human health and surrogates of human health.**

Finding 2: Phased Approach to New MRTPs. Many novel MRTPs are likely to be developed for marketing in the near future. There are inherent uncertainties and risks with new products that should be addressed. Risks should be minimized before new products are tested in humans. To address the risk of new products, a phased approach, similar to the New Drug Application framework for the regulation and control of new drugs, is appropriate for the evaluation of new MRTPs. A phased approach will help the FDA ensure that only products that are unlikely to be unsafe and have a reasonable expectation of reducing harm relative to conventional tobacco products will be used in human studies.

Recommendation 2: The FDA should establish guidance that conveys an expected sequencing of studies, such that preclinical work is com-

pleted and submitted to the FDA before clinical (human subjects) work commences, and that there is a reasonable expectation based on preclinical work that a reduction or lack of harm will be seen in humans.

Finding 3: Clinical Trial Studies. Although the use of randomized controlled trial methods will be constrained for a number of reasons (including the practical limitations of study cost, size, and follow-up, and ethical constraints on randomizing study participants to harmful exposures), they will continue to play an essential role in creating an evidence base on the public health effects of MRTPs. Randomized controlled trial methods can provide highly reliable data on the likelihood of addiction and initiation or cessation of product use. Also, these methods can provide reliable evidence on human exposure.

Recommendation 3: The FDA should require randomized controlled trials in the following domains:

- **Exposure reduction**
- **Self-administration of the MRTP**
- **Effects on use of conventional tobacco products**

These randomized controlled trials should include multiple comparison products (such as nicotine replacement products, conventional cigarettes or smokeless tobacco, placebo preparations, and alternative nicotine delivery systems). These trials should also assess the effect of the MRTP on human exposure and on human health and surrogates of human health.

Finding 4: Requirement for Postmarket, Prospective Epidemiologic Studies. Postmarket studies of MRTPs will be critical to evaluating the effect of MRTPs on both individuals and the public's health. In particular, the prospective cohort design will be an essential tool to validating anticipated or claimed effects of marketed MRTPs. These studies have several important strengths: (1) biochemical tobacco and MRTP exposure can be assessed at baseline, offering "unbiased" exposure assessment before health outcomes occur; (2) there is less of a problem with retrospective recall of product use, because this information can be summarized at the start of the study and followed prospectively; (3) changing product use habits can be monitored as the study progresses; (4) outcomes can be documented as they occur, and verification is more efficient; and (5) a wide variety of outcomes can be evaluated in the same study, particularly outcomes that are more common. Furthermore, cohort studies allow assessment of overall health status and outcomes.

Recommendation 4: The FDA should require prospective epidemiologic studies to commence upon issuance of a marketing order to confirm reduced exposure and reduced risk claims, and to examine effects of MRTP availability on the population as a whole, including the likelihood of initiation and cessation. The FDA should issue guidance on the design, conduct, and analysis of such studies.

Finding 5: Modeling of Public Health Outcomes. Mathematical modeling and simulation analysis provides a complementary approach to the conduct of empirical studies that can be useful at each stage of the regulatory process for MRTPs. Model-based analyses can (1) synthesize the available information from empirical studies of MRTPs; (2) enable researchers and decision makers to explore complex interactions and systems that may be impractical to evaluate in empirical studies; (3) allow researchers and decisions makers to explore "what if" questions relevant to decision making, which would not be practical to assess in empirical studies; and (4) be used to make projections about the short- and long-term effects of the introduction of MRTPs.

Recommendation 5: The FDA should issue guidance on the development and use of simulation and modeling approaches to predict public health impact through the systematic integration of information about relevant assumptions and influences. Such approaches should be tested for robustness with regard to results and assumptions, they should be public and transparent, and they should be validated against postmarketing epidemiologic research.

Design and Integration of Studies

Finding 6: Standards for Sampling in MRTP Studies. To have regulatory usefulness, studies of MRTPs must be generalizable to the overall population of interest and to specific populations, including populations at high risk for tobacco use. Failure to include relevant populations in studies will result in incomplete evidence on the effect of an MRTP on the public's health and, therefore, will be inadequate to support regulatory decisions about the marketing of MRTPs.

Recommendation 6: The FDA should require studies to include populations of special relevance, including (but not limited to)

- **users of tobacco products, including users who are and are not interested in quitting;**
- **in certain circumstances, nonusers of tobacco products;**

- **former smokers;**
- **beginning smokers;**
- **adolescents; and**
- **populations at a high risk for tobacco use, including, but not limited to, those low in socioeconomic status and educational attainment, and certain ethnic minorities.**

Finding 7: Quality of Studies. The usefulness of a study to inform a regulatory decision hinges on the quality and appropriateness of the design. In many cases, complementary studies might be needed to provide a breadth of evidence for an informed regulatory decision with appropriate control of confounders and internal and external validity.

Recommendation 7: For all studies of the effects of MRTPs on human health and behavior, the FDA should require a range of designs that are properly powered, balance internal and external validity, and comprise multiple populations appropriate to the experimental questions being addressed.

Finding 8: Standards for Good Research Practice. A significant amount of guidance on minimum standards for scientific studies directly relevant to the evaluation of MRTPs has already been developed. Guidelines for formatting, design, conduct, and reporting of science are articulated in consensus statements, such as the Consolidated Standards of Reporting Trials (CONSORT) reporting criteria for clinical trials, the Strengthening the Reporting of Observational Studies in Epidemiology (STROBE) statement for observational studies, the publication criteria of the International Council of Medical Journal Editors, and the reporting criteria of the International Conference on Harmonization. These existing guidelines represent robust standards for the conduct of science across many of the research domains relevant to the evaluation of MRTPs.

Recommendation 8: The FDA should issue guidance to the industry regarding the format, design, conduct, and reporting of studies in support of MRTP applications that is based upon current generally accepted principles for scientific investigation.

Finding 9: Standards for Integration of Evidence. Regulatory decisions regarding MRTPs will be based on a wide range and variety of scientific evidence, and the integration of scientific evidence will play a pivotal role in that decision making. The assessment of MRTPs will typically require the evaluation and integration of evidence on risks and benefits across multiple diverse outcomes, such as measures of toxicity, biomarkers, addictiveness, and disease endpoints. Modeling and simulation approaches are relevant to estimating public health effects of tobacco and, therefore, the FDA will likely engage in various methods of data integration, synthesis, and analysis, including, but not limited to, simulation and modeling. It is critical that these approaches are transparent and reproducible.

Recommendation 9: The FDA should develop and use an approach to data integration that is explicit and transparent with regard to the importance of the different outcomes, that uses optimal available evidence, and that employs objective and reproducible methods for data integration.

Governance of Studies

Finding 10: Independent Oversight and Conduct of Studies. It has been established in public records and as a matter of law that the tobacco industry has engaged in illegal and improper practices, including the destruction and manipulation of scientific data. As a result, the tobacco industry is profoundly isolated from the mainstream scientific community. Many major universities have policies against acceptance of tobacco funding, and many high-impact scientific and medical journals will not accept tobacco industry-supported manuscripts. The consequence of this isolation is a lack of the expertise and the resources necessary to produce high-quality science across the range of disciplines to support an application to market an MRTP. Use of a trusted third party, particularly for products developed by the tobacco industry, could provide an avenue for the production of credible evidence needed by the FDA to evaluate tobacco products. Ultimately, such a research structure could encourage and support the production and dissemination of credible and reliable evidence about the effects of tobacco products on the public's health.

Recommendation 10: MRTP sponsors should consider use of independent third parties to undertake one or more key functions, including the design and conduct of research, the oversight of specific studies, and the distribution of sponsor funds for research. Such independent third parties should be approved by the FDA in advance of the research.

Finding 11: Public Disclosure of Research. Public availability of data not only builds credibility and public trust, but also benefits the public because it allows for independent analysis of study methods and data. The model of Clinicaltrials.gov is particularly compelling and relevant, and a similar model of public accounting and open disclosure should be expected of the tobacco industry.

> **Recommendation 11: The FDA should require all MRTP sponsors to place all data generated in the development and marketing of the MRTP in a public repository selected by the FDA.**

Finding 12: Proper Conduct of Research. Standards for the conduct of science and the protection of human research participants have been established for biomedical research enterprises not only in academics but also in commercial research. The FDA has the tools to ensure studies adhere to established standards in the drug development framework, which can be applied to the development of MRTPs. Those standards not only protect human participants, but they also build credibility into any data that are provided to the FDA, particularly by the tobacco industry. Institutional credibility and trustworthiness is particularly relevant in this context, given the history of unethical and illegal practices of the tobacco industry.

> **Recommendation 12: The FDA should require studies offered in support of an MRTP application to adhere to established standards and principles of good research governance, including appropriately qualified investigators, transparency, independent institutional review board or ethical review, and adherence to the Common Rule (21 CFR parts 50 and 56).**

REFERENCES

CDC (Centers for Disease Control and Prevention). 2002. Cigarette smoking among adults—United States, 2000. *Morbidity and Mortality Weekly Report* 51(29):642.

CDC. 2008. Smoking-attributable mortality, years of potential life lost, and productivity losses—United States, 2000–2004. *Morbidity and Mortality Weekly Report* 57(45):1226-1228.

CDC. 2009a. *Tobacco use: Targeting the nation's leading killer—at a glance 2009.* National Center for Chronic Disease Prevention and Health Promotion, Office on Smoking and Health, Department of Health and Human Services.

CDC. 2009b. Cigarette smoking among adults and trends in smoking cessation—United States, 2008. *Morbidity and Mortality Weekly Report* 58(44):1227-1232.

CDC. 2010. Vital signs: Current cigarette smoking among adults aged ≥ 18 years—United States, 2009. *Morbidity and Mortality Weekly Report* 59(35):1135-1140.

CDC. 2011. Ten great public health achievements—United States, 2001–2010. *Morbidity and Mortality Weekly Report* 60(19):619-623.

HHS (U.S. Department of Health and Human Services). 2000. *Reducing tobacco use: A report of the surgeon general.* Atlanta, GA: U.S. Department of Health and Human Services, Centers for Disease Control and Prevention, National Center for Chronic Disease Prevention and Health Promotion, Office on Smoking and Health.

IOM (Institute of Medicine). 2001. *Clearing the smoke: Assessing the science base for tobacco harm reduction*. Washington, DC: National Academy Press.

Marlatt, G. A. 2002. *Harm reduction: Pragmatic strategies for managing high-risk behaviors.* New York: The Guilford Press.

Mokdad, A. H., J. S. Marks, D. F. Stroup, and J. L. Gerberding. 2004. Actual causes of death in the United States, 2000. *JAMA* 291(10):1238-1245.

1

Introduction

TOBACCO HARM IN THE UNITED STATES

Smoking is the leading cause of preventable morbidity and mortality in the United States, contributing to approximately 443,000 premature deaths each year nationally (CDC, 2008). Cigarette smoking is the most common form of tobacco use; there are an estimated 46.6 million current adult smokers, and initial use of tobacco typically occurs in youth (CDC, 2010; HHS, 1994). Many of the approximately 1,000 adolescents who initiate smoking each day continue use to become regular adult smokers (SAMHSA, 2011).

Smoking-related disease causes more deaths than alcohol, illicit drug use, homicide, and suicide combined (Mokdad et al., 2004). In total, tobacco-related mortality amounts to approximately 5.1 million years of potential life lost per year (CDC, 2008). Lung cancer, ischemic heart disease, and chronic obstructive pulmonary disease (COPD) are the leading causes of smoking-attributable deaths (CDC, 2008). Smoking can also increase a person's risk for stroke and other forms of cancers. A list of documented diseases and conditions that are caused by cigarette smoking can be found in Table 1-1. In total, cigarette smoking accounts for approximately 80 percent of deaths from COPD, and at least 30 percent of deaths due to cancers (CDC, 2008). Another 8.6 million smokers in the United States live with a smoking-attributable illness (CDC, 2009a). Smoking also imposes enormous costs on the U.S. health care system and economy, with an estimated $193 billion in losses due to health care costs and productivity losses per year (CDC, 2008).

TABLE 1-1 Diseases and Conditions Caused by Active Cigarette Smoking[a]

Disease	Effects
Malignant neoplasms[b]	*Tumor sites for which there is sufficient evidence:* • Oral cavity • Oropharynx, nasopharynx, and hypopharynx • Oesophagus (adenocarcinoma and squamous-cell carcinoma) • Stomach • Colorectum • Liver • Pancreas • Nasal cavity and paranasal sinuses • Larynx • Lung • Uterine cervix • Ovary (mucinous) • Urinary bladder • Kidney (body and pelvis) • Ureter • Bone marrow (myeloid leukaemia) *Tumor site for which there is limited evidence:* • Female breast
Cardiovascular diseases	• Coronary heart disease • Cerebrovascular disease • Atherosclerosis • Aortic aneurysm
Respiratory diseases in adults	• Chronic obstructive pulmonary disease (bronchitis, emphysema, chronic airway obstruction) • Pneumonia • Premature onset of and an accelerated age-related decline in lung function • All major respiratory symptoms (e.g., coughing, phlegm, wheezing, dyspnea) • Poor asthma control
Respiratory diseases in young people	• Impaired lung growth • Respiratory symptoms and asthma-related symptoms (e.g., wheezing) in childhood and adolescence • Early onset of lung function decline during late adolescence and early adulthood

TABLE 1-1 Continued

Disease	Effects
Reproductive and perinatal conditions	• Sudden infant death syndrome • Reduced fertility in women • Fetal growth restriction • Low birth weight • Premature rupture of the membranes • Placenta previa • Placental abruption • Preterm delivery and shortened gestation • Respiratory distress syndrome
Miscellaneous	• Cataracts • Hip fractures • Low bone density • Peptic ulcer disease in persons who are *Helicobacter pylori* positive • Diminished health status (i.e., increased absenteeism from work, increased use of medical care services) • Adverse surgical outcomes related to wound healing and respiratory complications

SOURCE:
[a] Modified from Giovino (2007). Reprinted from *American Journal of Preventive Medicine,* Vol 6, Supplement 1, Gary A. Giovino, The Tobacco Epidemic in the United States, S318-S326, Copyright 2007, with permission from Elsevier.
[b] Data from Secretan et al. (2009).

Cigarette smoking is harmful for nonusers as well. Secondhand smoke (also called environmental tobacco smoke, involuntary smoke, or passive smoke) is responsible for approximately 50,000 annual deaths, largely due to lung cancer, coronary heart disease, and sudden infant death syndrome (HHS, 2006).

Other forms of tobacco use, such as smokeless tobacco products and cigars, also have harmful consequences. Smokeless tobacco use—primarily in the form of chewing tobacco and snuff—causes cancer of the oral cavity and pancreas and other oral health problems such as gum recession and leukoplakia (IARC, 2007).

The United States faces particular challenges with youth smoking. Of the approximately 1 million people who start smoking regularly each year, 39.5 percent are under the age of 18 (SAMHSA, 2011). Like adult smokers, quit rates remain low in adolescent smokers (Mermelstein, 2003). It is estimated that 50 percent of those who start smoking in adolescence go on to smoke for 16 to 20 years (Pierce and Gilpin, 1996).

Since the mid 1960s when smoking rates among adults in the United States peaked at more than 42 percent, efforts to reduce cigarette use have made significant progress (CDC, 1999). The current prevalence of cigarette use is 20.6 percent among adults and 19.5 percent in youth (CDC, 2010). These tobacco control efforts have been recognized by the Centers for Disease Control and Prevention (CDC) as a major public health achievement, both of the 20th century and of the first decade of the 21st century (CDC, 2011a). Despite this improvement in the past half-century, however, progress has stalled in recent years. After substantial declines in adult smoking rates through the 1980s and 1990s, the rate of U.S. adult smokers has remained relatively static, from 20.9 percent in 2004 to 20.6 percent in 2009 (CDC, 2010). This lack of continued progress is also seen in youth smoking rates. Between 1997 and 2003, smoking prevalence among high school students declined substantially from 36.4 percent to 21.9 percent; this decline slowed from a youth smoking rate of 21.9 percent in 2003 to 19.5 percent in 2009 (CDC, 2011a).

Current disparities in tobacco use patterns are another cause for concern; trends in smoking show higher tobacco use rates among certain ethnic and racial minority groups, persons with low socioeconomic status, sexual minorities, and people in the South and Midwest of the United States (CDC, 2011b; HHS, 1998).

In spite of the well-established health risks for smoking and other forms of tobacco use, some individuals still choose to use tobacco for a number of potentially desired effects or outcomes. These effects or outcomes may range from perceived social benefits (Halpern-Felsher et al., 2004), to discrete physiological effects (Vezina et al., 2007), or even the mere enjoyment of the flavor and aroma of tobacco. Although the ultimate decision of whether the benefits outweigh the risks is up to the individual, the committee firmly maintains that the individual and public health hazards of tobacco use far outweigh any potentially desired effects of tobacco use.

HARM REDUCTION

The concept of harm reduction informs the public health rationale for permitting the development and potential marketing of modified risk tobacco products (MRTPs). The basic premise of harm reduction is the continuation of a potentially hazardous or dangerous behavior, with the aim of decreasing the potentially adverse consequences of these behaviors (Marlatt, 2002). The reduced risk can be for either participants or nonparticipants of the potentially harmful activity. Harm reduction is most typically associated with illicit substance use, including opioid substitution therapies, needle exchange programs, and supervised injecting sites. Harm reduction strategies can incorporate a wide spectrum of individual tactics, from safer use, to managed use, to complete abstinence from the risk behavior. In the context of tobacco harm reduction, "a product is harm reducing if it lowers total tobacco-related mortality and morbidity, even though use of that product may involve continued exposure to tobacco-related toxicants" (IOM, 2001).

Modification of the risk profiles of tobacco products is only one of several potential tactics to reduce the harm of tobacco. In addition to preventing initiation of tobacco use and promoting cessation of tobacco use, MRTPs with reduced risk profiles may potentially lessen the harm of tobacco for the substantial portion of U.S. smokers who are unable or unwilling to abstain. Of the 46 million adult smokers in the United States, an estimated 70 percent of smokers wish to quit completely, and approximately 45 percent of smokers attempt to quit each year (CDC, 2002, 2009b). However, only approximately 6 percent of the smokers who attempt to quit are successful for 1 month or more (HHS, 2000). Tobacco harm reduction efforts target users that are unwilling or unable to quit as one component of a comprehensive, multifaceted strategy to minimize the negative health effects of tobacco use.

In 2001 the Institute of Medicine (IOM) specifically addressed the potential of tobacco harm reduction in *Clearing the Smoke: Assessing the Science Base for Tobacco Harm Reduction*. The report articulated six principle conclusions:

1. For many diseases attributable to tobacco use, reducing risk of disease by reducing exposure to tobacco toxicants is feasible.
2. PREPs[1] have not yet been evaluated comprehensively enough (including for a sufficient time) to provide a scientific basis for concluding that they are associated with a reduced risk of disease compared to conventional tobacco use.
3. Surrogate biological markers that are associated with tobacco-related diseases could be used to offer guidance as to whether or not PREPs are likely to be risk-reducing.
4. Currently available PREPs have been or could be demonstrated to reduce exposure to some of the toxicants in most conventional tobacco products.
5. Regulation of all tobacco products, including conventional tobacco products as recommended in IOM (1994), as well as all other PREPs is a necessary precondition for assuring a scientific basis for judging the effects of using PREPs and for assuring that the health of the public is protected.
6. The public health impact of PREPs is unknown. They are potentially beneficial, but the net impact on population health could, in fact, be negative.

[1] PREPs or potential reduced-exposure products are defined in Box 1-2.

In the report, the IOM also suggests a research agenda for evaluating the potential for harm reduction of PREPs (IOM, 2001). Since the publication of *Clearing the Smoke*, significant advances have been made in the science of evaluating tobacco products (Carter et al., 2009; Cohen et al., 2009; Hatsukami et al., 2005, 2006, 2007, 2009; O'Connor et al., 2009; Shields, 2002). Also, in following the report's conclusion that harm reducing products might be possible, tobacco companies have developed a large number of PREPs. Several products remain on the market; however, no products have been proven to reduce harm or risk.

HISTORY OF TOBACCO REGULATION IN THE UNITED STATES

The Family Smoking Prevention and Tobacco Control Act of 2009 (FSPTCA)[2] grants the Food and Drug Administration (FDA) broad authority to regulate the manufacturing, distribution, and marketing of tobacco products. This law marks an important turning point in the history of tobacco control in the United States. In fewer than 70 years, the profile of tobacco has changed from that of a popular luxury item associated with athletes, movie stars, and celebrities to that of a highly regulated public health hazard. However, despite decades of heightened concern and public awareness, smoking is still the leading cause of preventable death in the United States.

Modern tobacco control began in the 1940s and 1950s with scientific and public health authorities establishing through emerging evidence that smoking causes disease. Prior to the 1940s, little existing evidence linked tobacco use to disease. As late as the 1950s, certain cigarettes were still marketed with explicitly positive health claims. The 2001 IOM report chronicled the early health claims of tobacco manufacturers in response to the health concerns of smokers (IOM, 2001).

Evidence collected in the 1940s and 1950s confirmed the link between smoking and lung cancer. Landmark publications from Richard Doll and Austin Hill, and from Ernst Wynder and Evarts Ambrose Graham, provided convincing evidence that smoking caused lung cancer (Doll and Hill, 1950, 1956; Wynder and Graham, 1950). In 1962, a report from the Royal College of Physicians of London reaffirmed findings that smoking significantly increased the risk of death from lung and heart disease (Royal College of Physicians, 1962). In 1964, the Surgeon General released a report that authoritatively linked smoking to lung cancer, cardiovascular disease, and emphysema (U.S. Public Health Service, 1964).

---

[2] *Family Smoking Prevention and Tobacco Control Act of 2009*, Public Law 111-31, 123 Stat. 1776 (June 22, 2009).

In addition to authoritatively linking smoking to disease, the Royal College of Physicians and the Surgeon General reports also directed the evidence to policy makers and the public with recommendations to reduce the harm of smoking. The Royal College report specifically recommended preventive measures including filtration of smoke, modifications of tobacco, and discouragement of smoking. The report also made specific recommendations for government action, including public education, taxation, and restrictions on advertisement and smoking in public places.

In response to the Surgeon General's report, which stated that "cigarette smoking is a health hazard of sufficient importance in the United States to warrant appropriate remedial action" (U.S. Public Health Service, 1964), Congress passed the Federal Cigarette Labeling and Advertising Act of 1965 requiring that the statement "Caution: cigarette smoking may be hazardous to your health" be placed on every cigarette package. In 1970, Congress strengthened the regulation through the Public Health Cigarette Smoking Act, which banned cigarette advertisements on any medium of electronic communication subject to the jurisdiction of the Federal Trade Commission. In total between 1965 and 2000, Congress passed six pieces of legislation to address the harm of tobacco use, including

- the Federal Cigarette Labeling and Advertising Act of 1965;[3]
- the Public Health Cigarette Smoking Act of 1970;[4]
- the Alcohol and Drug Abuse Amendments of 1983;[5]
- the Comprehensive Smoking Education Act of 1984;[6]
- the Comprehensive Smokeless Tobacco Health Education Act of 1986;[7] and
- the Alcohol, Drug Abuse, and Mental Health Administration Reorganization Act of 1992.[8]

Additional smoking-related regulations and interventions have followed, with the recognition of secondhand smoke as a serious public health hazard (IOM, 2010), including the restriction of smoking in government facilities and on all commercial U.S. airline flights. (See Table 1-2 for

---

[3] *Federal Cigarette Labeling and Advertising Act of 1965*, Public Law 89-92, 79 Stat. 282 (July 27, 1965).

[4] *Public Health Cigarette Smoking Act of 1970*, Public Law 91-222, 84 Stat. 87 (April 1, 1970).

[5] *Alcohol and Drug Abuse Amendments of 1983*, Public Law 98-24, 97 Stat. 175 (April 26, 1983).

[6] *Comprehensive Smoking Education Act of 1984*, Public Law 98-474, 98 Stat. 2200, 98th Congress (October 12, 1984).

[7] *Comprehensive Smokeless Tobacco Health Education Act of 1986*, Public Law 99-252, 100 Stat. 30, 99th Congress (February 27, 1986).

[8] *Alcohol, Drug Abuse, and Mental Health Administration Reorganization Act of 1992*, Public Law 102-321, 106 Stat. 323, 102nd Congress (July 10, 1992).

TABLE 1-2 Summary of Milestones in Decreasing Indoor Tobacco Smoke in the United States

Year	Event
1971	The Surgeon General proposes a federal smoking ban in public places.
1972	The first report of the Surgeon General to identify secondhand smoke as posing a health risk is released.
1973	Arizona becomes the first state to restrict smoking in several public places. The Civil Aeronautics Board requires no-smoking sections on all commercial airline flights.
1974	Connecticut passes the first state law to apply smoking restrictions in restaurants.
1975	Minnesota passes a statewide law restricting smoking in public places.
1977	Berkeley, California, becomes the first community to limit smoking in restaurants and other public places.
1983	San Francisco passes a law to place private workplaces under smoking restrictions.
1986	A report of the Surgeon General focuses entirely on the health consequences of involuntary smoking, proclaiming secondhand smoke a cause of lung cancer in healthy nonsmokers. The National Research Council issues a report on the health consequences of involuntary smoking. Americans for Nonsmokers' Rights becomes a national group; it had formed as the California Group Against Smoking Pollution.
1987	The U.S. Department of Health and Human Services establishes a smoke-free environment in all its buildings, affecting 120,000 employees nationwide. Minnesota passes a law requiring all hospitals in the state to prohibit smoking by 1990. A Gallup poll finds, for the first time, that a majority (55 percent) of U.S. adults favor a complete ban on smoking in all public places.
1988	A congressionally mandated smoking ban takes effect on all domestic airline flights of 2 hours or less. New York City's ordinance for clean indoor air takes effect; the ordinance bans or severely limits smoking in various public places and affects 7 million people. California implements a statewide ban on smoking aboard all commercial intrastate airplanes, trains, and buses.
1990	A congressionally mandated smoking ban takes effect on all domestic airline flights of 6 hours or less. The U.S. Environmental Protection Agency (EPA) issues a draft risk assessment of secondhand smoke.

TABLE 1-2 Continued

Year	Event
1991	The National Institute for Occupational Safety and Health issues a bulletin recommending that secondhand smoke be reduced to the lowest feasible concentration in the workplace.
1992	Hospitals applying to the Joint Commission on Accreditation of Healthcare Organizations for accreditation are required to develop a policy prohibiting smoking by patients, visitors, employees, volunteers, and medical staff. The EPA releases its report classifying secondhand smoke as a group A carcinogen (known to be harmful to humans), placing secondhand smoke in the same category as asbestos, benzene, and radon.
1993	Los Angeles passes a ban on smoking in all restaurants. The U.S. Postal Service eliminates smoking in all facilities. Congress enacts a smoke-free policy for Special Supplemental Food Program for Women, Infants, and Children clinics. A working group of 16 state Attorneys General releases recommendations for establishing smoke-free policies in fast-food restaurants. Vermont bans smoking in all public buildings and in many private buildings open to the public.
1994	The U.S. Department of Defense prohibits smoking in all indoor military facilities. The Occupational Safety and Health Administration proposes a rule that would ban smoking in most U.S. workplaces. San Francisco passes a ban on smoking in all restaurants and workplaces. The Pro-Children Act requires persons who provide federally funded children's services to prohibit smoking in their facilities. Utah enacts a law restricting smoking in most workplaces.
1995	New York City passes a comprehensive ordinance effectively banning smoking in most workplaces. Maryland enacts a smoke-free policy for all workplaces except hotels, bars, some restaurants, and private clubs. California passes comprehensive legislation that prohibits smoking in most enclosed workplaces. Vermont's smoking ban is extended to include restaurants, bars, hotels, and motels except establishments holding a cabaret license.
1996	The U.S. Department of Transportation reports that about 80 percent of nonstop scheduled U.S. airline flights between the United States and foreign points will be smoke free by June 1, 1996.
1997	President Clinton signs an executive order establishing a smoke-free environment for federal employees and all members of the public visiting federally owned facilities. The California EPA issues a report determining that secondhand smoke is a toxic air contaminant. Settlement is reached in the class-action lawsuit brought by flight attendants exposed to secondhand smoke.

continued

TABLE 1-2 Continued

Year	Event
1998	The U.S. Senate ends smoking in the Senate's public spaces. California law takes effect banning smoking in bars that do not have a separately ventilated smoking area. The Minnesota tobacco-document depository is created as a result of a tobacco-industry settlement with Minnesota and Blue Cross and Blue Shield of Minnesota. U.S. tobacco companies are required to maintain a public depository to house more than 32 million pages of previously secret internal tobacco industry documents.
2000	The New Jersey Supreme Court strikes down a local clean-indoor-air ordinance adopted by the city of Princeton on the grounds that state law preempts local smoking restrictions. A congressionally mandated smoking ban takes effect on all international flights departing from or arriving in the United States.
2002	New York City holds its first hearing on an indoor smoking ban that would include all bars and restaurants. The amended Clean Indoor Air Act enacted by the state of New York (Public Health Law, Article 13-E), which took effect July 24, 2003, prohibits smoking in virtually all workplaces, including restaurants and bars. The Michigan Supreme Court refuses to hear an appeal of lower-court rulings striking down a local clean-indoor-air ordinance enacted by the city of Marquette on the grounds that state law preempts local communities from adopting smoking restrictions in restaurants and bars that are more stringent than the state standard. Delaware enacts a comprehensive smoke-free law and repeals a preemption provision precluding communities from adopting local smoking restrictions that are more stringent than state law. Florida voters approve a ballot measure that amends the state constitution to require most workplaces and public places—with some exceptions, such as bars—to be smoke free.
2003	Dozens of U.S. airports—including airline clubs, passenger terminals, and nonpublic work areas—are designated as smoke free. Connecticut and New York enact comprehensive smoke-free laws. Maine enacts a law requiring bars, pool halls, and bingo venues to be smoke free. State supreme courts in Iowa and New Hampshire strike down local smoke-free ordinances, ruling that they are preempted by state law.
2004	Massachusetts and Rhode Island enact comprehensive smoke-free laws. The International Agency for Research on Cancer issues a new monograph identifying secondhand smoke as "carcinogenic to humans."
2005	North Dakota, Vermont, Montana, and Washington enact 100 percent smoke-free workplace and/or restaurant and/or bar regulations.
2006	New Jersey, Colorado, Hawaii, Ohio, and Nevada enact 100 percent smoke-free workplace and/or restaurant and/or bar regulations.
2007	Louisiana, Arizona, New Mexico, New Hampshire, and Minnesota enact 100 percent smoke-free workplace and/or restaurant and/or bar regulations.

TABLE 1-2 Continued

Year	Event
2008	Illinois, Maryland, Iowa, and Pennsylvania enact 100 percent smoke-free workplace and/or restaurant and/or bar regulations.
2009	Oregon and Nebraska enact 100 percent smoke-free workplace and/or restaurant and/or bar regulations.
2010	Wisconsin, Michigan, North Carolina, and Kansas enact 100 percent smoke-free workplace and/or restaurant and/or bar regulations.
As of Oct. 7, 2011	Across the United States, 21,875 municipalities are covered by a 100 percent smoke-free provision in workplaces and/or restaurants and/or bars by a state, commonwealth, or local law; this represents 79.6 percent of the U.S. population.
	The District of Columbia and 39 states have local laws in effect that require 100 percent smoke-free workplaces and/or restaurants and/or bars.

SOURCE: Derived from IOM (2010) with additional information from the American Nonsmokers' Rights Foundation (2011) and the Robert Wood Johnson Foundation (2011).

a summary of significant milestones in addressing indoor tobacco smoke exposure in the United States.)

In 1996, 13 years before the passage of the FSPTCA, the FDA unsuccessfully attempted to assert regulatory authority over tobacco. The FDA argued that tobacco products fell within its purview, claiming that nicotine is a drug, and that cigarettes and smokeless tobacco are combination products that consist of the drug (nicotine) and device components intended to deliver nicotine to the body.[9] Based on this framework, the FDA issued a regulation that included

- prohibiting the sale of tobacco products to individuals under age 18;
- prohibiting free samples and the sale of tobacco products through vending machines and self-service displays, except in facilities that ensure that no person under age 18 is present;
- restricting advertising exposed to children and adolescents to black-and-white, text-only displays;
- prohibiting billboards and outdoor advertising within 1,000 feet of schools and public playgrounds;
- prohibiting the sale or distribution of brand identified promotional nontobacco items; and
- prohibiting sponsorship of sporting and other events in the name of a product.

[9] 61 *Federal Register* 44396-45318; 21 CFR Parts 801, 803, 804, 807, 820, and 897.

The authority of the FDA to regulate tobacco and to enforce these rules was challenged by the tobacco industry, and in 2000 the Supreme Court held that Congress did not intend for the FDA to have the authority to regulate tobacco products.[10] This created a major barrier to the federal regulation of tobacco products. Without explicit authorization from Congress, the FDA would not have power to regulate tobacco.

Individual states still retained independent authority to regulate tobacco, and they have made significant reforms through both litigation and legislation. For example, most states have enacted laws banning smoking in public spaces. Table 1-2 displays significant scientific reports and clean-air policies enacted in the United States since 1971, when the Surgeon General first proposed a federal smoking ban for public spaces (IOM, 2010). Detailed information on these milestones are outlined in the Surgeon's General 2006 report, *The Health Consequences of Involuntary Exposure to Tobacco Smoke* (HHS, 2006).

Lawsuits filed by state attorneys seeking reimbursement for the costs of tobacco-related disease have had a profound effect, with the resulting settlements leading to significant restrictions on the tobacco industry that were previously not attainable. Many of these restrictions were a result of the Master Settlement Agreement (MSA), which was signed in November 1998. The MSA was a contractual agreement between the Attorneys General of 46 states and the largest U.S. cigarette companies. The companies accepted limitations and restrictions on certain marketing practices, and agreed to annual payments over a 25-year period to the states to contribute toward the economic costs of tobacco-related disease. A list of the restrictions on the advertising practices from the MSA can be found in Box 1-1.

In 1999, the federal government followed the successes of state Attorneys General and filed a lawsuit to reclaim health care expenses caused by tobacco-related disease. Specifically, the Department of Justice filed a lawsuit against the tobacco industry for violating the Racketeer Influenced and Corrupt Organizations (RICO) Act, seeking to "disgorge" billions of dollars in profits from past unlawful activities, and to prevent the tobacco industry from engaging in future fraudulent and unlawful conduct. The court eventually ruled that the government could not recover monetary damages, but did require the tobacco industry to engage in corrective advertising, to stop deceptive labels including the terms "low tar" or "light," and to submit to judicial oversight.[11] The RICO rulings are discussed in further detail in Chapter 2.

As a result, prior to the passing of the FSPTCA, tobacco regulation

---

[10] *FDA v. Brown and Williamson*, 529 U.S. 120 (2000).

[11] *United States v. Philip Morris USA, Inc., et al.*, 449 F. Supp. 2d 1, 26 (D.D.C. 2006).

BOX 1-1
Advertising, Promotion, and Marketing Restrictions Resulting from the Master Settlement Agreement

Entered on November 23, 1998, the Tobacco Master Settlement Agreement (MSA) placed restrictions on tobacco product advertising, marketing, and promotion. The MSA

- prohibited tobacco companies from targeting youth in the advertising, promotion, or marketing of their products;
- banned the use of cartoons in advertising;
- limited each company to brand-name sponsorship of one sporting or cultural event a year, excluding concerts, team sports, events with a significant youth audience, or events with underage contestants;
- banned public transit advertising;
- banned outdoor billboard advertising, excluding billboard advertising for brand-name sponsored events;
- limited advertising outside retail stores to signs no bigger than 14 square feet;
- banned company payments to promote tobacco products in various media, including movies and TV;
- banned non-tobacco apparel with brand-name logos except at brand-name sponsored events;
- banned gifts of non-tobacco items to youth in exchange for tobacco products;
- restricted the use of nationally recognized non-tobacco brand names for tobacco products; and
- limited free samples of tobacco products to adult-only facilities.

SOURCE: Adapted from Congressional Research Service. 2009. *FDA tobacco regulation: The Family Smoking Prevention and Tobacco Control Act of 2009.*

consisted of a patchwork of policy and regulation from various sources of authority: federal law, state law, the tobacco MSA, and RICO lawsuit rulings. These pieces of regulation were inadequate to address major systemic causes of tobacco use and tobacco-related disease in American society.

The need for comprehensive and systemic regulation of tobacco at a national level has been recognized for many years. In 1994, the IOM stated that a comprehensive national tobacco control strategy should include a regulatory component. The IOM recommended that "Congress should enact legislation that delegates to an appropriate agency authority to regulate tobacco products for the dual purposes of discouraging consumption and reducing the morbidity and mortality associated with use of tobacco" (IOM, 1994). In 2001, the IOM specifically addressed the question of PREPs and recommended the development of a surveillance system and a strengthened federal regulation of all modified tobacco products (IOM, 2001). In 2007, an IOM report gave extensive and detailed recommendations for federal regulation, including the specific recommendation that "Congress should confer upon the FDA broad regulatory authority over the manufacture, distribution, marketing, and use of tobacco products" (IOM, 2007).

FSPTCA OVERVIEW

The FSPTCA codified the recommendation from the 2007 IOM report, amending the Food, Drug and Cosmetic Act (FD&C Act) to grant the FDA broad authority to regulate tobacco products. Under the FSPTCA, a tobacco product is defined as "any product made or derived from tobacco that is intended for human consumption, including any component, part, or accessory of a tobacco product (except for raw materials other than tobacco used in manufacturing a component, part, or accessory of a tobacco product)." Under Section 301 of the FD&C Act, any adulterated or misbranded product is prohibited from being introduced into interstate commerce. Chapter IX articulates the definitions of adulterated or misbranded tobacco products. Among other things, Chapter IX gives the FDA authority to regulate the sale, distribution, labeling, and advertising of tobacco products, to set product standards, and to require reporting and disclosure of tobacco product ingredients and components.

SECTION 911

This report provides advice to the FDA on the minimum scientific standards for studies on MRTPs. Under Section 911 of the FSPTCA, the FDA is required to develop regulations or guidance, in consultation with the IOM, on the evidence that product sponsors should submit in filing an application to market an MRTP. This committee and the report adopt the definitions used in the legislation (see Box 1-2) and also take into account the regulatory framework established by the act in making any conclusions or recommendations.

Pursuant to the FSPTCA, no MRTP may be marketed without an order for sale from the Secretary of the U.S. Department of Health and Human Services (HHS). Under the FSPTCA, an MRTP is defined as any tobacco product that is sold or distributed for use to reduce harm or the risk of tobacco-related disease. The FSPTCA also specifies that a product will be categorized as an MRTP, and therefore be regulated under Section 911, if any labeling or advertising represents implicitly or explicitly that a product is reduced in risk or harm, or is free of or contains a reduced amount of a substance. Furthermore, any action that a tobacco product sponsor takes directed toward consumers through the media that would be reasonably expected to result in consumers believing a product is reduced in risk or exposure, or contains a reduced amount of a substance, renders the product subject to regulation under Section 911. A product must also obtain FDA approval to market or advertise the product using descriptors such as *light*, *low*, *mild*, or other similar descriptors. Smokeless tobacco products are not considered an MRTP solely because they use the word *smokeless* or other similar descriptors; if the smokeless tobacco product sponsor wishes to make additional modified risk claims, then the product must first apply for this claim. A product that is intended for tobacco cessation that is regulated by Chapter V of the FD&C Act cannot be considered an MRTP (for further detail, see Box 1-2).

To obtain an order for the sale of any new product, or for a new modified risk claim on an existing product, the manufacturer is required to submit an application to the FDA, which must include comprehensive documentation about the intended advertising and labeling, conditions of use, formulation, all documents relating to the product's effect on health, and information on how consumers actually use the product.

Under the FSPTCA the Secretary may only issue an order for the sale of an MRTP if the Secretary finds that the product that is the subject of the application meets one of two public health standards: either (1) an empirically demonstrated Modified Risk claim or (2) a Special Rule for Certain Products claim, specifying a reduced exposure product. The determination of whether an order is granted under either the Modified Risk standard or the Special Rule for Certain Products must be based on scientific evidence submitted by the applicant.

To issue an order pursuant to the Modified Risk standard, the Secretary must find that the applicant has demonstrated that the product, as actually used by consumers, will (1) significantly reduce harm and the risk of tobacco-related disease to individual tobacco users and (2) benefit the health of the population as a whole, taking into account both users and nonusers of the product. Under the Modified Risk standard, the orders expire after a time specified within the order.

Under the Special Rule for Certain Products, the Secretary may issue an order for the sale of a reduced-exposure product for which there is inadequate long-term epidemiologic data to support a finding under the Modified Risk standard. The specific conditions that must be met for approval under the Special Rule for Certain Products are outlined in Section 911(g)(2). Orders granted under the Special Rule for Certain Products cannot last longer than 5 years.

The law also specifies public health considerations that the Secretary must take into account when evaluating whether to issue an order, either under the Modified Risk standard or the Special Rule for Certain Products. These considerations are outlined in Section 911(g)(4) and include

BOX 1-2
Definitions and Historical Comparisons

Definition of MRTP from FSPTCA Section 911, Subsection (b), "Definitions":

(1) MODIFIED RISK TOBACCO PRODUCT- The term "modified risk tobacco product" means any tobacco product that is sold or distributed for use to reduce harm or the risk of tobacco-related disease associated with commercially marketed tobacco products.

(2) SOLD OR DISTRIBUTED-

(A) IN GENERAL- With respect to a tobacco product, the term "sold or distributed for use to reduce harm or the risk of tobacco-related disease associated with commercially marketed tobacco products" means a tobacco product—

(i) the label, labeling, or advertising of which represents explicitly or implicitly that—
(I) the tobacco product presents a lower risk of tobacco-related disease or is less harmful than one or more other commercially marketed tobacco products;
(II) the tobacco product or its smoke contains a reduced level of a substance or presents a reduced exposure to a substance; or
(III) the tobacco product or its smoke does not contain or is free of a substance;

(ii) the label, labeling, or advertising of which uses the descriptors "light," "mild," or "low" or similar descriptors; or
(iii) the tobacco product manufacturer of which has taken any action directed to consumers through the media or otherwise, other than by means of the tobacco product's label, labeling, or advertising, after the date of enactment of the Family Smoking Prevention and Tobacco Control Act, respecting the product that would be reasonably expected to result in consumers believing that the tobacco product or its smoke may present a lower risk of disease or is less harmful than one or more commercially marketed tobacco products, or presents a reduced exposure to, or does not contain or is free of, a substance or substances.

(B) LIMITATION- No tobacco product shall be considered to be "sold or distributed for use to reduce harm or the risk of tobacco-related disease associated with commercially marketed tobacco products," except as described in subparagraph (A).

(C) SMOKELESS TOBACCO PRODUCT- No smokeless tobacco product shall be considered to be "sold or distributed for use to reduce harm or the risk of tobacco-related disease associated with commercially marketed tobacco products" solely because its label, labeling, or advertising uses the following phrases to describe such product and its use: "smokeless tobacco," "smokeless tobacco product," "not consumed by smoking," "does not produce smoke," "smokefree," "smoke-free," "without smoke," "no smoke," or "not smoke."*

From Subsection (c), "Tobacco Dependence Products":

A product that is intended to be used for the treatment of tobacco dependence, including smoking cessation, is not a modified risk tobacco product under this section if it has been approved as a drug or device by the Food and Drug Administration and is subject to the requirements of Chapter V.

From *Clearing the Smoke: Assessing the Science Base for Tobacco Harm Reduction:*

The etymology of the term "Potential reduced-exposure products" or "PREP": Many tobacco and cigarette-like products have been introduced in the distant and recent past that do, under measurement systems such as the smoking machine, result in decreased emission of some toxicants compared to conventional products. These products could, therefore, *potentially*, results in reduced *exposure* to toxicants. The committee uses "potentially," because whether exposure to tobacco toxicants is reduced depends on the user's behavior, such as frequency and intensity of use. Reduced exposure, however, does not necessarily assure reduced risk to the user or reduced harm to the population. Therefore, and in order to avoid misinterpretation, the committee will use the generic phrase "potential reduced-exposure products," or PREPs, when discussing modified tobacco products, cigarette-like products (whether tobacco containing or not) developed for their tobacco harm reduction potential (IOM, 2001).

Comparison of "PREPs" and "MRTPs":

The term PREPs is not used by the committee and will not be included in this report. The term "PREPs" was coined in *Clearing the Smoke* to describe a category of products that theoretically could be used for tobacco harm reduction. While the term "PREPs" has been adopted within the academic literature following *Clearing the Smoke*, the passing of the FSPTCA has created a statutory definition that relates to PREPs. PREPs might be understood as a category of potential MRTPs that have not yet been shown to reduce exposure or risk. The task of this committee is essentially to advise the FDA in identifying scientific standards for studies to produce evidence showing that a PREP actually does reduce exposure, harm to users, and can protect public health. An MRTP may be thought of as a PREP that has been validated according to standards set in the FSPTCA and by the judgment of the FDA.

---

* It should be noted that some products, including some existing smokeless tobacco products, may be marketed without an order if they do not make a reduced risk or exposure claim. Products that do not make a claim or representation of reduced risk or exposure are not subject to regulation under section 911, and are not addressed in this report.

the health risks to individual current users, the likelihood that users will quit or that nonusers will start using the product, and the risks compared to cessation products.

If an order is approved, product sponsors are required to comply with certain conditions for marketing, as well as requirements to conduct postmarket surveillance and studies of the product to determine the impact of the order on consumer perception, behavior, and health. Results of the postmarket surveillance and studies must be submitted to the Secretary annually. Additionally, the Secretary can withdraw authorization, after an opportunity for hearing, if the product sponsors either fail to fulfill their obligations under the law, or if new evidence demonstrates that marketing of the product is not consistent with protecting the public's health.

COMPARISON OF REGULATORY FRAMEWORKS

The statutory framework established by the FSPTCA for regulation of MRTPs shares certain similarities with the existing regulatory frameworks for pharmaceuticals, biologics, and devices. For instance, the power of the FDA to inspect facilities and records, to require record keeping and reporting of data on health effects, and to require good manufacturing practices are generally consistent. In particular, all potential MRTPs must undergo a premarket approval process similar to new drugs and high-risk devices. However, tobacco products, including MRTPs, are fundamentally different than other products regulated by the FDA: tobacco products are inherently hazardous, addictive products. As a result, there are several significant differences in the regulatory standards and requirements established by the FSPTCA that are worth mention.

First, the standard to issue an order for the marketing of an MRTP is a public health standard. According to the FSPTCA, the Secretary's actions regarding tobacco products, such as setting product standards or restricting sales of certain products and advertising, must be based on a finding that the action is "appropriate for the protection of public health." In Section 911 the law specifies that the Secretary must take into account the effect of the product on nonusers when evaluating whether to issue an order for the sale of an MRTP.[12] The law also requires the Secretary take into account the increased or decreased likelihood of tobacco users quitting and nonusers initiating, and the risks and benefits compared to other smoking cessation products.

This public health standard is a significant departure from the standards for drugs and devices. Generally, to obtain approval, drugs and devices must be shown to be "safe and effective" for the individual

[12] 911(g)(1)(b) and 911(g)(2)(B)(4).

research participant or consumer. In contrast, MRTPs are potentially hazardous to the user and are never truly "safe" or "effective" in the sense that the product will improve the health of the user. The rationale for issuing an order for the sale of an MRTP is not that the products are safe and may improve health for individual users, but rather, that marketing the products may reduce the overall negative health effects in a population. To obtain an order to sell an MRTP, an MRTP candidate sponsor must demonstrate to the FDA that the overall burden of tobacco-related disease and death will be lower with the MRTP on the market than if it were not on the market.

Second, product sponsors are required as a condition of obtaining an order to market an MRTP to conduct postmarket studies and surveillance, and to report the data to the FDA annually. In contrast, at this point in time, drug and device manufacturers are generally not required to conduct postmarket studies of their products as an obligatory condition of approval.

Third, orders for the marketing of an MRTP must expire, and manufacturers must reapply for additional orders to allow for the ongoing sale of a product. MRTPs granted orders to market under the Special Rule for Certain Products are given a maximum of 5 years, while the terms of the orders granted under the Modified Risk standard are unspecified by the law. Following the expiration of an order, MRTP sponsors may obtain a renewal based on the filing of a new application. In contrast, drug and device sponsors do not have to reapply and essentially repeat the premarket approval process after a specified period of time.

BURDEN OF PROOF

The Secretary's decisions on whether an action will be appropriate for the protection of public health must be based on scientific evidence. According to the FSPTCA, a determination to allow for the marketing of an MRTP can be based on scientific evidence submitted by the product sponsor and on other scientific evidence made available to the Secretary.

The committee maintains that the burden of proof rests on the applicant. That is, the product sponsor that is applying for an order to market an MRTP bears the responsibility of producing evidence in support of an application, including the evidence necessary to demonstrate that an order is appropriate for the protection of public health. Without evidence, the Secretary cannot determine that issuing an order for the sale of an MRTP is appropriate for the protection of public health.

COMMITTEE CHARGE AND STATEMENT OF TASK

Origin of Task

In Section 911(l)(2) of the FSPTCA, Congress specifies that regulations and guidance on the scientific evidence required for the assessment and review of applications for a modified risk claim for a tobacco product must be developed in consultation with the IOM. The law states:

> The regulations or guidance issued under paragraph (1) shall be developed in consultation with the Institute of Medicine, and with input of other appropriate scientific and medical experts, on the design and conduct of such studies and surveillance.

Statement of Task

Pursuant to the legislation, the FDA Center for Tobacco Products (CTP) engaged the IOM to advise the agency. In discussing the scope of the report, the CTP and the IOM agreed that the committee would address premarket and postmarket studies in support of MRTPs and that postmarket surveillance would be excluded from the report to maintain appropriate focus and depth within the study time frame (see Box 1-3).

Scope of Task

The committee's interpretation of the task was informed by the text of the legislation and by input from the project sponsor in a public meeting. The law specifically states that regulation or guidance shall be developed in consultation with the IOM on both the design and conduct of studies. As such, the committee found authority to address issues not only concerning research methods and scientific standards, but also research conduct and governance.

The CTP also provided direction about the task to the committee during an open session of the first committee meeting. In a presentation

BOX 1-3
Statement of Task

The Institute of Medicine will establish a committee of 15 public health and medical experts to advise the Food and Drug Administration on the minimum standards for scientific studies to support the marketing of modified risk tobacco products and for post market studies of approved products.

to the committee, CTP Director Dr. Lawrence Deyton, emphasized that the committee should advise the CTP on the "types and characteristics" of evidence needed to evaluate an application for an order to market an MRTP. The director also specifically reminded the committee that it must take into account the population health regulatory standard described in the law. Finally, Dr. Deyton described tasks that he regarded as outside the committee's charge. Specifically, he indicated that he did not expect the committee to

- assess the evidence for any particular product;
- opine on whether any tobacco product or class of products is a potential candidate for modified risk determination;
- offer regulatory principles;
- define or recommend a conceptual or regulatory framework for modified risk assessment; and
- define terms from the act, such as "reasonably likely" or "measurable and substantial reduction."

The statutory mandate and the limits articulated by the sponsor presented a challenge to the committee. On the one hand, the law clearly placed the conduct of scientific studies within the scope of the IOM's task; on the other hand, the director indicated that the committee should not address regulatory principles. Furthermore, the state of tobacco science and regulation was well developed, with a significant existing body of literature on the evaluation and regulation of tobacco. As such, the committee concluded it would be inefficient and unproductive to simply reiterate basic scientific principles and review existing literature on tobacco. Accordingly, the committee sought to provide insight and direction to the CTP without usurping the judgments that should be left to the agency and the Secretary.

COMMITTEE PROCESS

With this task in mind, the IOM convened a committee of 15 experts. The committee conducted five meetings between February and September 2011. During these meetings, committee members heard from a wide variety of experts and stakeholders, including individuals representing the tobacco industry, researchers, public health experts and advocates, and government. The committee also extensively reviewed literature including original peer-reviewed research articles and published reports. In the course of deliberations, the committee commissioned work from and consulted with external experts to gain additional expertise.

OVERVIEW OF THE REPORT

This report is organized into six chapters. Chapter 2 discusses the standards for governance and the conduct of studies that are necessary to produce reliable and credible data to support a tobacco product's application for a modified risk claim. Chapters 3 through 5 then address the types and designs of studies available to assess these products' impact on an individual's health and on the population's health. Chapter 3 discusses studies on the health effects of the products, such as product composition, biomarkers, preclinical studies, clinical studies, and epidemiologic studies. Chapter 4 reviews the research methods to study the addictive potential of the product. This information provides evidence on the implications for human behavior patterns and public health risks. Chapter 5 outlines the standards for studies needed to address both consumer and nonconsumer risk perceptions and communication surrounding modified products. Issues of participant recruitment, measurement, and data analysis are also detailed in Chapter 5. In the final chapter, Chapter 6, the committee discusses the integration of these various classes of evidence required for informed decision making by the FDA. The committee presents its overarching findings and recommendations at the end of Chapter 6.

REFERENCES

American Nonsmokers' Rights Foundation. 2011. *Overview list—how many smokefree laws?* http://no-smoke.org/pdf/mediaordlist.pdf (accessed October 17, 2011).

Carter, L. P., M. L. Stitzer, J. E. Henningfield, R. J. O'Connor, K. M. Cummings, and D. K. Hatsukami. 2009. Abuse liability assessment of tobacco products including potential reduced exposure products. *Cancer Epidemiology, Biomarkers & Prevention* 18(12):3241-3262.

CDC (Centers for Disease Control and Prevention). 1999. Tobacco use—United States, 1990–1999. *Morbidity and Mortality Weekly Report* 48(43):986-993.

CDC. 2002. Cigarette smoking among adults—United States, 2000. *Morbidity and Mortality Weekly Report* 51(29):642.

CDC. 2008. Smoking-attributable mortality, years of potential life lost, and productivity losses—United States, 2000–2004. *Morbidity and Mortality Weekly Report* 57(45):1226-1228.

CDC. 2009a. *Tobacco use: Targeting the nation's leading killer—at a glance 2009.* National Center for Chronic Disease Prevention and Health Promotion, Office on Smoking and Health, Department of Health and Human Services.

CDC. 2009b. Cigarette smoking among adults and trends in smoking cessation—United States, 2008. *Morbidity and Mortality Weekly Report* 58(44):1227-1232.

CDC. 2010. Vital signs: Current cigarette smoking among adults aged ≥ 18 years—United States, 2009. *Morbidity and Mortality Weekly Report* 59(35):1135-1140.

CDC. 2011a. Ten great public health achievements—United States, 2001–2010. *Morbidity and Mortality Weekly Report* 60(19):619-623.

CDC. 2011b. Cigarette smoking—United States, 1965–2008. *Morbidity and Mortality Weekly Report* 60(1):109-113.

Cohen, J. E., M. Zeller, T. Eissenberg, M. Parascandola, R. O'Keefe, L. Planinac, and S. Leischow. 2009. Criteria for evaluating tobacco control research funding programs and their application to models that include financial support from the tobacco industry. *Tobacco Control* 18(3):228-234.

Doll, R., and A. B. Hill. 1950. Smoking and carcinoma of the lung; preliminary report. *British Medical Journal* 2(4682):739-748.

Doll, R., and A. B. Hill. 1956. Lung cancer and other causes of death in relation to smoking. *British Medical Journal* 2(5001):1071.

Giovino, G. A. 2007. The tobacco epidemic in the United States. *American Journal of Preventive Medicine* 33(6, Suppl. 1):S318-S326.

Halpern-Felsher, B. L., M. Biehl, R. Y. Kropp, and M. L. Rubinstein. 2004. Perceived risks and benefits of smoking: Differences among adolescents with different smoking experiences and intentions. *Preventive Medicine* 39(3):559-567.

Hatsukami, D. K., G. A. Giovino, T. Eissenberg, P. I. Clark, D. Lawrence, and S. Leischow. 2005. Methods to assess potential reduced exposure products. *Nicotine & Tobacco Research* 7(6):827-844.

Hatsukami, D. K., N. L. Benowitz, S. I. Rennard, C. Oncken, and S. S. Hecht. 2006. Biomarkers to assess the utility of potential reduced exposure tobacco products. *Nicotine & Tobacco Research* 8(2):169-191.

Hatsukami, D. K., R. M. Feuer, J. O. Ebbert, I. Stepanov, and S. S. Hecht. 2007. Changing smokeless tobacco products: New tobacco delivery systems. *American Journal of Preventive Medicine* 33(6S):S368-S378.

Hatsukami, D. K., K. Hanson, A. Briggs, M. Parascandola, J. M. Genkinger, R. O'Connor, and P. G. Shields. 2009. Clinical trials methods for evaluation of potential reduced exposure products. *Cancer Epidemiology, Biomarkers & Prevention* 18(12):3143-3195.

HHS (U.S. Department of Health and Human Services). 1994. *Preventing tobacco use among young people: A report of the Surgeon General*. Rockville, MD: U.S. Department of Health and Human Services, Centers for Disease Control and Prevention, National Center for Chronic Disease Prevention and Health Promotion, Office on Smoking and Health.

HHS. 1998. *Tobacco use among U.S. Ethnic/racial and minority groups—African Americans, American Indians and Alaska Natives, Asian Americans and Pacific Islanders, and Hispanics: A report of the Surgeon General*. Atlanta, GA: U.S. Department of Health and Human Services, Centers for Disease Control and Prevention, National Center for Chronic Disease Prevention and Health Promotion, Office on Smoking and Health.

HHS. 2000. *Reducing tobacco use: A report of the Surgeon General*. Atlanta, GA: U.S. Department of Health and Human Services, Centers for Disease Control and Prevention, National Center for Chronic Disease Prevention and Health Promotion, Office on Smoking and Health.

HHS. 2006. *The health consequences of involuntary exposure to tobacco smoke: A report of the Surgeon General*. Washington, DC: U.S. Department of Health and Human Services, Centers for Disease Control and Prevention, National Center for Chronic Disease Prevention and Health Promotion, Office on Smoking and Health.

IARC (International Agency for Research on Cancer). 2007. *Smokeless tobacco and tobacco-specific nitrosamines*. Lyon, FR: IARC.

IOM (Institute of Medicine). 1994. *Growing up tobacco free: Preventing nicotine addiction in children and youths*. Washington, DC: National Academy Press.

IOM. 2001. *Clearing the smoke: Assessing the science base for tobacco harm reduction*. Washington, DC: National Academy Press.

IOM. 2007. *Ending the tobacco problem: A blueprint for the nation*. Washington, DC: The National Academies Press.

IOM. 2010. *Secondhand smoke exposure and cardiovascular effects: Making sense of the evidence*. Washington, DC: The National Academies Press.

Marlatt, G. A. 2002. *Harm reduction: Pragmatic strategies for managing high-risk behaviors*. New York: The Guilford Press.

Mermelstein, R. 2003. Teen smoking cessation. *Tobacco Control* 12(Suppl. 1):i25-i34.

Mokdad, A. H., J. S. Marks, D. F. Stroup, and J. L. Gerberding. 2004. Actual causes of death in the United States, 2000. *JAMA* 291(10):1238-1245.

O'Connor, R. J., K. M. Cummings, V. W. Rees, G. N. Connolly, K. J. Norton, D. Sweanor, M. Parascandola, D. K. Hatsukami, and P. G. Shields. 2009. Surveillance methods for identifying, characterizing, and monitoring tobacco products: Potential reduced exposure products as an example. *Cancer Epidemiology, Biomarkers & Prevention* 18(12):3334-3348.

Pierce, J. P., and E. Gilpin. 1996. How long will today's new adolescent smoker be addicted to cigarettes? *American Journal of Public Health* 86(2):253-256.

Robert Wood Johnson Foundation. 2011. *Interactive tobacco map provides latest data on state smoking laws*. http://www.rwjf.org/publichealth/product.jsp?id=56548 (accessed October 17, 2011).

Royal College of Physicians. 1962. *Smoking and health. Summary and report of the Royal College of Physicians of London on smoking in relation to cancer of the lung and other diseases*. New York: Pitman Publishing.

SAMHSA (Substance Abuse and Mental Health Services Administration). 2011. *Results from the 2010 National Survey on Drug Use and Health: National findings*. NSDUH Series H-41, HHS Publication No. (SMA) 11-4658. Rockville, MD: Substance Abuse and Mental Health Services Administration.

Secretan, B., K. Straif, R. Baan, Y. Grosse, F. El Ghissassi, V. Bouvard, L. Benbrahim-Tallaa, N. Guha, C. Freeman, L. Galichet, and V. Cogliano. 2009. A review of human carcinogens—part E: Tobacco, areca nut, alcohol, coal smoke, and salted fish. *Lancet Oncology* 10(11):1033-1034.

Shields, P. G. 2002. Tobacco smoking, harm reduction, and biomarkers. *Journal of the National Cancer Institute* 94(19):1435-1444.

U.S. Public Health Service. 1964. *Smoking and health: Report of the Advisory Committee to the Surgeon General of the Public Health Service*. Washington, DC: U.S. Department of Health, Education, and Welfare, Public Health Service, Centers for Disease Control.

Vezina, P., D. S. McGehee, and W. N. Green. 2007. Exposure to nicotine and sensitization of nicotine-induced behaviors. *Progress in Neuro-Psychopharmacology and Biological Psychiatry* 31(8):1625-1638.

Wynder, E. L., and E. A. Graham. 1950. Tobacco smoking as a possible etiologic factor in bronchiogenic carcinoma—a study of 684 proved cases. *JAMA* 143(4):329-336.

2

Governance and Conduct of Studies

Under subsection (l) paragraph (2), the Family Smoking Prevention and Tobacco Control Act of 2009 (FSPTCA)[1] specifically directs the Food and Drug Administration (FDA) to develop regulations or guidance in consultation with the Institute of Medicine (IOM) "on the *design and conduct* of such studies and surveillance" (emphasis added). The specific requirement to advise the FDA on guidance and regulation for the *conduct* of studies is significant, as *conduct* encompasses more than sound study design and research methods.

A proper study design can produce meaningful results, while an improper study design produces meaningless data. In contrast, the improper *conduct* of scientific studies may encompass not only poor study design and execution, but also unethical or illegal activity. Consequences of improper conduct, such as the falsification, manipulation, or destruction of research findings, not only result in a loss of trust and credibility, but also can result in significant harm. It is critical that all data submitted in support of modified risk tobacco product (MRTP) applications are developed, generated, analyzed, and presented in a way that protects and maximizes credibility, scientific rigor, and public trust.

The mandate to advise the FDA on the conduct of studies was viewed as particularly important by the committee, given the history of the tobacco industry's efforts to obscure the true health effects of smoking.

---

[1] *Family Smoking Prevention and Tobacco Control Act of 2009*, Public Law 111-31, 123 Stat. 1776 (June 22, 2009).

While the industry currently acknowledges the health risks of smoking, this history continues to affect the legitimacy of self-sponsored research associated with their products. To provide confidence in the face of the history of tobacco industry–sponsored and tobacco industry–conducted research, additional measures may be required beyond what otherwise might be expected of industries.

The mandate to advise the FDA on the conduct of studies presented a unique challenge to the committee. The committee concluded that it would be neither helpful nor adequate to simply rearticulate minimum standards for research conduct; the basic standards for the ethically and socially responsible conduct of science are well established. The committee felt strongly that mechanisms to enforce or otherwise affirm minimum standards for the conduct of studies should be addressed, and would be of much greater relevance to the FDA. As such, in this chapter the committee addresses not only the principles for ethical and proper conduct of research, but also the governance mechanisms to ensure the ethical and proper conduct of research as well.

This chapter begins with a brief retelling of the history of tobacco research. The next section explores how the absence of governance and a history of improper conduct have resulted in a situation where the tobacco industry currently lacks the ability to independently produce and disseminate comprehensive and credible data about tobacco products. The chapter concludes with a discussion of one or more independent organizations that may be needed for the governance of tobacco industry studies in support of applications to market MRTPs.

HISTORY OF SCIENTIFIC RESEARCH FUNDED OR CONDUCTED BY THE TOBACCO INDUSTRY

To provide proper context for the committee's recommendations regarding the design and conduct of studies to support the marketing of MRTPs, it is necessary to briefly review the history of, and lessons learned from, research conducted, funded, or supported by the tobacco industry and its affiliate organizations. An earlier report from the IOM provides a more thorough review of the history of tobacco harm reduction approaches and products (IOM, 2001), so the current section is designed to briefly review the major issues.

Historical Overview of Tobacco Harm Reduction

The issue of reducing the harm associated with tobacco use emerged very early in the growth of the cigarette market in the United States. In the 1930s and 1940s, before smoking-related health effects began to be widely publicized, a prominent focus of advertising campaigns was irritation, which served as a proxy for health concerns as it was linked to prevalent theories of cancer (Kozlowski and O'Connor, 2010). The mentholated cigarette brand Kool was marketed in ways that highlighted the "soothing" properties and claimed to help ease cold symptoms (Sutton and Robinson, 2004). After the publication of epidemiologic evidence of the harms of cigarette smoking (Doll and Hill, 1950, 1952, 1954; Wynder and Graham, 1950), filtered cigarettes were heavily promoted to smokers to allay health concerns. This resulted in the so called "tar derby" where manufacturers competed to win customers on the basis of lower reported tar and nicotine in cigarettes (Hoffmann and Hoffmann, 1997). On July 18, 1957, John Blatnik led 6 days of Congressional hearings on filtered cigarette advertising,[2] the first of its kind in exploring the marketing of tobacco products (Harris, 2011). These hearings revealed that much of this marketing was fallacious, in that filters were largely ineffective, and that tar and nicotine numbers were largely incomparable between brands because manufacturers used different testing methods. The Federal Trade Commission (FTC) and the industry came to an agreement to not use tar and nicotine numbers in advertising in 1960, and themes in cigarette advertising turned more toward lifestyle and imagery (Kozlowski and O'Connor, 2010). By the 1960s, the cigarette market had shifted toward filtered brands.

The demonstration that cigarette tar could induce cancer in animal models resulted in the identification of tar as the primary aspect of concern (Wynder et al., 1953). This led to a widespread belief that reducing exposure to "tars" and nicotine would mitigate some of the associated health risks. Early epidemiologic findings appeared to support this view, inasmuch as those who used filtered brands were somewhat less likely to develop lung cancer (Wynder and Stellman, 1979). At the time, however, it was not broadly accepted that smoking was driven by nicotine addiction, nor that smokers might adjust their smoking behaviors to maintain their accustomed nicotine doses (NCI, 2001). In the early 1960s, the FTC began working with industry to refine a test method to compare brands, and this was implemented beginning in 1964 (NCI, 1997). Many public health advocates and institutions, including the National Cancer Institute (NCI), believed that publicizing tar values and switching to lower tar cigarettes would generate a public health gain (Hoffmann and Hoffmann, 1997). Tragically, rather than decreasing disease risk, the use of these products likely had a profoundly negative effect on the public's health (NCI, 2001).

---

[2] U.S. Congress, House of Representatives. Subcommittee of the Committee on Government Operations. *False and misleading advertising (filter-tip cigarettes)*. 85th Cong., 1st Sess. July 18–26, 1957.

In 1968, in response to calls for greater attention to the problem of lung cancer associated with smoking, the NCI established the Tobacco Working Group, an advisory group to establish a research agenda on the development of less hazardous cigarettes (Parascandola, 2005a, 2005b). This working group comprised members from government, academia, and the tobacco industry. From 1968 through about 1980, the Smoking and Health Program spent more than $50 million in research, 74 percent of which was directed toward chemical and biological assays of cigarette prototypes (Parascandola, 2005a). Documents revealed that the industry members of the working group helped to set research priorities and steer research findings and reporting (Parascandola, 2005b).

Internal tobacco industry documents provided insight into the industry's knowledge of the relationships between cigarette smoking behaviors and cigarette design. The industry's greater knowledge of human smoking behavior allowed for the design of "elastic" products, from which different amounts of smoke (and nicotine) could be extracted to satisfy consumer needs (Hammond et al., 2006; Kozlowski and O'Connor, 2002). Tobacco companies used many techniques to continue the appearance of relatively healthier cigarettes (Pollay and Dewhirst, 2002). Product features intending to lower toxin yields, as measured by the FTC machine, included air dilution and the reduction of tobacco density (NCI, 2001). The dilution of mainstream smoke by air could be accomplished in a number of ways, including increased paper porosity and diffusivity, porous tipping, and the inclusion of ventilation holes in the filter. These features acted to increase burn rate and to reduce the concentration of smoke taken at the tip. The reduction of tobacco density was achieved through engineered tobaccos, such as "expanded" tobacco, which was essentially "puffed" using gases to decrease density. This modification was advantageous for tobacco companies because less dense cigarettes burn more quickly when left in smoking machines, meaning that the measured tar yields were reduced by virtue of decreasing the number of puffs taken. Filter ventilation, however, was the key feature that drove cigarette elasticity. Ventilation holes were often placed in locations that are likely covered by the smoker's lips or fingers. Because they also acted to cool smoke and reduce the puff concentration, they also served to make the smoke taste and feel lighter to the smoker (Kozlowski and O'Connor, 2002). Finally, the inclusion of ventilation holes reduced resistance to draw, which in turn made it easier for smokers to draw more smoke from the cigarette for a given amount of puffing effort (Kozlowski and O'Connor, 2002; NCI, 2001). Some products were explicitly designed to be highly elastic, including the product Barclay, introduced by Brown & Williamson in the early 1980s. The filter design used grooved air channels that made it very easy for smokers to compensate, while giving very low yield for tar, nicotine, and carbon monoxide. The FTC eventually ruled the FTC method did not accurately measure the brand's delivery (Kozlowski et al., 2005).

Initially, lower-tar cigarettes were marketed as distinct brands. However, this changed when Philip Morris introduced Marlboro Lights in 1972, beginning a broad trend toward product line extensions (Pollay and Dewhirst, 2002). Line extensions carried associations with the parent brand (e.g., taste, quality) and likely attracted more smokers to switch to lower-tar cigarettes. The history of light and low-tar cigarettes shows that tobacco harm reduction research—despite its potential to reduce risk—deserves careful and unique consideration because the migration of smokers toward lower-yield cigarettes has not improved either the health of individuals or the public (Harris et al., 2004; NCI, 2001; Thun and Burns, 2001).

In the 1980s, industry research and development turned toward more radical reengineering of products, in part as a result of rising litigation risk and in response to a growing smoke-free environments movement. In 1989, R.J. Reynolds introduced Premier, which it claimed as a smokeless cigarette. This product was withdrawn and later reengineered as Eclipse, which continued to be sold until 2008. Philip Morris tested various versions of an electrically heated cigarette smoking system, which used an external heating element to heat tobacco on specially designed cigarettes to produce smoke. The Accord, the first such system, was introduced in 1990, and the most recent incarnation was the Heatbar, tested by Philip Morris International in Switzerland from 2006 to 2009. Other approaches focused on applying technology to selectively reduce toxicants in traditional cigarette designs (e.g., Advance, Marlboro UltraSmooth, and Omni).

Early in the 2000s, evidence began to emerge from Sweden that showed dramatic reductions in smoking-related disease coincident with a rise in the use of snus, a form of moist smokeless tobacco (Foulds et al., 2003; Henningfield and Fagerstrom, 2001). Snus, as produced in Sweden, was regulated as a food product and thus subject to quality controls that led manufacturers to reduce levels of toxicants such as nitrosamines and heavy metals. These data encouraged some in tobacco control that harm reduction was possible if smokers could be convinced to adopt use of smokeless tobacco (Kozlowski, 2007), while others raised serious concerns about unintended consequences (Tomar et al., 2009). This message was seized upon in the United States, where smoke-free restrictions were growing, and existing smokeless tobacco companies began to aggressively court smokers. By 2009, the two major smokeless tobacco companies—Conwood Sales Company, LLC, and U.S. Smokeless Tobacco Company—had been purchased by the leading cigarette manufacturers (R.J. Reynolds Tobacco Company and Philip Morris, respectively), horizontally integrating the tobacco market. Both companies introduced forms of snus into the U.S. market, carrying cigarette brand names (Marlboro and Camel).

Other tobacco products have also been promoted as having potential for harm reduction. In 2001, Star Scientific introduced dissolvable tobacco products (Ariva and later Stonewall), lozenges made from powered tobacco that would be used orally and disintegrate. In 2009, R.J. Reynolds followed suit with Camel Strips, Sticks, and Orbs, all different configurations of dissolvable tobacco. In 2006, electronic nicotine delivery systems (commonly referred to as e-cigarettes) emerged; these products have a physical form that resemble a traditional cigarette, but they use electrical heating elements to vaporize a nicotine containing glycerol solution. Some scientists have suggested these products hold promise for harm reduction, if subject to proper testing, regulation, and quality control (Etter et al., 2011).

Scientific Research Conducted, Funded, or Supported by the Tobacco Industry

In examining scientific standards for the design and conduct of studies related to MRTPs, an additional relevant consideration is the past behavior of the newly regulated industry. Cigarette manufacturers stated for more than six decades, either implicitly or explicitly, that cigarettes were not dangerous to health (Cummings, 2003; Cummings et al., 2002). However, industry officials and tobacco scientists were aware of smoking's relationship to cancer risks as early as the 1940s, with broad internal acceptance seen by the late 1950s (Cummings et al., 2007). The wide discrepancy between internal knowledge and public posturing required efforts to maintain a perception among the general public and policy makers that scientific controversy still surrounded the relationship between smoking and health, and scientific research was essential to this.

Like most industries, tobacco manufacturers have maintained significant research and development arms, with a significant portion focused on product development and testing. Much of this was directed toward optimizing products in terms of taste and nicotine delivery (Carpenter et al., 2007; Cook et al., 2003; Harris, 2011; Hurt and Robertson, 1998; Megerdichian et al., 2007; Wayne et al., 2004). However, the tobacco industry has also engaged in health-relevant research on its products, including nicotine self-administration (DeNoble and Mele, 2006), mental illness (Hirshbein, 2011), and the composition and toxicity of secondhand smoke (Schick and Glantz, 2007a). Philip Morris determined that cigarette filters released inhalable fibers, yet never reported this to consumers (Pauly et al., 2002). Documents reveal that lawyers exerted considerable control over internal research, primarily to guard against product liability lawsuits

(Hanauer et al., 1995). Industry scientists did publish selected internal research, sometimes in the form of monographs or conference proceedings (Dunn, 1973; International Smoking Behaviour Conference and Raymond E. Thornton, 1978), and toxicological and chemical research was often published over the years. In addition to the recognized tobacco-specific journals *Tobacco Science* and *Beitraige zur Tabakforschung [Contributions to Tobacco Research]*, tobacco industry scientists and consultants served on editorial boards of a number of scientific journals, including *Indoor and Built Environment*, *Inhalation Toxicology*, *Regulatory Toxicology and Pharmacology*, *Mutagenesis*, and the *Journal of Clinical Epidemiology* (Bitton et al., 2005; Garne et al., 2005).

A number of companies also sponsored external research, and a review of documents found that tobacco industry lawyers, rather than scientific merit, heavily influenced the selection of external research programs with the intent to improve public relations, divert public focus away from the negative health consequences of tobacco use, and influence policy (Bero et al., 1995). The tobacco industry's scientific consulting program on secondhand smoke was largely attorney managed and intended to sway public opinion, but it also influenced funded scientists in terms of how they could express their research in public debates and conferences (Muggli et al., 2003). Further analyses of documents show a 40-year effort by Philip Morris USA to fund and influence the work of Dr. Ernst Wynder, a highly respected researcher on smoking and health (Fields and Chapman, 2003). The industry sought to fund research into alternative explanations of smoking-health links, including genetics (Gundle et al., 2010), stress (Landman et al., 2008; Petticrew and Lee, 2011), personality factors (Eysenck, 1991), and environmental pollution. Cataldo et al. (2010) describe industry efforts to gain control of the Framingham heart health cohort study by funding its principal investigator, so as to gain access to the dataset to produce favorable reanalyses questioning the link between smoking and heart disease. Research on secondhand smoke and health was designed and analyzed so as to achieve favorable conclusions (Barnes et al., 2006; Neilsen and Glantz, 2004; Ong and Glantz, 2000; Schick and Glantz, 2005; Tong et al., 2005; Yano, 2005). Other studies have shown that research funded by industry tended to come to different conclusions about secondhand smoke health effects (Barnes and Bero, 1998) and the economic impacts of smoking restrictions (Scollo et al., 2003).

The Tobacco Industry Research Committee, which was created in 1954 in response to emerging evidence of smoking-related cancer risks, later subdivided into the Council for Tobacco Research (CTR), which funded research, and the Tobacco Institute (TI), which focused on lobbying and communications. While the CTR existed, nominally, to fund independent research into smoking and health, it was part of the broader public relations approach to questioning the validity of smoking-health links. Internal documents show that the organization was controlled by industry lawyers and funded special projects to favored scientists who would reliably cast doubt on smoking-disease claims (Bero et al., 1995). Similarly, the Center for Indoor Air Research (CIAR), organized in 1988, funded external peer-reviewed research as well as special projects. The CIAR was formed in response to growing calls to limit indoor smoking, and in particular the 1986 Surgeon General's report on involuntary smoking. Barnes and Bero (1996) examined the CIAR's project portfolio and showed that while 70 percent of the peer-reviewed projects were on topics not related to secondhand smoke, 63 percent of the special projects were related to secondhand smoke. Furthermore, while only 2 percent of the peer-reviewed projects had what could be termed "pro-industry" conclusions, the special projects showed 29 percent in favor of industry (Barnes and Bero, 1996). The industry also supported the Associates for Research into the Science of Enjoyment (ARISE), an organization created in 1988 in direct response to the classification of nicotine as an addictive drug by the U.S. Surgeon General (Landman et al., 2008; Smith, 2006). ARISE aimed to tout the health benefits of the use of legal substances such as tobacco in terms of stress relief and performance enhancement, and received more than 90 percent of its support from the tobacco industry. While ARISE did not sponsor research, it did organize symposia, conferences, and publications that served to disseminate its members' research (many of whom were funded by the tobacco industry). The Master Settlement Agreement dissolved the TI, CTR, and CIAR in 1998.

Racketeer Influenced and Corrupt Organizations (RICO) Findings

In 1999, the federal government filed against the tobacco industry (Phillip Morris USA, R.J. Reynolds, Brown & Williamson, British American Tobacco, Lorillard, and Liggett) for violating the Racketeer Influenced and Corrupt Organizations (RICO) Act. District Judge Gladys Kessler stated in her 2006 findings of fact that the tobacco companies "conspired together to violate the substantive provisions of RICO."[3] A key element furthering the conspiracy was the coordination of research activities (such as those described above) designed to cast doubt on the health risks of smoking. Kessler noted in the findings of fact that

> Defendants attempted to and, at times, did prevent/stop ongoing research, hide existing research, and destroy sensitive documents in order to protect their public positions on smoking and health, avoid or limit liability for smoking and health-related claims in litigation, and prevent regulatory limitations on the cigarette industry.

---

[3] *United States v. Philip Morris USA, Inc., et al.*, 449 F. Supp. 2d 1, 26 (D.D.C. 2006).

Although the government was not permitted to recover monetary damages (disgorgement of illegal profits, which were estimated at $280 billion), the defendants were ordered to engage in corrective advertising, remove misleading labels from products, and submit to judicial oversight. The ruling has survived several levels of appeal. Most recently, Philip Morris et al. have argued that FDA regulation is a sufficient deterrent to future violations, and thus the RICO case should be vacated. In a recent opinion, Judge Kessler noted that FDA regulation was unlikely to deter defendants' future bad acts because they were vigorously fighting the regulations via other court cases and regulatory challenges.[4]

Conclusions on Scientific Research Funded or Conducted by the Tobacco Industry

The history of research conducted, funded, or supported by the tobacco industry is not raised to be retributive or punitive, but simply to acknowledge that past actions reflect on the credibility of the industry's current research, which may pose a problem for regulators, particularly in the contentious area of MRTPs.

An additional concern is that any perception of cavalier attitudes to tobacco research may tarnish the reputation of the FDA itself. The tobacco control statute places a high-capacity and historically well-trusted agency in the practice of regulating a commodity quite different from the products historically under its purview. The FDA carries a near-unique stature in the degree of public trust it has received, and there are plausible reasons to believe that this reservoir of public trust has imparted stability to the agency and has rendered its difficult combination of tasks easier (Carpenter, 2010).

If data generated for the FDA by tobacco companies is perceived to lack credibility, the FDA could in general, and the Center for Tobacco Products (CTP) could in particular, find its reputation, its scientific credibility, and its public trust severely compromised and perhaps irreversibly damaged. This reputational damage to the FDA and to public health institutions is a critical issue. Concerns about problematic data have surfaced occasionally in the past with the pharmaceutical industry; there is little reason to think that such scandals will not arise with the tobacco industry. Yet given any scandal, the consequences of the perception that the FDA and the CTP wrongly trusted tobacco industry claims will be far worse in terms of public, scientific, and legislative credibility.

These concerns are not isolated; similar concerns have been raised by the National Advisory Council on Drug Abuse (NACDA). In providing guidance to the National Institute on Drug Abuse (NIDA) about provid-

[4] *United States v. Phillip Morris, Inc.* 2011 U.S. Dist. (D.D.C., June 22, 2011).

ing research funding to potential grantees who also receive money from the tobacco industry, the NACDA made several points to consider, including that receiving funding from the tobacco industry could compromise perceived objectivity and credibility of research, and that "any connection between tobacco industry supported research (or tobacco industry scientists) and NIDA could negatively impact NIDA's credibility and the public's trust in NIDA funded research" (NIDA, 2011).

RELEVANCE OF THIRD-PARTY GOVERNANCE

The conduct of tobacco product research presents a case that is unique from other FDA-regulated commodities. First, there is profound public distrust in both the tobacco industry and in the research it sponsors. Since the 1960s, public trust in tobacco companies has been at historical lows compared to virtually all other institutions or industries (Ashley and Cohen, 2003; Harris Interactive, 2003; NCI, 2008), and these patterns have not abated in recent years (Harris Interactive, 2010). Prior to the FSPTCA, commercial tobacco products were not regulated by the FDA (White et al., 2007). As a result, compared to other industries that develop products also requiring premarket approval (the drug and device industries), the tobacco industry does not possess, and will not possess for some time to come, the same degree of organization; accepted measures, methods, and models; and routine involvement and consultation of qualified experts.

The fundamental problem that confronts the FDA is a critical shortage of credible and reliable evidence about the effects of MRTPs. The history of public distrust and the absence of governance in the tobacco industry have created an isolated industry that lacks not only the expertise to produce the necessary range of credible and reliable data, but also the trustworthiness to acquire external expertise and avenues to disseminate acquired data. The committee also recognizes that other industries, including the pharmaceutical and device industries, may develop and sponsor MRTP candidates, and although these institutions should also be held to high standards for the design and conduct of studies, they may not have to overcome the same hurdles in maintaining or restoring credibility to their research.

Role of Governance in Sustaining Credibility in Tobacco Industry Research

The idea that research on commercial products that carry public health risks should be supervised, funded, or structured by independent entities has important precedents and models (Marks, 1997). When pharmaceutical and medical product companies engage universities, medical schools, or research hospitals to conduct research, the institutions conducting the research studies contractually embed research autonomy into the funding arrangements, and all such studies in human subjects are approved by institutional review boards (IRBs) before they commence. So too, human subjects research is overseen by the National Institutes of Health, the Office of Human Subjects Research of the Public Health Service, and the FDA itself. This is true of all researchers receiving public funds and those conducting research on an FDA-regulated product (FDA, 2010; HHS, 2009; White et al., 2007). Academic and medical journals also exercise a gatekeeping and oversight role for clinical research with human subjects.

The production of reliable and credible data depends upon building rigor, oversight, and independence into the entire research process. It is well recognized that data problems often cannot be detected after study completion; therefore, integrity and accountability need to be built into the research throughout the study's execution. For balanced and rigorous evaluation of data in support of any marketing application, the FDA has traditionally expected or required independent oversight. Unlike the tobacco industry, clinical research models in the pharmaceutical industry were developed in academic medicine and pharmacology circles in the 20th century, with significant input from pharmaceutical industry partnerships, which had from the 1940s onward sought pharmacological, statistical, and other medical expertise for the improvement of their experimental methods (Marks, 1997). Various officials and bureaus of the FDA itself also participated in the modernization of the research paradigm in pharmaceuticals, either through regulation or through advisory or participatory roles (Carpenter, 2010). The credibility of data in support of new pharmaceutical products or medical devices is, in other words, supported by a national and global infrastructure of research that has taken decades to evolve, and even now it is not free of problems.

The FSPTCA places the CTP in a difficult position. The center will now be regulating tobacco companies as product sponsors, without the long-run institutional knowledge of these companies that is gained through decades of regulation and oversight. There is not an established set of regulatory practices for the review of MRTPs, nor is there an established set of federal research standards for the design, conduct, analysis, monitoring, and completion of studies in support of MRTPs. Development of the "clinical trial" industry for MRTPs is, in a sense, being initiated in the next few years, because current tobacco industry practices suggest a degree of immaturity in the development of methods, measures, and standards (Rees et al., 2009).

Although industry- and company-sponsored studies are very common, they have been largely unregulated in the way pharmaceutical trials have been, and they lack the same level of oversight, governance, and rigor. A related point is that as major academic journals increasingly refuse to publish tobacco industry–funded research, they do not provide their traditional gatekeeping or oversight role via peer review. This hypothesis gains credence from studies of tobacco industry research, including research done by some of the largest and most established companies, where independent researchers have found significant problems with governance. In one examination of more than 73 different studies with human subjects conducted by R.J. Reynolds from 1985 to 2000, White et al. (2007) report that "in *all* 73 studies, [informed] consent procedures failed to meet *five or more human subjects research standards*" (emphasis added). Although R.J. Reynolds formed a human subjects review committee in 1985, the authors conclude that "the committee's structure and procedures did not meet generally accepted practices of the time regarding community representation, written procedures for adverse events, and other factors" (White et al., 2007).

Similarly, in a December 2009 review of industry research on potential reduced-exposure products (PREPs), Rees et al. (2009) suggest that the industry is catching up to clinical methodology standards now broadly accepted in the academic and medical realms. Basic good research practices such as documentation of data and analysis appear to be lacking from internal industry records, as well as cutting-edge methods of trial design, adaptation of design to hypothesis, and statistical analysis (Rees et al., 2009). Furthermore, switching paradigms that accommodate dual use of a PREP and conventional product, and switching to nicotine replacement therapy or cessation were not observed (Rees et al., 2009). As PREP assessment methods continue to be refined, such methods have become increasingly important to independent investigators. Clinical trial methods need to reflect real-life use patterns within the context of a research study, including *ad libitum* use of a PREP alone or in combination with conventional products, as well as employment of rigorous controls such as nicotine replacement therapy or forced switching conditions. Perhaps the narrow objective of demonstrating reduced exposure risk compared with a conventional product in support of product claims has constrained the scope of clinical research methods used by the industry.

Because the tobacco industry is currently limited in its ability to produce credible and comprehensive data, at least part of the research base in support of an MRTP may need to be generated by researchers and organizations independent of the sponsors of the MRTP in question. Rees et al. (2009) conclude that "research independent of the tobacco industry is essential to provide an effective and unbiased evaluation of industry claims" and note that "claims for PREPs, both implied and explicit, must ultimately be evaluated independently, by the broader scientific community, using validated assessment strategies and accepted clinical methodology." Such independent research oversight would support the generation of credible and scientifically rigorous data that meet the unique challenges that tobacco product research presents.

Conduct and Publication of Tobacco Industry Research

The FSPTCA requires product sponsors to provide evidence that the issuance of an order for the sale of an MRTP will benefit public health, including the effect of marketing the product on users and nonusers of tobacco products. As discussed in later chapters, an essential element in establishing the public health benefit of an MRTP is assessing the effect of the product on high-risk populations, in particular adolescent populations. As such, it is inevitable that product sponsors will need to collect extensive data on the effect of products in these populations in both pre- and postmarket studies. This poses a significant problem to product sponsors, because the tobacco industry currently lacks expertise and experience in conducting behavioral and addiction research in high-risk populations. Recognizing the risks involved, some tobacco companies appear resistant to the notion of conducting the research themselves. This issue was specifically discussed with industry representatives during an open meeting of the committee in May 2011. Representatives from multiple tobacco companies acknowledged that while research on adolescent populations is relevant to support an MRTP application, the companies were at that time reluctant to commence such research and were seeking guidance from the FDA on how best to proceed. In a personal communication, Lars Erik Rutqvist, Senior Vice President of Scientific Affairs of Swedish Match, indicates that industry is very unlikely to conduct research on "sensitive subpopulations such as adolescents . . ." because of ". . . ethical and product stewardship concerns."[5] If the position of Swedish Match is generally representative of the tobacco industry, then the risks and issues inherent in research on adolescent and other high-risk populations seem likely to dissuade most tobacco companies from conducting the research themselves. Without a framework that allows the industry to fund independent investigation on adolescents and other high-risk populations, it is likely that major gaps in knowledge about MRTPs will remain.

To assess the health impacts of an MRTP, product sponsors may have universities and research hospitals conduct the requisite studies with tobacco-industry and MRTP-sponsor funding. There are at least two problems with a university- or hospital-based research model with which the FDA and the scientific community may need to grapple. Firstly, many

[5] Personal communication, Dr. Lars E. Rutqvist, Swedish Match AB, Stockholm Sweden, August 5, 2011.

universities disallow tobacco industry funds in support of research on tobacco or tobacco products. As of March 2007, more than 20 academic institutions in the United States instituted policies banning tobacco industry funding of tobacco research. Secondly, even if a university permits tobacco-funded research on its campus, it does not ensure the resulting research will be widely trusted or considered valid by the broader scientific community. The CTP will wish to avoid a regime where product sponsors simply "forum shop" or "venue shop" for those institutions—characterized by a least common denominator of standards—that will permit industry-funded research on tobacco claims. One way of doing so would be to prescribe that whenever a tobacco company contracts with a university to conduct tobacco industry–funded research in support of an MRTP application, it must include in the contract certain essential conditions designed to assure the independence, integrity, and transparency of research.

Similarly, many medical and scientific journals have refused, and will continue to refuse, to publish research funded by tobacco companies or affiliated foundations or institutes. The passage of the FSPTCA will not alone change this fact. If the research supporting MRTP claims is of sufficient academic quality for publication in an academic journal, this refusal of journals to publish tobacco-sponsored research may be de facto prohibitive. Alternatively, if the CTP requires any substantial part of the portfolio of research supporting an MRTP application to consist of actual published research, it may be difficult for sponsors to meet this standard. If a governance framework is created that fosters credible and trustworthy tobacco research, journals may be willing to reconsider acceptance of tobacco industry–sponsored manuscripts.

An additional concern relates to the experience and qualifications of investigators conducting tobacco research. Use of unqualified or inexperienced investigators not only increases the risk for poorly conducted research, but also undermines the credibility of the research findings and research sponsor. Furthermore, use of unqualified and inexperienced investigators may expose research participants to greater risks for harm. It is in the best interest of all stakeholders involved in the evaluation of an MRTP to maintain high standards for the qualifications of investigators. This is embodied in the FSPTCA: according to Section 911(i)(2), the qualifications and experience of investigators conducting postmarket surveillance of MRTPs must be reviewed and approved by the U.S. Department of Health and Human Services (HHS) Secretary. The credibility of the investigators is equally important. Investigators should be free from real or perceived conflicts of interest and biases. It is critical that the investigators involved in research in support of MRTPs and potential MRTPs have adequate qualifications, experience, and credibility.

Another critical component of the FSPTCA assures that MRTP sponsors make and follow through with commitments to design, conduct, and report on postmarketing studies with thoroughness and diligence. The commencement and completion of postmarketing studies has long been a difficult area of regulation for the FDA, especially for Phase IV studies in the area of prescription drug regulation. These studies have often been slow to be completed and in some cases tardy to commence, and a number of independent entities have expressed their concern about the FDA's ability to commit product sponsors to finish these studies with the due diligence the law requires (Glasser et al., 2007; HHS, 1996; U.S. Government Accountability Office, 2009; Wood et al., 1998). This has been partially addressed by the Food Drug Administration Amendments Act,[6] which grants the FDA authority to regulate Phase IV studies and apply penalties if they are not conducted in a timely fashion. Like other features of FDA regulation (e.g., drugs with accelerated approval based upon studies using surrogate endpoint measures) the marketing approval for an MRTP claim under Section 911 of the FSPTCA is, according to law, a conditional and provisional approval. In accordance with the FSPTCA, when the FDA approves an MRTP, it will plan a series of postmarketing studies designed to address questions that were not fully answerable at the premarket stage. An independent tobacco research governance entity (TRGE) can play an important role in the design of these studies and in the monitoring of their completion.

Ethical Considerations of Tobacco Research

Robust standards for the ethical conduct of research have been developed to guide studies that involve human participants. Prominent examples include the Nuremburg Code, the Belmont Report, the Declaration of Helsinki, and the International Conference on Harmonisation Guideline for Good Clinical Practice. These documents, and in particular the Belmont Report, inform the federal regulations for the protection of human research participants, collectively known as the Common Rule (specifically 21 Code of Federal Regulations [CFR] parts 50 and 56 for FDA regulations). The committee affirms the protections enforced by the Common Rule as requisite in all tobacco studies that involve human participants. In addition to the basic protections afforded by the Common Rule, the committee identified several ethical issues in tobacco research worth particular attention.

The first issue is the risk of conducting clinical trials of MRTPs or other tobacco products in populations with a high risk for tobacco initia-

[6] *Food and Drug Administration Amendments Act of 2007*, Public Law 110-85, 121 Stat. 823 (September 27, 2007).

tion and addiction, including but not limited to adolescents, certain ethnic minorities, and individuals with mental health disorders. Randomization of participants to a product known to be potentially addictive and hazardous is ethically problematic. The committee maintains that the only circumstances under which an MRTP candidate should be provided to high-risk individuals is when (1) the individual is a current user of conventional tobacco products, (2) the individual does not want to quit using tobacco products or the individual wants to quit using tobacco products, but is unable to quit, (3) the MRTP candidate is not more hazardous than conventional tobacco products, and (4) at the end of the trial the individual is offered the best available treatment option for tobacco cessation.

A related issue is research involving individuals who do not use tobacco products or tobacco product users who are on the verge of quitting. There are certain groups of people, such as adolescents or individuals who are tobacco naive, who are at risk for starting to use an MRTP, and who may be especially vulnerable to developing nicotine dependence. Data on their initial or early reactions to the use of such products are relevant to an estimation of public health risk. In fact, a comprehensive analysis of potential public health impact demands that their vulnerability to chronic MRTP use (and, subsequently, other tobacco use) be empirically or experimentally addressed. However, there are clear potential risks to providing an MRTP or other tobacco products to such populations: e.g., experimental use might foster addiction and life-long use, with all its negative consequences. A decision to engage in research with such populations would, therefore, require the careful consideration of the potential risks and benefits. Generally, the committee concludes that

a. Research involving users of tobacco products is ethically permissible as long as the exposures in the study are not more risky than the hazards from their current tobacco use (i.e., the MRTP being tested is less dangerous than a cigarette for a smoker). Also, standard of care cessation treatment should be made available.

b. Survey research or perception/messaging research among nonsmokers is acceptable where the nonsmokers are not being exposed to the product.

c. Experimental research that exposes nonusers to products is ethically problematic; but such research cannot completely be ruled out because it could provide critically valuable information. The ethics, risks, and benefits need to be determined on a case-by-case basis.

Although practically challenging and ethically problematic, research involving high-risk populations is essential to ascertain the characteristics and mechanisms that make them more susceptible to tobacco use. Understanding these characteristics and mechanisms can help estimate the effect of marketing MRTPs, and can inform interventions to reduce the rates of tobacco use in these populations.

The third issue is the risk of improperly disclosing the substance abuse of a minor to the minor's parents or guardians in the process of obtaining parental consent for research. Generally, it is critical that a minor's assent and parental consent be obtained prior to any research involving the minor. However, there are circumstances where obtaining parental consent for the minor's participation in research will disclose information about the minor's behavior, including potentially illegal behavior. Disclosure of this information is problematic because it may result in a number of unwanted consequences for the minor. While the assent of minors is always necessary, investigators should also be cognizant of circumstances where obtaining parental consent will violate the confidentiality of the study participant, and where waiver of parental consent is warranted.

A last issue is the inclusion of individuals from high-risk groups with reduced decision-making capacity. Some populations at a high risk for tobacco use, such as adolescents and populations with mental health issues, may have a higher prevalence of individuals with reduced decision-making capacity. When investigators are conducting research involving these high-risk groups, they should be particularly cautious about the inclusion of individuals who lack the capacity to provide meaningful consent.

TOWARD A TOBACCO RESEARCH GOVERNANCE ENTITY

To improve the credibility of the studies in support of an MRTP application under Section 911, tobacco product sponsors and the CTP should consider facilitating the creation of a third party or third parties for the conduct and oversight of these studies. The committee will not recommend a specific model for adoption, but it will instead discuss existing arrangements in other fields and the general properties of a governance entity that would be desirable or appropriate in the MRTP field.

Health Effects Institute Model and Other Potential Organizational Models

The idea for an independent research entity in a contentious area of research on public health risks is not new. In 1980, the Environmental Protection Agency (EPA) and the automobile industry agreed to create a new governance and research organization to resolve conflicts over research on health and air quality. The EPA and automakers had clashed over standards on which the federal government wrote rules enforcing the National Ambient Air Quality Standards and the tailpipe emissions standards of the late 1970s. Because there was little agreement on the science supporting air quality regulations, it was increasingly difficult for dialogue between industry and regulators to proceed. Led by the efforts of Cummins Engine executive Charles Powers and EPA official Michael Walsh, the Health Effects Institute (HEI) was created in 1980 (Jasanoff, 1990; Keating, 2001).

The HEI is a nonprofit corporation with approximately one-half of its funding coming from the automobile industry and the other half coming from the federal government and other government sources. HEI is based in Boston, Massachusetts, and is well situated among the top research universities and hospitals. A number of scholarly analyses have described the HEI as having successfully managed the boundary between industry and government, as well as between the research community in health effects and the research community in air quality (Keating, 2001).

The HEI has multiple roles, including the funding of research through competitive requests for applications (RFAs). These RFAs serve, like those developed by a grant agency, to create open competition. Such open competition ensures that research funds will not be directed consistently or privately to those recipients most likely to produce certain outcomes, and it also promotes implicit competition among researchers on the basis of research quality and rigor rather than upon loyalty to the financial sponsor.

The relevant organizational structure of the HEI includes a board of directors and three committees. The board of directors is independent of the sponsors of the institute, and it "acts as the principal guardian of the HEI's objectivity" (quoted in Keating, 2001). The board monitors potential conflicts of interest and oversees the institute's staff, checks appointments to its expert committees, and monitors sponsored research projects. Below the board rest three committees. The Health Research Committee develops 5-year plans for research and awards research funds to investigators. The Health Research Committee also oversees research investigators and their work. Independently of the Health Research Committee, the Health Review Committee evaluates research produced by HEI-funded investigators and interprets the meanings of such research for policy makers. A third committee, the special Committee on Emerging Technologies, examines new fuels and their potential environmental and health impacts. A key feature of this structure is the independence of the board of directors from the sponsors and from the staff, and the independence of the two principal committees from one another (Keating, 2001).

The Health Research Committee develops and publishes the RFAs through which competitive research is applied for and ultimately funded. Project selection is undertaken by expert panels that convene under the authority of the committee and review and rank applications. The committee and HEI staff often work with the sponsor of the successful application to refine the scope and methods of the research project, examining research design, methods of analysis, and data. When research commences, the research committee oversees the research, reviewing progress reports from the investigators, overseeing quality audits of the project research, and visiting the investigators' research sites.

There are important limits to the HEI model that must be considered when thinking about it as a possible prototype for a TRGE. Perhaps the most important difference between the HEI and any TRGE is that the HEI does not fund projects in support of marketing applications; rather, it funds projects that contribute to general knowledge. Hence the commercial stakes of the research funded by the HEI may be somewhat less than the kind of research that could be funded by a TRGE. In particular, it may be problematic for individual tobacco companies to contribute funds to a TRGE if those funds will be used to fund research that potentially benefits a competitor's product more than its own product. As such, it will be important to distinguish between two different types of research: (1) individual product testing and (2) research that contributes to general knowledge, including research on better methods for product testing. Institutions like the HEI may be better suited to develop study methods or standards, rather than individual product evaluation. It should also be noted that the public health standard articulated in the FSPTCA is different from any other existing premarket approval standard. Additionally, the level of public, medical, and academic distrust in the tobacco industry and its research is much greater than any that has ever buffeted the automobile industry. Important issues of trust would need to be confronted in order for any such model to be entertained.

Another possibility for an organizational model lies in the Reagan-Udall Foundation (RUF), which advises the FDA on modernizing regulatory science. It conducts and oversees studies on regulatory science, particularly in the emerging fields of pharmacogenomics and genomic-based prediction of drug response and adverse event risk. The RUF receives grants from independent foundations for its work in advancing regulatory science, ranging from work in systems toxicology funded by the Komen Foundation to work on antitubercular drugs in the critical path to tuberculosis drug regimens. The RUF has a board of directors composed of a diverse mix of consumer representatives, industry representatives, scientific and medical authorities, and government officials; none of these groups accounts for a majority of the board's members. The foundation has implemented a number of strategies to attempt to ward off conflict of interest and undue industry influence. The RUF prohibits board members from participating in any activity or matter in which they have a financial interest. RUF board members must also openly disclose any financial interest they may have, or have had in the past, in entities doing business with the RUF and in any FDA-regulated entity. Additionally, the board requires conflict-of-interest measures be undertaken for each individual project the foundation undertakes. All projects undertaken by the RUF are reviewed by its board of directors and are subject to an independent review. While the RUF has some features, it has not existed for as long as the HEI, and thus it has far less of a track record. It also has no experience in funding projects. Still, the attempted controls for bias and conflict of interest are potentially noteworthy in thinking about a TRGE.

It should be noted that several third-party institutions have been engaged in independent tobacco research, including the Life Sciences Research Office and the Institute for Science and Health. However, the credibility and independence of these organizations have been questioned, which illustrates the importance of oversight, transparency, and governance (Schick and Glantz, 2007b).

Possible Design and Structure of a TRGE

Funding of a TRGE

A TRGE could receive funding from a mix of public and private sources. Independent organizations and foundations would also provide potential sources of funds, especially those foundations specializing in health research and risk reduction. The consideration of industry funding would need to be cautious. Unregulated or unstructured industry funding could potentially contribute to a perception of bias, so it is quite possible that the funding from tobacco companies and potential MRTP sponsors could be structured in a tax-like manner. The HEI model of regular, equalized contributions from members of the industry—with expected contributions independent of research and no bargaining over HEI governance between contributions—would be a useful model for consideration.

Oversight Board

A board of directors or general oversight board for any TRGE should be placed in a position of responsibility for maintaining the credibility and objectivity of the organization. It would be critical to ensure that a TRGE board be independent both of the FDA and the tobacco/MRTP industry. It would be prudent to institute a conflict-of-interest policy with prohibitions on participation in any matter where the board member may have a financial or other conflict of interest or plausible bias. An oversight board would need to be composed of a diverse membership—

nontobacco-related businesses, medicine and academics, consumers—with each individual openly disclosing any potential conflicts of interest. A board could assist the entity in selecting research contractors in any research competition.

Research Protocol Advice

An important feature of the entity would be in ensuring the independent design of research protocols by researchers. Independence of research design from the study sponsor is a critical feature of rigorous research, as the design of a study (its measures, its statistical methods, even the particulars of the hypothesis tested, duration of treatment, and other features) can deeply shape the research outcome. If research funding were provided through a TRGE, the competition might create additional incentives to cleave to robust research design models.

Organization, Oversight, and Training

As with the HEI, a TRGE could monitor contractors' research performance, provide staff members and/or training for IRB members at universities and contract research organizations, and set up Data Monitoring Committees (DMCs), including Data and Safety Monitoring Boards (DSMBs) and Observational Study Monitoring Boards (OSMBs). Given the nascent character of research in the MRTP field, it would be important for any research team to receive and consider advice midstream on study conduct. Independent monitoring of IRBs and DMCs/DSMBs/OSMBs would also be important, given the lack of broad university- and hospital-based experience in conducting research on MRTPs. It is doubtful that the CTP would be able to handle these responsibilities on its own.

Contract Mechanisms

A TRGE could fund research in several ways. It could issue contracts to independent investigators or contract research organizations, including commercial laboratories as well as hospital or academic contract research organizations. Another possibility and one well worth considering would be an RFA model, not unlike the model of the HEI or the funding models used by government granting agencies. Upon the preparation of an MRTP application that would involve premarket or postmarket clinical studies, the TRGE could develop and post an RFA for each study—or suite of studies—that the sponsor would need to have performed in order for an MRTP application to be considered complete by the CTP. Some (if not many) of the details of these studies would be left unspecified at the time of the RFA, so that upon the award of an investigator contract the TRGE could participate in the design of the study.

Quality Control

It would be critical for any TRGE to ensure adherence to rigorous quality control measures on the part of researchers conducting studies for an MRTP application to the FDA. Keating identifies a

> zealous approach to quality control on the part of HEI-funded investigators. Adherence to quality control guidelines and favorable reports from quality assurance audits, along with rigorous peer review, are the first line of defense against attacks on the credibility of the research (Keating, 2001).

The TRGE could promulgate good research practices for MRTPs in conjunction with academic specialists (e.g., Rees et al., 2009). Consistent with the HEI model and with other forms of research governance at universities and hospitals, the TRGE could perform scheduled or random data audits and other forms of site-specific research investigation. The TRGE could also assist the CTP in ensuring that postmarket studies are being launched, monitored, and completed in a timely fashion.

CONCLUSIONS

After extensive consideration of both the unique nature of the tobacco industry, the FSPTCA, and other relevant precedents, the committee identified a number of potential considerations that should guide the governance of tobacco industry studies.

1. **Research Funding.** Although the funding of such research will usually originate with the company developing the product, there may be cases where sponsors themselves may wish to have the research overseen or conducted independently. The FDA should expect that some of the research performed for MRTP applications that it reviews will be performed or conducted independently, by choice of sponsor. This raises the issue of how such third-party research can be governed.

2. **Research with Special Populations.** In some cases, especially that of experimentation with adolescents or populations vulnerable to high use rates, the FDA may wish to require or expect that research should be overseen by an independent third party who would be the recipient of tobacco industry study funding but would be responsible for

- choice of investigators,
- funding of investigators,

- oversight of studies,
- data collection,
- analysis of results, and
- publication of results.

3. **Data Transparency.** It is critical that the public have access to the totality of the data on MRTPs; therefore, all trials should be registered on the National Library of Medicine website Clinicaltrials.gov with the same time limits defined in the Food and Drug Administration Amendments Act (FDAAA) of 2007.

- In addition, for the same reasons, all trial results should be posted at Clinicaltrials.gov within 6 months of the last research participant completing the trial, the trial being terminated, or there being no further activity in the trial.
- Companies may attest annually to their posting of results and trials. The FDAAA penalties for nonposting should apply to tobacco studies.

4. **Engaging Academic Researchers**. Where a third-party entity carries out some or all of the research, such an entity should work with representatives of academic medical centers and scientific journals to develop a transparent funding process for tobacco studies that will allow academic medical centers to accept such funding and will satisfy the journal editors' requirements regarding independence from tobacco funding.
5. **Communicating Risks and Benefits.** Marketing materials for MRTPs should only be allowed to use the conclusions from studies reached by the analysis of the independent entity described above.
6. **Research Oversight.** Where independent research entities are used, any independent institute

- should have as its mission the performance of high-quality studies to determine the risks of modified tobacco products;
- should be governed by individuals appointed by an organization independent of the tobacco industry and with sufficient scientific stature to inspire public confidence; and
- should receive "core funding" from a tax on tobacco products that will maintain its basic functions, while individual studies will be funded by the interested companies.

REFERENCES

Ashley, M. J., and J. E. Cohen. 2003. What the public thinks about the tobacco industry and its products. *Tobacco Control* 12(4):396-400.

Barnes, D. E., and L. A. Bero. 1996. Industry-funded research and conflict of interest: An analysis of research sponsored by the tobacco industry through the center for indoor air research. *Journal of Health Politics Policy and Law* 21(3):515-542.

Barnes, D. E., and L. A. Bero. 1998. Why review articles on the health effects of passive smoking reach different conclusions. *JAMA* 279(19):1566-1570.

Barnes, R. L., S. K. Hammond, and S. A. Glantz. 2006. The tobacco industry's role in the 16 cities study of secondhand tobacco smoke: Do the data support the stated conclusions? *Environmental Health Perspectives* 114(12):1890-1897.

Bero, L., D. E. Barnes, P. Hanauer, J. Slade, and S. A. Glantz. 1995. Lawyer control in tobacco industry's external research program: The Brown and Williamson documents. *JAMA* 274(3):241-247.

Bitton, A., M. D. Neuman, J. Barnoya, and S. A. Glantz. 2005. The p53 tumour suppressor gene and the tobacco industry: Research, debate, and conflict of interest. *Lancet* 365(9458):531-540.

Carpenter, C. M., G. F. Wayne, and G. N. Connolly. 2007. The role of sensory perception in the development and targeting of tobacco products. *Addiction* 102(1):136-147.

Carpenter, D. P. 2010. *Reputation and power: Organizational image and pharmaceutical regulation at the FDA*. Princeton: Princeton University Press.

Cataldo, J. K., L. A. Bero, and R. E. Malone. 2010. "A delicate diplomatic situation": Tobacco industry efforts to gain control of the Framingham study. *Journal of Clinical Epidemiology* 63(8):841-853.

Cook, B. L., G. F. Wayne, L. Keithly, and G. Connolly. 2003. One size does not fit all: How the tobacco industry has altered cigarette design to target consumer groups with specific psychological and psychosocial needs. *Addiction* 98(11):1547-1561.

Cummings, K. M. 2003. A promise is a promise. *Tobacco Control* 12(2):117-118.

Cummings, K. M., C. Morley, and A. Hyland. 2002. Failed promises of the cigarette industry and its effect on consumer misperceptions about the health risks of smoking. *Tobacco Control* 11(Suppl. 1):i110-i117.

Cummings, K. M., A. Brown, and R. O'Connor. 2007. The cigarette controversy. *Cancer Epidemiology, Biomarkers & Prevention* 16(6):1070-1076.

DeNoble, V. J., and P. C. Mele. 2006. Intravenous nicotine self-administration in rats: Effects of mecamylamine, hexamethonium and naloxone. *Psychopharmacology* 184(3-4):266-272.

Doll, R., and A. B. Hill. 1950. Smoking and carcinoma of the lung; preliminary report. *British Medical Journal* 2(4682):739-748.

Doll, R., and A. B. Hill. 1952. A study of the aetiology of carcinoma of the lung. *British Medical Journal* 2(4797):1271-1286.

Doll, R., and A. B. Hill. 1954. The mortality of doctors in relation to their smoking habits. *British Medical Journal* 1(4877):1451-1455.

Dunn, W. L. 1973. *Smoking behavior: Motives and incentives*. Washington: V. H. Winston; distributed by the Halsted Press Division, Wiley.

Etter, J. F., C. Bullen, A. D. Flouris, M. Laugesen, and T. Eissenberg. 2011. Electronic nicotine delivery systems: A research agenda. *Tobacco Control* 20(3):243-248.

Eysenck, H. J. 1991. *Smoking, personality, and stress: Psychosocial factors in the prevention of cancer and coronary heart disease*. New York: Springer.

FDA (Food and Drug Administration). 2010. Code of Federal Regulations Title 21. In *Chapter 1—Food and Drug Administration, Part 50—Protection of Human Subjects*.

Fields, N., and S. Chapman. 2003. Chasing Ernst L Wynder: 40 years of Philip Morris' efforts to influence a leading scientist. *Journal of Epidemiology and Community Health* 57(8):571-578.

Foulds, J., L. Ramstrom, M. Burke, and K. Fagerstrom. 2003. Effect of smokeless tobacco (snus) on smoking and public health in Sweden. *Tobacco Control* 12(4):349-359.

Garne, D., M. Watson, S. Chapman, and F. Byrne. 2005. Environmental tobacco smoke research published in the journal *Indoor and Built Environment* and associations with the tobacco industry. *Lancet* 365(9461):804-809.

Glasser, S. P., M. Salas, and E. Delzell. 2007. Importance and challenges of studying marketed drugs: What is a phase IV study? Common clinical research designs, registries, and self-reporting systems. *The Journal of Clinical Pharmacology* 47(9):1074.

Gundle, K. R., M. J. Dingel, and B. A. Koenig. 2010. "To prove this is the industry's best hope": Big tobacco's support of research on the genetics of nicotine addiction. *Addiction* 105(6):974-983.

Hammond, D., N. E. Collishaw, and C. Callard. 2006. Secret science: Tobacco industry research on smoking behaviour and cigarette toxicity. *Lancet* 367(9512):781-787.

Hanauer, P., J. Slade, D. E. Barnes, L. Bero, and S. A. Glantz. 1995. Lawyer control of internal scientific-research to protect against products-liability lawsuits—the Brown-and-Williamson documents. *JAMA* 274(3):234-240.

Harris, B. 2011. The intractable cigarette "filter problem." *Tobacco Control* 20(Suppl. 1):i10-i16.

Harris, J. E., M. J. Thun, A. M. Mondul, and E. E. Calle. 2004. Cigarette tar yields in relation to mortality from lung cancer in the cancer prevention study II prospective cohort, 1982-8. *BMJ* 328(7431):72.

Harris Interactive. 2003. *Attitudes to government regulation vary greatly for different industries; The Harris Poll #19*. http://www.harrisinteractive.com/vault/Harris-Interactive-Poll-Research-Attitudes-to-Government-Regulation-Vary-Greatly-for-Different-Industries-2003-04.pdf (accessed December 5, 2011).

Harris Interactive. 2010. *Oil, pharmaceutical, health insurance, and tobacco top the list of industries that people think should be more regulated; The Harris Poll #149*. http://www.harrisinteractive.com/vault/HI-Harris-Poll-Industry-Regulation-2010-12-02.pdf (accessed December 5, 2011).

Henningfield, J. E., and K. O. Fagerstrom. 2001. Swedish match company, Swedish snus and public health: A harm reduction experiment in progress? *Tobacco Control* 10(3):253-257.

HHS (U.S. Department of Health and Human Services). 1996. *Postmarketing studies of prescription drugs*. Philadelphia, PA: Office of Inspector General.

HHS. 2009. *U.S. Code of Federal Regulations. Title 45, public welfare. Part 46, protection of human subjects*. http://www.hhs.gov/ohrp/humansubjects/guidance/45cfr46.html (accessed September 30, 2011).

Hirshbein, L. 2011. Scientific research and corporate influence: Smoking, mental illness, and the tobacco industry. *Journal of the History of Medicine and Allied Sciences* published online May 19, 2011:1-24.

Hoffmann, D., and I. Hoffmann. 1997. The changing cigarette, 1950-1995. *Journal of Toxicology and Environmental Health* 50(4):307-364.

Hurt, R. D., and C. R. Robertson. 1998. Prying open the door to the tobacco industry's secrets about nicotine: The Minnesota Tobacco Trial. *JAMA* 280(13):1173-1181.

International Smoking Behaviour Conference and R. E. Thornton. 1978. *Smoking behavior: Physiological and psychological influences*. Edinburgh: Churchill Livingstone.

IOM (Institute of Medicine). 2001. *Clearing the smoke: Assessing the science base for tobacco harm reduction*. Washington, DC: National Academy Press.

Jasanoff, S. 1990. The fifth branch: Science advisors as policy makers. Cambridge, Massachusetts: Harvard University Press.

Keating, T. J. 2001. Lessons from the recent history of the Health Effects Institute. *Science, Technology, & Human Values* 26(4):409-430.

Kozlowski, L. T. 2007. Effect of smokeless tobacco product marketing and use on population harm from tobacco use: Policy perspective for tobacco-risk reduction. *American Journal of Preventive Medicine* 33(6):S379-S386.

Kozlowski, L. T., and R. J. O'Connor. 2002. Cigarette filter ventilation is a defective design because of misleading taste, bigger puffs, and blocked vents. *Tobacco Control* 11(Suppl. 1):i40-i50.

Kozlowski, L. T., and R. J. O'Connor. 2010. Dealing with health fears: Cigarette advertising in the United States in the twentieth century. In *Tobacco: Science, policy, and public health*. 2nd ed, edited by P. Boyle, N. Gray, J. Henningfield, J. Seffrin, and W. Zatonski. Oxford: Oxford University Press. Pp. 31-45.

Kozlowski, L. T., N. A. Dreschel, S. D. Stellman, J. Wilkenfeld, E. B. Weiss, and M. E. Goldberg. 2005. An extremely compensatible cigarette by design: Documentary evidence on industry awareness and reactions to the barclay filter design cheating the tar testing system. *Tobacco Control* 14(1):64-70.

Landman, A., D. K. Cortese, and S. Glantz. 2008. Tobacco industry sociological programs to influence public beliefs about smoking. *Social Science and Medicine* 66(4):970-981.

Marks, H. M. 1997. *The progress of experiment: Science and therapeutic reform in the United States, 1900-1990*. Cambridge, England: Cambridge University Press.

Megerdichian, C. L., V. W. Rees, G. F. Wayne, and G. N. Connolly. 2007. Internal tobacco industry research on olfactory and trigeminal nerve response to nicotine and other smoke components. *Nicotine & Tobacco Research* 9(11):1119-1129.

Muggli, M. E., R. D. Hurt, and D. D. Blanke. 2003. Science for hire: A tobacco industry strategy to influence public opinion on secondhand smoke. *Nicotine & Tobacco Research* 5(3):303-314.

NCI (National Cancer Institute). 1997. *The FTC cigarette test method for detemining tar, nicotine, and carbon monoxide yields of U.S. cigarettes, Smoking and tobacco control monograph no. 7*. Bethesda, MD: U.S. Department of Health and Human Services, National Institutes of Health. NIH Pub. No. 96-4028.

NCI. 2001. *Risks associated with smoking cigarettes with low machine-measured yields of tar and nicotine, Smoking and tobacco control monograph no. 13*. Bethesda, MD: U.S. Department of Health and Human Services, National Institutes of Health, National Cancer Institute. NIH Pub. No. 99-4645.

NCI. 2008. *The role of the media in promoting and reducing tobacco use*. Edited by R. M. Davis, *NCI tobacco control monograph series*. Bethesda, MD: U.S. Dept. of Health and Human Services, National Institutes of Health, National Cancer Institute.

Neilsen, K., and S. A. Glantz. 2004. A tobacco industry study of airline cabin air quality: Dropping inconvenient findings. *Tobacco Control* 13(Suppl. 1):i20-i29.

NIDA (National Institute on Drug Abuse). 2011. *Points to consider regarding tobacco industry funding of NIDA applications*. http://www.nida.nih.gov/about/organization/nacda/points-to-consider.html (accessed September 30, 2011).

Ong, E. K., and S. A. Glantz. 2000. Tobacco industry efforts subverting international agency for research on cancer's second-hand smoke study. *Lancet* 355(9211):1253-1259.

Parascandola, M. 2005a. Lessons from the history of tobacco harm reduction: The National Cancer Institute's smoking and health program and the "less hazardous cigarette." *Nicotine & Tobacco Research* 7(5):779-789.

Parascandola, M. 2005b. Science, industry, and tobacco harm reduction: A case study of tobacco industry scientists' involvement in the National Cancer Institute's smoking and health program, 1964-1980. *Public Health Reports* 120(3):338-349.

Pauly, J. L., A. B. Mepani, J. D. Lesses, K. M. Cummings, and R. J. Streck. 2002. Cigarettes with defective filters marketed for 40 years: What Philip Morris never told smokers. *Tobacco Control* 11(Suppl. 1):i51-i61.

Petticrew, M. P., and K. Lee. 2011. The "father of stress" meets "big tobacco": Hans Selye and the tobacco industry. *American Journal of Public Health* 101(3):411-418.

Pollay, R. W., and T. Dewhirst. 2002. The dark side of marketing seemingly "light" cigarettes: Successful images and failed fact. *Tobacco Control* 11(Suppl. 1):i18-i31.

Rees, V. W., J. M. Kreslake, R. J. O'Connor, K. M. Cummings, M. Parascandola, D. Hatsukami, P. G. Shields, and G. N. Connolly. 2009. Methods used in internal industry clinical trials to assess tobacco risk reduction. *Cancer Epidemiology, Biomarkers & Prevention* 18(12):3196-3208.

Schick, S., and S. Glantz. 2005. Philip Morris toxicological experiments with fresh sidestream smoke: More toxic than mainstream smoke. *Tobacco Control* 14(6):396-404.

Schick, S. F., and S. Glantz. 2007a. Concentrations of the carcinogen 4-(methylnitrosamino)-1-(3-pyridyl)-1-butanone in sidestream cigarette smoke increase after release into indoor air: Results from unpublished tobacco industry research. *Cancer Epidemiology, Biomarkers & Prevention* 16(8):1547-1553.

Schick, S. F., and S. A. Glantz. 2007b. Old ways, new means: Tobacco industry funding of academic and private sector scientists since the Master Settlement Agreement. *Tobacco Control* 16(3):157-164.

Scollo, M., A. Lal, A. Hyland, and S. Glantz. 2003. Review of the quality of studies on the economic effects of smoke-free policies on the hospitality industry. *Tobacco Control* 12(1):13-20.

Smith, E. A. 2006. "It's interesting how few people die from smoking": Tobacco industry efforts to minimize risk and discredit health promotion. *The European Journal of Public Health* 17(2):162.

Sutton, C. D., and R. G. Robinson. 2004. The marketing of menthol cigarettes in the United States: Populations, messages, and channels. *Nicotine & Tobacco Research* 6(Suppl. 1):S83-S91.

Thun, M. J., and D. M. Burns. 2001. Health impact of "reduced yield" cigarettes: A critical assessment of the epidemiological evidence. *Tobacco Control* 10(Suppl. 1):i4-i11.

Tomar, S. L., B. J. Fox, and H. H. Severson. 2009. Is smokeless tobacco use an appropriate public health strategy for reducing societal harm from cigarette smoking? *International Journal of Environmental Research and Public Health* 6(1):10-24.

Tong, E. K., L. England, and S. A. Glantz. 2005. Changing conclusions on secondhand smoke in a sudden infant death syndrome review funded by the tobacco industry. *Pediatrics* 115(3):e356-e366.

U.S. Government Accountability Office. 2009. *New drug approval: FDA needs to enhance its oversight of drugs approved on the basis of surrogate endpoints*. (Publication No. GAO-09-866). Washington, DC: Government Printing Office.

Wayne, G. F., G. N. Connolly, and J. E. Henningfield. 2004. Assessing internal tobacco industry knowledge of the neurobiology of tobacco dependence. *Nicotine & Tobacco Research* 6(6):927-940.

White, J., M. Parascandola, and L. Bero. 2007. Tobacco industry research and protection of human subjects: A case study of R. J. Reynolds. *Nicotine & Tobacco Research* 9(11):1213-1225.

Wood, A. J. J., C. M. Stein, and R. Woosley. 1998. Making medicines safer—the need for an independent drug safety board. *NEJM* 339(25):1851-1854.

Wynder, E. L., and E. A. Graham. 1950. Tobacco smoking as a possible etiologic factor in bronchiogenic carcinoma—a study of 684 proved cases. *JAMA* 143(4):329-336.

Wynder, E. L., and S. D. Stellman. 1979. Impact of long-term filter cigarette usage on lung and larynx cancer risk: A case-control study. *Journal of the National Cancer Institute* 62(3):471-477.

Wynder, E. L., E. A. Graham, and A. B. Croninger. 1953. Experimental production of carcinoma with cigarette tar. *Cancer Research* 13(12):855-864.

Yano, E. 2005. Japanese spousal smoking study revisited: How a tobacco industry funded paper reached erroneous conclusions. *Tobacco Control* 14(4):227-233.

3

Evidence Base and Methods for Studying Health Effects

Decades of research on the health effects of tobacco use have generated overwhelming evidence to support the conclusion that tobacco use causes disease. An inference of causality requires evidence along the causal pathway from exposure to disease, including evidence on the effects of tobacco from experimental and observational study designs, and from investigations into the biological mechanisms of disease. A widely cited criteria for making a causal inference in epidemiology and public health are the Hill Criteria (Weed, 2000). The judgment that tobacco use causes diseases such as lung cancer and heart disease has been based on evidence from a wide range of investigations that fulfill the requirements of the Hill Criteria. This has been thoroughly reviewed and documented in reports of the Surgeon General on tobacco, such as the 2004 and 2010 reports (HHS, 2004a, 2010).

The evaluation of the health effects and mechanisms of modified risk tobacco products (MRTPs) is a closely related enterprise. Development of many MRTPs will be based on existing evidence and knowledge of the mechanisms of tobacco-related disease. In general, MRTPs are designed to remove or block a step in the causal pathway between tobacco exposure and disease. As such, evidence about how an MRTP intervenes on the causal pathways for tobacco-related disease will be critical. However, inferences about the health effects of an MRTP based on prior knowledge of the causal pathways of tobacco disease, while relevant, will not be sufficient to inform regulatory decisions. Independent evidence on the health effects of the MRTP will be necessary. The study of the health effects of tobacco use can provide an illustrative precedent for the evaluation of MRTPs. The same range of research methods employed to establish a causal relationship between tobacco and disease will be needed to provide evidence on the health effects of MRTPs on both individual and public health. This chapter discusses that evidence and provides guidance on how the Food and Drug Administration (FDA) should consider different types of that evidence in its decision-making process. The chapter begins with a discussion of the composition of modified tobacco products. The committee then discusses biomarkers of MRTPs, including biomarkers of exposure and biomarkers of effects. Next, it discusses preclinical and clinical studies, including the advantages and disadvantages of those studies, and what evidence the various study types can provide to inform the FDA's decisions on MRTPs.

PRODUCT COMPOSITION

Smokeless tobacco products, such as oral snuff, and combusted tobacco products, such as cigarettes, are the main types of tobacco products used in the United States (SAMHSA, 2007). The composition of tobacco and tobacco smoke has been the subject of intense study for at least the past 60 years, and studies have identified more than 8,000 constituents of tobacco and tobacco smoke (Rodgman and Perfetti, 2009). Validated methods are available to quantify many constituents of tobacco and tobacco smoke (Borgerding and Klus, 2005; Rodgman and Perfetti, 2009), and the chemical composition can have a large effect on the potential health risks of a given product. Product composition, including how the constituents compare to other products, therefore, is an important aspect of any new product. Although different tobacco products continue to be introduced, this section discusses the types of tobacco products currently available, the methods for analyzing them, and the commonly reported constituents. Smokeless products are discussed first, followed by a discussion of combusted products.

Smokeless Tobacco Products

Types of Smokeless Products

Smokeless tobacco products used in the United States include moist snuff and chewing tobacco (for oral use), and dry snuff (for nasal use). Types of chewing tobacco include plug, twist, and loose leaf varieties. The use of chewing tobacco and dry snuff has declined over time. Oral moist snuff is by far the most popular kind of smokeless tobacco in the United States (Federal Trade Commission, 2007). Oral moist snuff is used by placing the tobacco—either loose or packaged in a tea bag–like sachet—

in the space between the cheek and gum, or lip and gum. Generally, oral moist snuff is not chewed. Brands such as Copenhagen and Skoal, manufactured by Altria Group, Inc., and Grizzly and Kodiak, marketed by Reynolds American, Inc., are common.

The use of any form of smokeless tobacco has declined substantially between 1986 and 2003 (Nelson et al., 2006); in this time period, there was an approximately 5 percent decrease in overall smokeless tobacco sales (in pounds) (Federal Trade Commission, 2007). However, the use of moist snuff or dip increased by approximately 87 percent over the same period (Nelson et al., 2006). In 2005, total dollar sales for moist snuff accounted for more than 80 percent of total sales for smokeless tobacco (Federal Trade Commission, 2007). In 2008, 3.5 percent of Americans aged 12 or older (0.4 percent of women aged 12 or older and 6.8 percent of men aged 12 or older) had used a smokeless tobacco product in the previous month (SAMHSA, 2011).

Moist snuff for oral use contains both high salt and high moisture content (Stepanov et al., 2010). When placed in the oral cavity, the product generates excess saliva, usually requiring spitting. Recently, the tobacco industry has introduced and promoted spit-free smokeless tobacco products. These new products, such as Camel Snus and Marlboro Snus, contain low moisture content and are distributed in small pouches of flavored tobacco. The products have been marketed to current cigarette smokers for situations where smoking is prohibited (Hatsukami et al., 2007a). These products have design features in common with snus products that have been used in Sweden for many years. Users of Swedish Snus place the product between the gum and upper lip; it does not usually stimulate salivation. Other new smokeless tobacco products continue to appear. These include dissolvable products such as Camel Orbs (a pellet), Camel Sticks (a twisted toothpick-size stick), and Camel Strips (a film strip placed on the tongue). All of those new products are made from finely ground flavored tobacco (Rainey et al., 2011).

Methods of Analysis

Methods of analysis of the components of smokeless tobacco are standardized (IARC, 2007; Richter and Spierto, 2003; Richter et al., 2008; Song and Ashley, 1999; Stepanov and Hecht, 2005; Stepanov et al., 2008, 2010). Smokeless tobacco analyses include analyses for moisture content, pH, and components. Moisture content can be determined by the difference in weight before and after drying. For measurement of pH, the tobacco is extracted with water and the pH is determined with a pH meter. Nicotine can be determined by extraction of the tobacco and analysis by combined gas chromatography-mass spectrometry (GC-MS) or high-performance liquid chromatography-mass spectrometry (LC-MS). Minor tobacco alkaloids such as nornicotine and anatabine are extracted, derivatized by reductive alkylation, and determined by gas chromatography-tandem mass spectrometry (GC-MS/MS). Tobacco-specific *N*-nitrosamines (TSNAs) are extracted and analyzed by either gas chromatography with nitrosamine selective detection or by liquid chromatography-tandem mass spectrometry (LC-MS/MS). Both conventional and supercritical fluid extractions have been used. Polycyclic aromatic hydrocarbons can be quantified by extraction with cyclohexane followed by solid-phase extraction and GC-MS. Aldehydes are measured by extraction, derivatization with 2,4-dinitrophenylhydrazine, and GC-MS. Anions such as nitrate, nitrite, and chloride are determined by anion exchange with conductivity detection.

Laboratory analysis of constituents in these products would be a standard first step in the initial evaluation of any new product. These analyses are generally quite straightforward involving standard methods of extraction, sample cleanup, analyte identification, and quantitation. Data from diverse laboratories involved in the analysis of various products give comparable results for most analytes. There are differences in the literature in the manner in which the analytical data are expressed. Some investigators have expressed their data per dry weight of product, while others use wet weight, or even portion size. Because traditional moist snuff products typically contain about 50 percent water, it is crucial to recognize the manner in which the data are being expressed and to take this into consideration when making judgments on constituent levels. The expression of constituent levels per dry weight of product, with inclusion of data on water content is standard (Stepanov et al., 2008). Because portion sizes are fixed in the products encased in tea bag–like sachets, it is also important to report constituent levels per portion size for these products.

Laboratory analysis of constituents, however, may not reflect constituent uptake under conditions of use. Biomarker of exposure studies, described below, provide a more realistic indication of exposure.

Commonly Reported Constituents

Thousands of compounds have been identified in unburned tobacco (Rodgman and Perfetti, 2009), but routine analyses of smokeless tobacco have focused on relatively few of these compounds thought to be critical in its biological activities (IARC, 2007; Richter and Spierto, 2003; Richter et al., 2008; Song and Ashley, 1999; Stepanov and Hecht, 2005; Stepanov et al., 2008, 2010). Commonly reported constituents include TSNAs, nicotine and minor tobacco alkaloids, nitrite, nitrate and other anions, metals, aldehydes, and polycyclic aromatic hydrocarbons. Nicotine is generally reported as protonated and unprotonated (determined by measuring pH of the product). This is important because unprotonated nicotine is absorbed more readily through the oral mucosa than protonated nicotine. Plasma nicotine levels are directly related to pH of the product: higher pH values lead to higher levels of plasma nicotine (IARC, 2007). Minor tobacco alkaloids might, along with nicotine, contribute to addiction. Unlike cigarette smoke, the most common strong carcinogens in smokeless tobacco products are TSNAs. Extensive data demonstrating their presence in parts per million quantities, greater than nitrosamine concentrations in any other consumer product intended for oral use, are available (IARC, 2007; Richter et al., 2008; Stepanov et al., 2008). Levels of polycyclic aromatic hydrocarbons and aldehydes have been less frequently reported (Stepanov et al., 2008, 2010).

There is solid evidence that nicotine is addictive, but little evidence of addictive potential for other constituents of smokeless tobacco products. With respect to the induction of cancer, it is suspected but not proven that TSNAs play a major role, while other compounds such as polycyclic aromatic hydrocarbons and aldehydes may also contribute. There may be other unidentified or unrecognized compounds in smokeless tobacco that contribute in important ways to its adverse health effects. Among the thousands of identified compounds in smokeless tobacco products, the 28 currently identified carcinogens represent only a small fraction (IARC, 2007; Rodgman and Perfetti, 2009). Furthermore, seemingly innocuous compounds such as sodium chloride, which occurs in amounts more than 5 percent in some smokeless tobacco products (IARC, 2007), could exacerbate the effects of carcinogens by leading to local irritation, among other effects (Stepanov et al., 2008).

Combusted Products

Types of Products

Cigarettes are by far the most used combusted tobacco product. In 2009, there were more than 46 million cigarette smokers in the United States, about 20.6 percent of the adult population (CDC, 2010). Between the mid-1960s and 2004, cigarette smoking among adults decreased from approximately 42 percent to 21 percent; however, prevalence has not changed substantially since then (CDC, 1999, 2011b). Additionally, after substantial declines (66 percent) in cigar consumption from 1964 to 1993, consumption rates for cigars increased by close to 50 percent from 1993 to 1997 (NCI, 1998). In 2010, 5.2 percent of Americans aged 12 or older had smoked cigars in the past month (SAMHSA, 2011). Other combusted products include pipes and water pipes.

Methods of Analysis

Because combusted products are burned, their constituents cannot simply be extracted as with smokeless tobacco products. Various machine methods attempt to simulate the smoking of tobacco products, and the smoke is collected and analyzed (IARC, 2004). Different organizations use different methods for generating smoke. For example, the International Organization for Standardization and the U.S. Federal Trade Commission smoking regimen uses a 35 mL puff every 60 seconds, and a puff duration of 2 seconds, with the filter ventilation holes (if present) open. Health Canada uses an intense smoking regimen with a 55 mL puff every 30 seconds, and a puff duration of 2 seconds, with the filter ventilation holes completely blocked. The Massachusetts Department of Health method has a 45 mL puff every 30 seconds, and a puff duration of 2 seconds, with the filter ventilation holes 50 percent blocked. It is widely recognized that none of these methods accurately reproduces the many ways smokers actually use cigarettes, but the methods can be used for comparison of one product to another (IARC, 2004).

Researchers can collect and analyze both mainstream smoke, which emanates from the filter end of the cigarette, and sidestream smoke, which emanates mainly from the burning tip of the product. For collection, a glass fiber filter separates arbitrarily gas phase constituents from total particulate matter, which collects on the filter (Adam et al., 2006). Once the combusted material is collected, the methods of analysis of the various constituents of cigarette smoke have some similarities to those used for smokeless tobacco. Because the products of combustion are generally more complex than those obtained by extraction of unburned tobacco, multiple extraction or purification steps are often necessary before the analysis can be completed, usually by GC-MS or LC-MS/MS techniques (IARC, 2004).

Laboratory analyses by machine smoking would be a standard first step in the initial evaluation of any new product, even though it is widely recognized that this approach has limitations. Machine smoking methods do not replicate human smoking conditions because smokers may vary their way of smoking a cigarette depending on many factors. Important among these is the well-established phenomenon of compensation, in which smokers may alter their method of smoking in order to compensate for lower machine measured amounts of nicotine and other constituents. They accomplish this in a number of different ways including increasing puff number or volume and blocking filter vents (NCI, 2001). Under a given set of machine smoking conditions, analyses of particular constituents are generally well standardized leading to reasonable agreement in constituent levels among different laboratories. However, formalized interlaboratory comparisons have only been carried out for a few constituents. When reporting constituent levels for any product, it is crucial to describe the type of smoking regimen that has been used.

There is no proven method to replicate the many ways humans smoke cigarettes. The World Health Organization, under the Framework Convention on Tobacco Control, has adopted the approach of expressing machine-measured constituents per mg of nicotine for use in regulation, because this would presumably mitigate some of the effects of compensation (Burns et al., 2008). However, this approach is untested in a regulatory setting.

The measurement of smoke constituents can be challenging. Even measurement of parameters seemingly as simple as pH and free nicotine have led to controversy (Chen and Pankow, 2009; Pankow et al., 2003).

Commonly Reported Constituents

The FDA has developed a list of "harmful and potentially harmful constituents in tobacco products and tobacco smoke" that includes more than 100 constituents from various classes of chemicals (FDA, 2011a, 2011c). These include "tar," nicotine and minor tobacco alkaloids, carbon monoxide (CO), nitrogen oxides, polycyclic aromatic hydrocarbons, TSNAs, volatile nitrosamines, aldehydes, aromatic amines, metals, phenols, ketones, volatile hydrocarbons such as benzene and butadiene, ethylene and propylene oxide, furan, hydrazine, hydrogen cyanide, heterocyclic aromatic amines, nitrogen compounds, pyridine, vinyl chloride, polonium-210, and others. The majority of these constituents have been routinely analyzed, and extensive data are available on their concentrations in tobacco smoke (Chen and Moldüveanu, 2003; Counts et al., 2004; Ding et al., 2006, 2007; Gregg et al., 2004; Hammond and O'Connor, 2008; IARC, 2004; Roemer et al., 2004).

Furthermore, the same considerations discussed above with respect to smokeless tobacco apply to combusted products. It is not certain that the current list of harmful and potentially harmful constituents is complete. There may be other constituents among the more than 8,000 in tobacco and tobacco smoke (Rodgman and Perfetti, 2009) that are important but currently unrecognized. It is also known that there are interactions between carcinogens and tumor promoters or cocarcinogens that may not be recognized when simply analyzing a list of compounds (HHS, 2010; IARC, 2004).

Summary of Product Composition

Analysis of smokeless tobacco products or combusted products can be achieved using standardized and validated methods for a variety of constituents. Although there could be some inter-laboratory differences in results of these analyses, most data are generally comparable for a given product. In the analysis of smokeless tobacco products, the method of extraction and the method of expressing the results need to be taken into account when comparing data. In the analysis of combusted products, the method of machine smoking is critical when comparisons are to be made. None of the standard machine smoking methods replicate human smoking conditions, but these methods can be useful for comparison of different products under comparable conditions.

BIOMARKERS

Studies of tobacco and tobacco-related diseases use a number of different biomarkers, and the validity of those biomarkers are key to the validity of the studies; biomarkers will continue to play an important role in the FDA's regulation of MRTPs. The FDA will be making regulatory decisions about the marketing of MRTPs in the immediate future, but the latency period between tobacco exposure and the development of major clinical adverse health consequences is usually quite long. Validated biomarkers and other surrogates of tobacco-related disease outcomes that can provide information over a shorter time frame, therefore, will play a critical role in the evaluation of MRTPs. The Family Smoking Prevention and Tobacco Control Act of 2009 (FSPTCA) highlights the importance of addressing biomarkers and surrogates when it specifies that regulations or guidance issued by the Agency shall "include validated biomarkers, intermediate clinical endpoints, and other feasible outcome measures, as appropriate."[1]

Terminology around biomarkers can be a controversial issue. Over the course of evaluating both the statutory language and the prevailing literature, the committee encountered inconsistencies in the definitions for terms central to this discussion, including the terms "biomarker," "surrogate," "intermediate endpoint," and "endpoint." The committee also found it important to differentiate between biomarkers of exposure and biomarkers of effect or risk. In this report, the committee broadly categorizes biomarkers as biomarkers of exposure and biomarkers of risk, and further distinguishes among specific types of biomarkers of risk. Specifically, the committee adopts the definitions articulated in the Institute

[1] *Family Smoking Prevention and Tobacco Control Act of 2009*, Public Law 111-31, 123 Stat. 1776 (June 22, 2009).

BOX 3-1
Definitions Related to Biomarkers, Clinical Endpoints, and Surrogate Endpoints

Biomarker: A characteristic that is objectively measured and evaluated as an indicator of normal biological responses, pathogenic processes, or pharmacologic responses to an intervention.

Biomarker of exposure: The chemical, or its metabolite, or the product of an interaction between a chemical and some target molecule or cell, that is measured in a compartment in an organism.

Biomarker of risk: A biomarker that indicates a risk factor for a disease.

Clinical endpoint: A characteristic or variable that reflects how a patient or consumer feels, functions, or survives.

Surrogate endpoint: A biomarker that is intended to substitute for a clinical endpoint. A surrogate endpoint is expected to predict clinical benefit (or harm or lack of benefit or harm) based on epidemiologic, therapeutic, pathophysiologic, or other scientific evidence.

SOURCE: Adapted from IOM (2010).

of Medicine's (IOM's) 2010 report, *Evaluation of Biomarkers and Surrogate Endpoints in Chronic Disease* (IOM, 2010). Relevant definitions from that report are presented in Box 3-1. Biomarkers of exposure and biomarkers of risks are discussed below.

Biomarkers of Exposure

Biomarkers of human exposure to specific constituents of tobacco products may be the constituents themselves; metabolites of the constituents in urine, blood, breath, saliva, nails, or hair; or protein- or DNA-binding products (adducts) of the constituents or their metabolites. These biomarkers have the potential to bypass many of the uncertainties in product analysis and provide a realistic and direct assessment of carcinogen and toxicant dose in an individual. It should be emphasized however that the biomarkers discussed here are virtually all biomarkers of exposure to specific tobacco or tobacco smoke constituents. In most cases, they have not been validated as biomarkers of risk. Furthermore, these biomarkers are derived from specific constituents of tobacco products thought to be harmful to the consumer, but there may be unknown or unmeasured constituents that are also harmful, or there may be combination effects of individual constituents that cannot be recognized by measurement of individual biomarkers of exposure. Presently, there is no single accepted biomarker that predicts the risk of disease in people who use tobacco products.

Analytical Validation

These biomarkers of exposure to tobacco toxicants and carcinogens are most frequently quantified by LC-MS/MS, GC-MS/MS, and related techniques. The first step in validation is analytical validation. This topic has been previously discussed in detail in a recent IOM report (IOM, 2010). Chapters of this 2010 report are provided in Appendix B.

Validation with Respect to Product Use

The second step in validation of a biomarker of exposure to tobacco toxicants and carcinogens is demonstrating that the biomarker is actually related to tobacco product exposure. The most reliable method of demonstrating this relationship is to assess levels of the biomarker after a research participant has stopped using the tobacco product. In one representative study, researchers assessed at various times (3, 7, 14, 21, 28, 42, and 56 days) the persistence of eight tobacco smoke carcinogens and toxicant biomarkers in the urine of 17 people who had stopped smoking. The biomarkers were metabolites of 1,3-butadiene, acrolein, crotonaldehyde, benzene, ethylene oxide, pyrene (a representative polycyclic aromatic hydrocarbon), and nicotine-derived nitrosamine ketone (NNK), a TSNA. These biomarkers, which are described in more detail below, include some of the major carcinogens and toxicants present in cigarette smoke. Levels of all these biomarkers—except for 1,3-butadiene metabolites (called dihydroxybutyl mercapturic acid)—decreased significantly after 3 days of smoking cessation ($P < .001$). The rates of decrease for most of the biomarkers were rapid, reaching nearly their ultimate levels (81–91 percent reduction) after 3 days, while that of the NNK metabolite (called 4-[methylnitrosamino]-1-[3-pyridyl]-1-butanol and its glucuronides [total NNAL]) was gradual, reaching a 92 percent reduction after 42 days. The decrease in the pyrene metabolite was variable among research participants, reaching about 50 percent of baseline, consistent with other common environmental sources of pyrene, such as diet. These results demonstrated that all biomarkers investigated in this study except dihydroxybutyl mercapturic acid were related to cigarette smoking (Carmella et al., 2009). A similar study carried out in smokeless tobacco users demonstrated the reduction of total NNAL after cessation of product use (Hecht, 2002).

Another method of validating tobacco carcinogen and toxicant biomarkers with respect to tobacco product exposure is to compare their levels in smokers and nonsmokers. Numerous studies of this type have been reported, and individual biomarkers are described in an upcoming section and presented in Table 3-1. Biomarkers of exposure of tobacco-specific compounds such as NNK, *N*-nitrosonornicotine (NNN), and nicotine are not found in non-tobacco users unless they have been exposed to secondhand tobacco smoke (see Table 3-1). Other biomarkers, such as those related to combustion products such as pyrene, are detected in both smokers and nonsmokers because they occur not only in tobacco products, but also in the diet and polluted air. Therefore, some of the ranges of values overlap between smokers and nonsmokers, as shown in Table 3-1. However, biomarker levels are consistently higher in smokers compared to those in nonsmokers in individual studies (Hecht et al., 2010). Biomarkers of the tobacco-specific compounds are similar in smokers and smokeless tobacco users, while those of some of the volatile organic combustion products are considerably lower in smokeless tobacco users (Hecht, 2002; Hecht et al., 2010).

Exposure to secondhand cigarette smoke can contribute to biomarker levels in nonsmokers, but the levels are generally small, about 1–5 percent of the levels in smokers (Hecht et al., 2010). Some biomarkers that are consistently elevated in nonsmokers exposed to secondhand tobacco smoke are cotinine, a major metabolite of nicotine, and NNAL and its glucuronides, metabolites of NNK (Hecht, 2002, 2003b; HHS, 2006). Cut points in these biomarkers for distinguishing light smokers from nonsmokers exposed to secondhand smoke have been discussed (Goniewicz et al., 2011).

Validation with Respect to Disease Risk

One approach to determining the relationship of exposure biomarkers to disease risk is to carry out prospective epidemiologic studies, or cohort studies. In these studies, samples from healthy research participants are collected and stored, and demographic and lifestyle data are obtained using questionnaires. The participants are then followed for years, and eventually diseases such as cancers will occur in some of them. The stored samples from these research participants are retrieved, along with samples from appropriately matched controls that remain disease free, to form a nested case-control study. These samples can be analyzed for the biomarkers to determine their relationship to disease. The magnitude of the relationship to disease risk for each biomarker or their combinations can be evaluated using standard statistical analysis methods. Although there are certain limitations of this approach, which have been discussed

TABLE 3-1 Representative Exposure Biomarkers Related to Tobacco Carcinogens and Toxicants

Biomarker	Source	Range of Recent Mean Values or Concentrations: Smokers	Nonsmokers	References (smokers)	References (nonsmokers)
Urinary biomarker[a]					
Nicotine equivalents	Nicotine	70.4–154 μmol/24hr	N/A[b]	(Lowe et al., 2009; Mendes et al., 2008; Roethig et al., 2007, 2009; Scherer et al., 2006, 2007a; Zedler et al., 2006)	
Total NNAL	NNK	1.1–2.9 nmol/24hr	N/A[b]	(Carmella et al., 2009; Kavvadias et al., 2009b; Lowe et al., 2009; Melikian et al., 2007; Mendes et al., 2008; Roethig et al., 2009; Sarkar et al., 2008; Scherer et al., 2007a; Stepanov and Hecht, 2005)	
Total NNN	NNN	0.049–0.24 nmol/24hr	N/A[b]	(Kavvadias et al., 2009a, 2009b; Stepanov and Hecht, 2005; Stepanov et al., 2009)	
1-HOP	Pyrene	0.50–1.45 nmol/24hr	0.18–0.50 nmol/24hr	(Carmella et al., 2009; Feng et al., 2006a; Mendes et al., 2008; Roethig et al., 2007, 2009; Sarkar et al., 2008; Scherer et al., 2007a; Suwan-ampai et al., 2009)	(Feng et al., 2006a; Roethig et al., 2007, 2009; Scherer et al., 2007a; Suwan-ampai et al., 2009)
MHBMA	1,3-butadiene	15.5–322 nmol/24hr	0.65–7.5 nmol/24hr	(Carmella et al., 2009; Roethig et al., 2009; Sarkar et al., 2008; Scherer et al., 2006)	(Carmella et al., 2009; Roethig et al., 2009; Sarkar et al., 2008)

SPMA	Benzene	3.2–32.1 nmol/24hr	0.17–3.14 nmol/24hr	(Carmella et al., 2009; Ding et al., 2009; Feng et al., 2006a; Mendes et al., 2008; Roethig et al., 2007; Sarkar et al., 2008; Scherer et al., 2006, 2007a)	(Carmella et al., 2009; Ding et al., 2009; Feng et al., 2006a; Roethig et al., 2007; Sarkar et al., 2008; Scherer et al., 2006, 2007a; Suwan-ampai et al., 2009)
HPMA	Acrolein	5,869–11,190 nmol/24hr	1,131–1,847 nmol/24hr	(Carmella et al., 2009; Ding et al., 2009; Mendes et al., 2008; Roethig et al., 2007, 2009; Sarkar et al., 2008; Scherer et al., 2006, 2007a)	(Carmella et al., 2009; Roethig et al., 2007, 2009; Sarkar et al., 2008; Scherer et al., 2006, 2007a)
HBMA	Crotonaldehyde	9,800–26,000 nmol/24hr	1,200–3,200 nmol/24hr	(Carmella et al., 2009; Scherer et al., 2006, 2007b)	(Carmella et al., 2009; Scherer et al., 2006, 2007b)
HEMA	Ethylene oxide	19.1–102 nmol/24hr	6.51–38.8 nmol/24hr	(Carmella et al., 2009; Ding et al., 2009)	(Carmella et al., 2009; Ding et al., 2009)
Cd	Cadmium	2.3–12.8 nmol/24hr	1.34–8.04 nmol/24hr	(Batariova et al., 2006; Hoffmann et al., 2000; McElroy et al., 2007; Paschal et al., 2000)	(Batariova et al., 2006; Hoffmann et al., 2000; McElroy et al., 2007; Paschal et al., 2000)
Hemoglobin adducts[c]					
Cyanoethylvaline	Acrylonitrile	112±81 pmol/g globin	6.5±6.4 pmol/g globin	(Scherer, 2005; Scherer et al., 2007a)	(Scherer, 2005; Scherer et al., 2007a)

continued

TABLE 3-1 Continued

		Range of Recent Mean Values or Concentrations			
Biomarker	Source	Smokers	Nonsmokers	References (smokers)	References (nonsmokers)
Carbamoylethylvaline	Acrylamide	84.1±41.8 pmol/g globin	27.8±7.1 pmol/g globin	(Scherer, 2005; Scherer et al., 2007a)	(Scherer, 2005; Scherer et al., 2007a)
Hydroxyethylvaline	Ethylene oxide	132±92 pmol/g globin	21.1±12.7 pmol/g globin	(Scherer, 2005; Scherer et al., 2007a)	(Scherer, 2005; Scherer et al., 2007a)
4-aminobiphenyl-globin	4-aminobiphenyl	0.26±0.006[d] pmol/g globin	0.067±0.009[d] pmol/g globin	(Roethig et al., 2009; Scherer, 2005)	(Roethig et al., 2009; Scherer, 2005)
Other[e]					
Exhaled CO	Carbon monoxide	17.4–34.4 ppm	2.6–6.5 ppm	(Scherer, 2006; Scherer et al., 2007a)	(Scherer, 2006; Scherer et al., 2007a)
Carboxyhemoglobin	Carbon monoxide	3.4–7.1%	0.35–1.45%	(Lowe et al., 2009; Roethig et al., 2009; Scherer, 2006; Scherer et al., 2007a)	(Lowe et al., 2009; Roethig et al., 2009; Scherer, 2006; Scherer et al., 2007a)

[a] Measured in nmol/24hr unless noted otherwise (based on 1.3g creatinine per 24hr in smokers and 1.5g creatinine per 24hr in nonsmokers, or 1.5 L urine per 24hr).

[b] N/A = Not applicable because these are not detected in the urine of nonsmokers unless they use other tobacco products, nicotine replacement products (for nicotine equivalents, and sometimes NNN [Stepanov et al., 2009]), or are exposed to secondhand smoke, in which case levels are usually less than 5 percent of smoker levels (IARC, 1999b, 2006).

[c] Measured in pmol/g globin; mean ± S.D.

[d] Weighted mean ± S.D.

[e] Mean concentrations

Abbreviations:

1-HOP = 1-hydroxypyrene and its glucuronides/sulfates; Cd = Cadmium; HBMA = 4-hydroxybut-2-yl mercapturic acid; HEMA = 2-hydroxyethyl mercapturic acid; HPMA = 3-hydroxypropyl mercapturic acid; hr = hour; MHBMA = the sum of 1-hydroxy-2-(*N*-acetylcysteinyl)-3-butene and 1-(*N*-acetylcysteinyl)-2-hydroxy-3-butene; nicotine equivalents = the sum of nicotine, cotinine, 3′-hydroxycotinine, and their glucuronides; SPMA = *S*-phenylmercapturic acid; total NNN = *N*′-nitrosonornicotine and its glucuronides.

NOTE: Adapted with permission from Hecht, S. S., J. M. Yuan, and D. Hatsukami. Applying tobacco carcinogen and toxicant biomarkers in product regulation and cancer prevention. *Chemical Research in Toxicology* 23(6):1001-1008. Copyright (2010) American Chemical Society.

(Rundle and Ahsan, 2008), such epidemiologic studies with prospective study design and objective measurements of biomarkers in biospecimens would provide a direct link of the disease of interest to the biomarker and its parent compound. The relationship of tobacco carcinogens and toxicant biomarkers to cancer and other diseases has been examined in only limited prospective studies to date. Examples are cotinine and total NNAL with respect to lung cancer. In one prospective study, serum cotinine was related linearly to lung cancer risk, with no suggestion of a plateau at high exposure levels (Boffetta et al., 2006). Two molecular epidemiologic studies related total NNAL to lung cancer risk. In the first study, researchers saw a dose-dependent association between urinary levels of total NNAL and risk of lung cancer (Yuan et al., 2009). In relation to the lowest quartile of total NNAL, the risk of lung cancer associated with the second and third tertiles were 1.43 (95% CI, 0.86–2.37) and 2.11 (95% CI, 1.25–3.54), respectively (P for trend = .005) after adjustment for number of cigarettes smoked per day, number of years of cigarette use, and total cotinine (cotinine plus its glucuronide). Smokers in the highest tertiles of total urinary NNAL and cotinine displayed an 8.5-fold increased risk for developing lung cancer as compared to smokers in the lowest tertiles of these measures but otherwise similar in smoking history. A second study also showed this association using prospective measurements of total NNAL in serum, although no relationship with cotinine was seen (Church et al., 2009). Prospective measures have also been used to evaluate the association between baseline cotinine and cardiovascular disease (Whincup et al., 2004).

Description of Some Widely Used Biomarkers of Exposure

This section provides greater discussion on the common biomarkers of exposure. "Nicotine equivalents," the combination of nicotine, cotinine, 3′-hydroxycotinine, and their glucuronides, represent 73–96 percent of the nicotine levels delivered to a user of tobacco products (Hukkanen et al., 2005). This combination is a widely accepted biomarker of nicotine uptake that directly measures, to a high percentage, the nicotine dose.

Total NNAL, the sum of free and glucuronidated NNAL, and total NNN, the sum of free and glucuronidated NNN, are biomarkers of uptake of the carcinogenic tobacco-specific nitrosamines NNK and NNN (Hecht, 2008). NNK and NNN always occur together in tobacco products, and they are potent carcinogens in laboratory animals (IARC, 2007). Nicotine equivalents, total NNAL, and total NNN are unique biomarkers because of their tobacco specificity. They are only detected in people exposed to tobacco products or (for nicotine equivalents and occasionally NNN) in people who use nicotine replacement products

(Stepanov et al., 2009). As indicated in Table 3-1, the levels of these biomarkers in nonusers of tobacco exposed to secondhand tobacco smoke are generally considerably low compared to those observed in users of tobacco products.

1-HOP is a biomarker of exposure to polycyclic aromatic hydrocarbons, tobacco smoke particulate phase constituents, and products of incomplete combustion. These compounds are also commonly found in polluted air and the diet. Many polycyclic aromatic hydrocarbons are potent carcinogens in laboratory animals. The most widely studied polycyclic aromatic hydrocarbon carcinogen is benzo[a]pyrene (BaP). Polycyclic aromatic hydrocarbons always occur as mixtures, and 1-HOP, which is a metabolite of the noncarcinogen pyrene, an ever-present component of these mixtures, is a widely accepted biomarker of exposure to this class of compounds.

The mercapturic acids MHBMA, SPMA, HPMA, HBMA, and HEMA are biomarkers of the tobacco smoke gas phase constituents 1,3-butadiene, benzene, acrolein, crotonaldehyde, and ethylene oxide, respectively (Carmella et al., 2009). 1,3-butadiene, benzene, and ethylene oxide cause tumors in multiple organs of mice and rats (HHS, 2004b; IARC, 2008). Both acrolein and crotonaldehyde are associated with lipid peroxidation and perhaps inflammation (Chung et al., 1999; Thompson and Burcham, 2008). Acrolein reacts with the p53 gene at codons associated with lung cancer, a phenomenon also observed in studies of polycyclic aromatic hydrocarbon diol epoxide metabolites (Feng et al., 2006b). Acrolein is an intense irritant and cilia-toxic compound (IARC, 1995). Acrylonitrile, acrylamide, and 4-aminobiphenyl are also well-established carcinogens (HHS, 2004b; IARC, 1987, 1999b; Klaunig, 2008).

CO competes with oxygen for binding to hemoglobin and hinders the ability of oxygen to be released from hemoglobin. Although smokers are unlikely to experience acute CO-related symptoms (Scherer, 2006), CO is believed to impair oxygen delivery and cause complications of atherosclerosis and other cardiovascular diseases in smokers (HHS, 2004a).

Among the compounds related to the biomarkers, NNK and NNN, BaP, 1,3-butadiene, benzene, ethylene oxide, cadmium, and 4-aminobiphenyl are considered "carcinogenic to humans" by the International Agency for Research on Cancer (IARC, 1987, 1999a, 2006, 2007, 2008; Straif et al., 2005) and potentially may be involved in causing different types of cancer in tobacco users (Hecht, 1999, 2003a, 2010). Many of these compounds also have considerable toxic effects. Additionally, NNK, NNN, BaP, 1,3-butadiene, benzene, acrolein, acetaldehyde, formaldehyde, and CO were recommended for regulation under the World Health Organization's Framework Convention on Tobacco Control (Burns et al., 2008).

These and other widely used biomarkers of exposure are presented in Table 3-1, which presents urinary biomarkers, hemoglobin adduct biomarkers, and others. Recent data on the range of values for these biomarkers of exposure are given for both smokers and nonsmokers. Although the ranges of values for smokers and nonsmokers overlap for certain biomarkers in the table, biomarker levels are consistently elevated in smokers in the individual studies referenced in the table.

Examples of Other Biomarkers

Examples of some other exposure biomarkers include urinary or plasma phenanthrene tetraol and phenanthrols (Church et al., 2010; Hecht et al., 2005), 3-hydroxybenzo[a]pyrene and BaP tetraols (Forster et al., 2008; Zhong et al., 2010), and hydroxyfluorenes (Jacob et al., 2007) for polycyclic aromatic hydrocarbons; DNA adducts of various compounds in white cells and various tissues (Phillips, 2002); 3-ethyladenine in urine (Feng et al., 2006a); and 2-cyanoethylmercapturic acid in urine for acrylonitrile (Scherer et al., 2010).

A group of biomarkers related to inflammation, oxidative stress, and other conditions that could be influenced by tobacco products have been termed "biomarkers of potential harm" by authors from Altria, and these might be considered as risk biomarkers. Some of these, such as markers of oxidative damage, straddle the border between exposure and effect markers because they are caused by exposure to tobacco products but do not directly result from a known measured constituent of these products. One example is 8-epi-prostaglandin $F_{2\alpha}$, an established biomarker of oxidative damage, which is significantly higher in smokers than in nonsmokers (Frost-Pineda et al., 2011).

"Biomarkers of potential harm" also include those biomarkers related to inflammation (such as white blood cell count, high-sensitivity C-reactive protein [CRP], fibrinogen, and von Willebrand's factor) and platelet activation (such as 11-dehydrothromboxane B_2) as well as triglycerides and alkaline phosphatase, all of which were significantly elevated in smokers, including in one recent study that examined the relationship between these biomarkers, machine-measured tar yields, and biomarkers of exposure to cigarette smoke constituents in more than 3,500 smokers and more than 1,000 nonsmokers (Frost-Pineda et al., 2011; Liu et al., 2011). Body mass index, smoking duration, cigarette tar category, and some biomarkers of exposure were significant factors in multiple regression models for the biomarkers of potential harm. Body mass index was the highest ranking factor in the models for white blood cell count, high-sensitivity CRP, fibrinogen, and 8-epi-prostaglandin $F_{2\alpha}$, while gender and smoking duration influenced 11-dehydrothromboxane B_2 and von Willebrand's factor. Overall, the relationship between cigarette smoking, biomarkers of exposure, other factors, and these biomarkers of potential harm was quite complex (Liu et al., 2011).

Analysis of spent filters has also been used to estimate exposure. Examples include the measurement of solanesol, nicotine, NNK, and acrolein in filters. In some studies, these measurements correlated with urinary exposure biomarkers (Mariner et al., 2010; Morin et al., 2010; Pauly et al., 2009).

Examples of Biomarker Application in Product Evaluation

Exposure biomarkers are useful in evaluating new products that, according to laboratory analyses, have lower levels of certain constituents. Studies of this type have been reviewed (Hatsukami et al., 2007b). Some typical results are presented here.

Omni cigarettes were advertised as having reduced carcinogens, including nitrosamines and polycyclic aromatic hydrocarbons. Decreases of 53 percent in levels of NNK and 20 percent in levels of pyrene in smoke were advertised, based on machine measurements. Smokers were randomized to use either the Omni cigarette or medicinal nicotine, and exposure biomarkers were assessed for a 4-week period. The reductions in total NNAL were only 21 percent in those who used the Omni cigarette compared to baseline levels with their usual brand, while there was no significant reduction in 1-HOP (Hatsukami et al., 2004).

Quest cigarettes were available with deliveries of 0.3 mg nicotine per cigarette or 0.05 mg nicotine per cigarette. When smokers switched from their customary brand to the 0.05 mg nicotine yield cigarette for 6 weeks, they experienced significant reductions in cotinine (96 percent), total NNAL (78 percent), total NNN (67 percent), 1-HOP (36 percent), HPMA (56 percent), and SPMA (69 percent). In addition to these reductions, the 0.05 mg cigarette was associated with relief from withdrawal symptoms from the users' usual cigarette (Hatsukami et al., 2010).

In a 4-week study of smokeless tobacco users who switched from their usual conventional brand of smokeless tobacco to either Swedish Snus or the nicotine patch, total NNAL levels decreased significantly, although the overall mean total NNAL level was significantly lower for research participants who switched to the nicotine patch than for research participants who switched to snus. These results are consistent with the lower levels of NNK in Swedish Snus than in conventional moist snuff products available in the United States (Hatsukami et al., 2004).

In a recent study, smokers were randomized to receive the smokeless tobacco products Camel Snus, Taboka, or medicinal nicotine over a 4-week period in which they quit smoking. Significant reductions of exhaled CO, urinary cotinine, and total NNAL were observed in all groups. A significant reduction of total NNN was also observed in the treatment groups, except for the Camel Snus group. Total NNAL levels were greater in the Camel Snus group than in those who used medicinal nicotine (Kotlyar et al., 2011). These results reflect the lower levels of NNK and NNN in these products compared to the amounts delivered in cigarette smoke.

Cross-sectional studies of biomarkers and product use have also been reported. In one study, data from the U.S. National Health and Nutrition Examination Survey (1999–2008) were used to evaluate levels of biomarkers of a variety of toxicants and carcinogens in smokers compared to smokeless tobacco users. Smokeless tobacco users had higher levels of several polycyclic aromatic hydrocarbon biomarkers, as well as higher levels of total NNAL, than did nonusers of tobacco. Of 33 biomarkers analyzed, 18 were significantly lower in smokeless tobacco users than in smokers, while 10 of the 33 biomarkers were not different. The levels of the other five biomarkers, including total NNAL, were higher in smokeless tobacco users than in smokers (Naufal et al., 2011).

In summary, biomarkers can provide a more realistic assessment of the consumer's exposure to carcinogens and toxicants in tobacco products than simple analyses of the products because laboratory analyses cannot fully duplicate human use conditions. In most cases, the general trend of laboratory results is reflected in the biomarker data.

Summary of Biomarkers of Exposure

Validated tobacco carcinogen and toxicant biomarkers of exposure for a variety of compounds are now available. Measurement of a panel of these biomarkers in an appropriately conducted study can provide a realistic assessment of human uptake of a variety of toxicants and carcinogens in tobacco products. Many studies of this type show a relationship between product constituent levels and biomarker levels, but the relationship is not always straightforward. If the panel of biomarkers presented were decreased to the levels found in nonsmokers, it is likely that there would be a beneficial effect on health, but this has not been proven. Some tobacco carcinogen and toxicant biomarkers such as cotinine and total NNAL have been related to cancer risk in molecular epidemiologic studies, but most of the biomarkers discussed here would still be best described as exposure biomarkers, pending the availability of more data.

In summary, the evaluation of new products would always include standard laboratory analyses of constituents as a first step. Whether differences in constituent levels translate to differences in exposure to tobacco carcinogen and toxicant biomarkers requires testing in an appropriately designed clinical study.

Although many studies have shown a relationship between individual constituents of tobacco products and chronic diseases, there is no proof that any individual constituent or group of constituents is responsible for a given disease. Therefore, it is possible that constituents that play a decisive role in disease causation are simply not being measured, or that there are interactive effects among constituents that are critical in disease etiology but are not taken into account in the analyses. There may also be interactions between particular constituents and biological processes such as inflammation that are not fully captured by biomarker analyses. There are also limited dose response data relating constituents such as TSNAs, polycyclic aromatic hydrocarbons, volatile organic compounds, or heavy metals to specific diseases, and therefore reductions in the levels of a particular chemical or class of chemicals cannot be reliably generalized to a reduction in disease.

A particularly important question is whether a given measurement has evolved from being a "biomarker of exposure" to a "biomarker of risk" or "surrogate endpoint for disease." The committee recognizes that this question could be critical in the design of studies on MRTPs. For example, studies on smokeless tobacco products would produce significantly lower biomarkers of volatile combustion products (such as CO, acrolein, or benzene) than studies on combusted products because smokeless tobacco products do not deliver significant quantities of these materials. Epidemiologic studies demonstrate that the risk for lung cancer is higher in smokers than in smokeless tobacco users. Furthermore, when smokers stop smoking, their risk for lung cancer gradually decreases over a period of years. Based only on these facts, one might propose for example that exhaled CO is a biomarker of risk because it would clearly decrease when one stopped smoking and presumably when a smoker switched to smokeless tobacco. But there is no biological rationale for this observation, because CO is not known to be involved as a causative agent for lung cancer. Therefore, the committee believes that, for a biomarker of exposure to be accepted as a biomarker of risk or a surrogate endpoint for disease, there should be a strong biological rationale as well as compelling data from clinical or epidemiologic studies. Presently, there are only limited data on the relationship of exposure biomarkers to chronic disease.

There is no standard approved design for clinical trials in which new products would be evaluated with respect to biomarker outcomes. This topic has been reviewed recently, and further studies are required (Hatsukami et al., 2009).

Biomarkers of Risk

The validity of a study that uses a biomarker of risk is only as good as the validity of the biomarker. The utility of biomarkers of risk ultimately hangs on the assumption that they not only *correlate* to the clinical endpoint of interest, but also that the biomarker will *fully capture* the complete effect of an intervention on the clinical endpoint (Prentice, 1989).

Biomarkers of risk can include blood, other bodily fluid, or tissue markers and risk factors that relate to the natural history and progression of specific diseases and conditions. However, they cannot be considered as markers of disease occurrence on their own. Further, a single biomarker could be a predictor of many diverse conditions, such as markers of systemic inflammation or other immune system dysfunction (e.g., cytokines or CRP, blood immunoglobulin A levels or eosinophil counts). Another example of a biomarker that could predict many conditions is high levels of oxidative stress. Other biomarkers are applied to particular conditions such as cardiovascular disease (e.g., high-density lipoprotein or low-density lipoprotein cholesterol) or adult-onset diabetes (e.g., glucose intolerance, intermediate fasting blood glucose levels, glycosylated hemoglobin levels), but again, they are not indicators of the disease *per se*. Some biomarkers of disease can be extremely complex at the cellular or molecular level, such as rates of nuclear DNA repair, which can predict the occurrence of various cancers or other systemic conditions, but do not necessarily indicate disease presence.

Although more speculative, another related issue relevant to future use of biomarkers is the concomitant use of pharmacological ("chemopreventive") interventions that may be used to prevent cancer or other conditions. There are currently, for example, several candidate pharmacological interventions for human cancer chemoprevention. These are based mostly on basic research, but are also subject to human testing, including certain vitamins, resveratrol, polyamines, and flavanoids. Although none of these are fully proven in humans at the present time, in the future it is conceivable that as these products emerge as proven preventive entities, tobacco products that contain some of these agents may emerge, and complicate the regulation of health claims. As is currently the case, some of these may be marketed as dietary supplements or under the umbrella of the "nutraceuticals" movement. Regulators should be alert to the emergence of such combination products, and refer to Section 201 (rr)(4) of the Federal Food, Drug, and Cosmetic Act (as amended by the FSPTCA), which states that a "tobacco product shall not be marketed in combination with any other article or product regulated under this Act (including a drug, biologic, food, cosmetic, medical device, or a dietary supplement)."[2]

---

[2] *Family Smoking Prevention and Tobacco Control Act of 2009*, Public Law 111-31, 123 Stat. 1776 (June 22, 2009).

Surrogate Endpoints in the Study of Disease Outcomes

Surrogate endpoints are a set of predisease measures that are not clinically overt conditions but nonetheless represent nascent or early pathological processes for many subsequent clinical conditions. The presumption, with varying amount of evidence, is that some portion of these early processes progress over time to produce overt clinical illness. For many years, there has been substantial concern about adopting surrogate endpoints as the sole measure of therapeutic efficacy in clinical trials, particularly because there are very important counterexamples in the history of drug regulation where surrogate endpoint control did not lead to disease prevention or amelioration; such intermediate endpoints included blood pressure control, antiarrhythmic treatments, and cholesterol-lowering agents.

The standards for using biomarkers of risk as surrogate endpoints are even more stringent as "the surrogate endpoints should be a perfect proxy for the effect of an intervention on the recipient's risk of important clinical outcomes" (IOM, 2010). It is not uncommon, however, for potential surrogate endpoints to fail to predict clinical outcomes. According to Fleming and DeMets (1996), such failures often occur because: (1) the surrogate endpoint does not affect the same pathophysiologic pathway that leads to the clinical outcome of interest; (2) there are multiple causal pathways linked to a particular clinical outcome, but the intervention in question only affects one pathway mediated through the surrogate among several causal pathways linked to the disease; (3) the surrogate under study is insensitive to or is not a part of the causal pathway of the intervention's effect, or is insensitive to its effect; or (4) the intervention results in additional mechanisms of action independent of the disease process. The pharmacologic suppression of ventricular arrhythmias in the post-myocardial infarction setting well illustrates this. Premature ventricular contractions in the presence of ongoing myocardial damage or ischemia confer a poor prognosis, and it was thought that the pharmacologic suppression of these would result in clinical benefit. In fact, however, the Cardiac Arrhythmia Suppression Trial demonstrated that the suppression was harmful (CAST II Investigators, 1992).

The ideal setting for the use of a surrogate endpoint is when the surrogate endpoint lies along the only causal pathway of the clinical endpoint's process, and the surrogate's effect mediates the intervention's entire effect on the clinical outcome. Ideally, one should have a thorough understanding of the disease process and causal pathways, as well as a deep appreciation of the intervention's mechanisms of action. Admittedly, that is unlikely to occur with MRTPs that have a multiplicity of biological and organ system effects. It is important, therefore, to validate a surrogate used to assess the health effects of MRTPs. To be validated, it is essential that its effect be simultaneously, prospectively, and directly assessed against the desired clinical endpoint. In 2010, the IOM published a comprehensive report on the evaluation of biomarkers and surrogates, *Evaluation of Biomarkers and Surrogate Endpoints in Chronic Disease*; the committee refers the reader to that report for a detailed description of standards for the evaluation of biomarkers and surrogates. Appendix B presents the framework developed by that committee.

Because most chronic disease progression occurs over a pathogenic continuum, it is possible that in some circumstances surrogate endpoints that are close to overt clinical illness may have value in MRTP assessment. There may be instances where endpoints, such as coronary calcification levels or abnormal bone architecture and density, may be adequate to be considered in product evaluation studies. Endpoints will require a thorough evidence review and explicit specification, and could possibly improve the MRTP evaluation process. Furthermore, some of the outcomes can only be obtained with invasive procedures, and may not be suitable for all research studies.

It should be noted that with respect to reflecting true disease outcomes, biomarkers have been controversial. In general, because there have been many documented instances where pharmacological alteration of biomarker levels has not led to disease progression in the predicted direction, biomarkers have received limited credibility as disease endpoints (Hatsukami et al., 2006; Hecht et al., 2010). In general, they are not acceptable alternatives to true disease endpoints.

PRECLINICAL STUDIES

Preclinical assessment is an established step in the evaluation of any new product. In the case of a potential MRTP, the first step would be the analysis of harmful and potentially harmful constituents, as discussed in previous sections. This would be followed by in vitro toxicity and genetic toxicology tests in bacterial and mammalian systems. In these tests, extracts or fractions of the MRTP would be compared to standard conventional products. Although all in vitro tests have limitations, the collective results can nevertheless provide potentially useful information. If the results of these tests signaled decreased activity compared to a standard conventional product, the evaluation would proceed to the next stage of studies with laboratory animals. The potential MRTP would again be compared to a standard conventional product using a suitable animal model system. Finally, the evaluation would proceed to short-term genetic toxicology tests in people who used the new or conventional product. The choice of the comparison product in all of these studies is clearly important. Generally, initial comparisons should be between products of the same class, either combusted or noncombusted. The committee discusses the selection of comparison products further in Chapter 6.

Preclinical studies of the effects of smokeless tobacco products and combusted tobacco products are discussed below.

Smokeless Tobacco Products

Reviews of in vitro assays (Johnson et al., 2009) and animal models (IARC, 1985, 2004, 2007; Secretan et al., 2009) for the evaluation of smokeless tobacco or smokeless tobacco extracts have been published. Table 3-2 summarizes preclinical studies for the evaluation of harms from smokeless tobacco products.

In Vitro *Studies*

In vitro laboratory assays include the Ames test, and tests on cytotoxicity, proliferation, and programmed cell death (apoptosis); these tests provide routine tandem toxicology analyses. Mutation induction by *Salmonella typhimurium* in the Ames test or toxicologic effects noted in various human or animal cells are evaluated after exposure to smokeless tobacco extracts. As depicted in Table 3-2, smokeless tobacco extracts are a product of physical (e.g., grinding, freeze drying) or chemical methods (e.g., organics: dimethylsulfoxide, methylene chloride, methanol, acetone, ethanol; or inorganics: buffered salt solutions [Hanks, phosphate buffered saline, saline], and water or artificial saliva) (Bernzweig et al., 1998; Lindemann and Park, 1988; Merne et al., 2004; Rohatgi et al., 2005; Shirname-More, 1991; Yildiz et al., 1999).

Further assessment of genotoxic activities of smokeless tobacco extracts requires standardization not only for the method of extraction, but also for levels of moisture and humectant content.

With widespread use of these aforementioned cell assays, molecular expression patterns for epithelial and mesenchyme cells are also routinely determined using multispectral cytometric instruments. However, this approach should not ignore normal physical adherence characteristics of experimental cell targets such as epithelial and mesenchymal cells that adhere to tissue culture surfaces; false positive or negative results may be obtained in comparison to nonadherent immune cells that are also examined using flow cytometry. Furthermore, introduction of new smokeless tobacco products will require incorporation of additional cell laboratory designs to evaluate genotoxic potential.

An important cell culture design improvement is a raft three-dimension assay. This design uses mimicry of human mucosa structure to assess genotoxic responses. In addition, commercial molecular kits are available

TABLE 3-2 Summary of Preclinical Studies for the Evaluation of Harms from Smokeless Tobacco Products

Type of Assay	Summary of Results	References for Assay
In Vitro Assays		
Ames Assay *S. typhimurium* mutagenesis S9 (+/−); strains TA98,TA100 Base change mutations (rfa, uvrB, pkM101)	Some carcinogens and ST products have produced a range of results due to experience and spontaneous revision of mutations in *S. typhimurium* strains and intralaboratory variability. In general STE use demonstrated a positive Ames test but use of S9, dual strains, and base change mutations are infrequent but they add to specificity.	(Hakura et al., 2005; Jansson et al., 1991; Johnson et al., 2009; Niphadkar et al., 1996; Stamm et al., 1994; Whong et al., 1984, 1985)
Cell Assays	An increase in cell assays that evaluate genotoxicity as determined by cytotoxicity, proliferation, and apoptosis has shown STE to function in a dose- and time-dependent manner in some cell targets. Variables in cell number, cell density, time of incubations, and form of STE create variability. A standardization of these features will aid in determinations. Critical for evaluation is the origin of the cell used and the type of cell: primary, immortalized, or transformed/immortalized.	
Cytotoxicity	This assay uses trypan blue dye exclusion, tertrazolium salt assays (MTT), nuclear identification using DAPI staining, and propidium iodide nuclear identification. Extraction methods also affect results in a dose-dependent manner to increase cytotoxicity with DMSO compared to aqueous extractions (DMEM, H_20, PBS, artificial saliva). Number of ST products tested included American moist snuffs, a commercial Swedish moist snuff, and 11 ST products. Other ST products used were Kentucky moist ST, a loose-leaf form of ST, and reference 1S3.	(Rickert et al., 2008)
Human cells Normal human keratinocytes	Kentucky moist ST extracted using PBS produced a time-dependent response. Assays used included trypan blue dye exclusion and a MTT assay with reduction in tetrazolium salt to formazan dye.	(Bagchi et al., 2001, 2002)

Human cell lines Lymphoblasts: • AHH-1,TK-6 • Fibroblasts • Oral leukoplakia (AMOL-III) • Colon adeno-carcinoma (HT-29)	The absence of normal cells and the use of only transformed or immortalized cells will not permit a cytotoxic association with a strong relationship to human exposure to ST. It is unfortunate that there is only limited analysis of ST products to determine validity of responses.	(Copp et al., 2009; Gregory and Gfell, 1996; Rohatgi et al., 2005; Shirname-More, 1991)
Rodent cell lines: • Golden hamster oral squamous cell carcinoma • Chinese hamster ovary cells • Rat or mouse • Macrophage J774A.1 • Spleen T cells and oral epithelial cells from rats	Cytotoxicity assays for rodent cell lines used a concentration and time dependence to assess activity. However, except for limited use of oral epithelial cells, no other primary cells were examined. The other cell targets are either immortalized or transformed. These assays used DAPI staining, a nuclear fluorescent dye; lactic dehydrogenase release as a function of membrane integrity; neutral red stain to observe lysosomes; tetrazolium salt to formazan to determine mitochondrion viability (MTT/MTS); and Trypan dye exclusion to test for viability.	(Bagchi et al., 1996; Hasséus et al., 1997; Mangipudy and Vishwanatha, 1999; Muns et al., 1994; Rickert et al., 2008; Yildiz et al., 1999)
Proliferation of human cells: • Normal human oral epithelial cells • Rodent cell lines • Chinese hamster ovary cells • Embryonal mouse tongue epithelial cells • Rat spleen T cells and oral epithelial cells	These studies used the proliferation marker BrdU, ^{3}HdT, or colony formation counts. MTT/MTS assays were also used. Exposures to assess this assay used STE prepared from HBSS or DMEM. The extract was derived from Kentucky loose-leaf ST, dry and moist snuffs, Kentucky reference chewing tobacco, commercial Swedish snuff, or commercial khaini. The results showed a dose- and time-dependent response for normal human oral epithelial cells, but other cells produced no effect as noted for hamster carcinoma cells, inhibition for rat cells, or variable responses for low versus high doses for human oral keratinocytes and fibroblasts. An extract of dichloromethane showed a reduction in proliferation of mouse tongue epithelial cells.	(Bagchi et al., 2001; Gijare et al., 1989; Hasséus et al., 1997; Wang et al., 2001; Yildiz et al., 1999)

continued

TABLE 3-2 Continued

Type of Assay	Summary of Results	References for Assay
Apoptosis: • Human cells • Normal human oral epithelial cells • Rodent cell lines • Golden Syrian hamster oral squamous cell carcinoma • Rat or mouse epithelial cells	Programmed cell death distinguishes cell death from an internal cell death and an inflammatory imposed cell death. Assays assess membrane blebbing; flipping of membrane phosphatidyl-serine, using Annexin V, which is a calcium-binding protein; activated receptors, TRADD or Fas; cytosol cysteine proteases designated caspases; or nuclear fragmentation by denoting nucleosome formation as noted by 180 to 200bp band ladder in an electrophoretic gel. Nuclear TUNEL of cleaved ends of DNA is another routinely used marker. Assays used a DMEM or PBS extraction and exposed cells to reference STs (2S1,2S3). A positive apoptotic result requires at least two assays, which has not been uniformly performed. Furthermore, STE produces apoptosis in normal human oral keratinocytes and hamster epidermoid carcinoma cells in a dose dependent manner.	(Bagchi et al., 1999; Banerjee et al , 2007; Mangipudy and Vishwanatha, 1999)
Other Genotoxic Assays		
Chromosome Aberrations *Sister Chromatid Exchange* *Micronuclei*	The number and types of cell targets have been limited, and these assays are not as often reported.	(Jaju et al., 1992; Jansson et al., 1991; Patel et al., 1994; Trivedi et al., 1993)

Animal Studies		
	Preclinical models use either ST or STE, but they are not applied at the frequency humans with which use an ST product per day. They are conducted on hamsters, rats, and mice to produce a few tumors on an inconsistent basis. Tumors are noted in the oral cavity, lip canal, and forestomach but not in a significant frequency. There is a need to evaluate levels of carcinogen before placement and after placement of ST product into animals. There is also a lack of examination of oral tissues for pathology changes similar to human users of ST. There is a concern that continual placement of ST will produce false positive results because of local irritation. Therefore improvement in preclinical animal evaluations of ST harms rests upon use of various tissues.	(Grasso and Mann, 1998; IARC, 2007; Johansson et al., 1989, 1991b)
Hamster Buccal Pouch Studies	The hamster buccal pouch permits placement of ST in close contact with mucosa. This tissue lacks lymphatic drainage and robust T lymphocyte responses, but Langerhan cell numbers can be increased as integrity of mucosa is reduced. ST either by placement, ligation, or with bees wax plugs for up to 2 years produced no tumors. Carcinogens, polycyclic aromatic hydrocarbon, 7,12 DMBA, MCA, or quinone, exemplified by 4-nitroquinoline-N-oxide or virus (herpes simplex virus), were required to facilitate tumorigenesis of ST. TSNA injections similar to rats were also used to produce tumors in nasal, lung, trachea, and adrenal tissues.	(Ashrafi et al., 1992; Correa et al., 1990; Dunham et al., 1966, 1975; Herrold and Dunham, 1962; Hoffmann et al., 1981; Homburger, 1971; Jorquera et al., 1992; Papageorge et al., 1996; Park et al., 1986; Peacock and Brawley, 1959; Peacock et al., 1960; Schuller et al., 1993, 1994; Schwartz and Gu, 2002; Shklar et al., 1985; Worawongvasu et al., 1991)

continued

TABLE 3-2 Continued

Type of Assay	Summary of Results	References for Assay
Rat Lip Canal Studies	ST and STE has been placed into surgically formed lip canals, swabbed into the oral cavity, and placed into the rat diet. Changes in mucosa are recorded (e.g., hyperplasia, hyperothokeratosis, hyperchromatic nuclei, nuclear and cytoplasmic reversal, and mitotic figures) with stroma fibrosis and hyperplasia of forestomach. Small and inconsistent numbers of tumors are noted. Exposure levels of TSNA derived from ST in the saliva are generally higher than a single application of ST by humans (e.g., 11–55 ppm compared to 0.2 ppm). The concentration of ST used in studies is five times the exposure by humans from a single application. ST products with a high TSNA produce more mucosa pathology than ST products with low TSNA levels. Combined exposure of ST with 4-NQO increased tumorigenesis, but HSV-1 inoculation did not increase tumors with snuff administration.	(Hirsch and Johansson, 1983; Hirsch and Thilander, 1981; Hirsch et al., 1984; Hoffmann and Adams, 1981; Hoffmann et al., 1992; Johansson et al., 1989, 1991a, 1991b; Larsson et al., 1989; Palladino et al., 1986; Schwartz et al., 2010)
Rat Dietary Studies	Dietary incorporation with TSNA produced several different types of malignant tumors: lung, liver, pancreas, prostate, mammary glands, leukemia, and lymphoma. Dried snuff fed to rats or mice produced a tumor of the kidney, and one rat and three mice developed leukemia, but these latter neoplasias are likely spontaneous.	(DiPaolo, 1962; Hecht et al., 1985)
Mouse Dietary Studies	Dietary exposure of ST or injection of TSNA in mice or transgenic mice (INS-GAS/FVB) produced respectively chronic pancreatitis, or lung adenomas.	(IARC, 2007; Stenstrom et al., 2007)

NOTE: ^{3}HdT = tritiated thymidine; 4-NQO = 4-nitroquinoline 1-oxide; bp = base pair; BrdU = 5′-bromodeoxy-uridine; DAPI = 4′,6-diamidino-2-phenylindole; DMBA = dimethylbenz[a] anthracene; DMEM = Dulbecco's Modified Eagle Medium; DMSO = dimethyl sulfoxide; HBSS = Hanks buffered salt solution; HSV-1= Herpes simplex virus 1; INS-GAS = insulin–gastrin; MCA = methylcoanthrene; MTS = 3-(4,5-dimethylthiazol-2-yl)-5-(3-carboxymethoxyphenyl)-2-(4-sulfophenyl)-2H-tetrazolium; MTT = 3-(4,5-dimethylthiazol-2-yl)-2,5-diphenyltetrazolium bromide; PBS = phosphate buffered saline; ppm = parts per million; ST = smokeless tobacco; STE = smokeless tobacco extract; TRADD = tumor necrosis factor receptor type 1-associated death domain protein; TUNEL = terminal deoxynucleotidyl transferase dUTP nick end labeling.

to facilitate examination for genotoxic change among target cells. Some of these kits permit tagging and identification of chemical substances in cell sites, silencing of specific RNAs, transfection of genetic material to modify specific cellular pathways, or immortalization of epithelial cells, which facilitates cell culture growth (Andrei, 2006; Andrei et al., 2010; Singh and Nalwa, 2011).

A consistency of cell number, type, and differentiation of cell type is achievable. However, specific attention to cell features of primary cells in comparison to immortalized (or transformed immortalized, malignant cell lines) is suggested. It is also a practical conclusion that persistent genotoxic cell harm will result in a redesign of the smokeless tobacco product.

Furthermore, it is also expected that assays will address loss of normal cell physiology as reflected in regards to not only cancer, but also infection, inflammation, respiratory, or cardiovascular disease processes. These latter pathologies are often neglected in cell studies, but they are suspected to be indirect targets for ST-derived substances.

Animal Studies

Animal models for the evaluation of harm from smokeless tobacco or smokeless tobacco extracts products have included Syrian hamster buccal pouch; various strains of rats that are exposed through the diet; or various strains of rats that have undergone surgery to produce a lip canal that allows placement of smokeless tobacco into a tube of mucosa (e.g., F344, Sprague Dawley, Wistar, and SD). Dietary exposure among transgenic mice has also been reported.

A concentration and use pattern consistent with human exposure to smokeless tobacco products should be employed in animal models, but this has not been achieved. Previous studies used smokeless tobacco extracts or derivative concentrations several fold above the single self-administered exposure by humans (Hoffmann and Adams, 1981; Palladino et al., 1986). Moreover, smokeless tobacco and smokeless tobacco extracts handling and storage under carefully controlled conditions are required to prevent inappropriate formation of TSNAs (Brunnemann et al., 2002; Djordjevic et al., 1993).

At best, animal models mimic human tissue responses. However, in our present situation with the introduction of new spit-less smokeless tobacco products (e.g., additives flavorings), there is an increased difficulty to achieve this goal and evaluate chemical and biologic interplay in animals. Furthermore, attention needs to be focused upon direct contact of pathology sites in the oral cavity, gingiva/periodontum, and in nondirect contact disease tissues in respiratory and cardiovascular sites, which have been reported to be under the oral tissue's influence (Fisher et al., 2005; Ismail et al., 1983).

Animal models provide avenues to assess direct tissue damage. Additionally, animal models also offer opportunities to determine—prior to tumor induction in the oral cavity—infection, inflammation, or major organ damages in locations other than the site of smokeless tobacco application.

Epithelial oral pathology changes, benign tumors, and malignant tumors are observed in animal models and humans after exposures to smokeless tobacco or smokeless tobacco extracts products under various study conditions. However, under identical experimental conditions, not every study produced tumors (described in Table 3-2). In response to this observation and to enhance tumorigenesis, a combination of smokeless tobacco exposure with chemical promoters (in comparison to only smokeless tobacco exposures) was used to produce more local and distant tumors. A reevaluation of smokeless tobacco with promoters is still required to include human carcinogens such as BaP or viral infection patterns similar to human exposure.

It is also recognized that differences between animal species as reflected by liver microsome activity, cytochrome P450 expressions, or disposition of smokeless tobacco- or smokeless tobacco extract-derived substances result in a variability of tumor induction. However, a consistent tissue response trend is expected to determine genotoxicity or tissue harm (Leslie et al., 2007; Wu et al., 2002).

Persistent observed formation of tumors or pathologies associated with increased infection, inflammation, or respiratory or cardiovascular harm will be causes for redesign of smokeless tobacco products.

Combusted Products

A number of in vitro and animal studies have investigated the effects of combusted tobacco products on cancer- and noncancer-related endpoints. Table 3-3 summarizes the key models used in those studies.

Cell-Based Models

Evaluation of Oxidative and Nitrosative Stress Oxidative and nitrosative stress is produced by combustible tobacco products, and these reactive oxygen species and reactive nitrogen species (ROS/RNS) lead to modifications of DNA, proteins, and lipids. Extract of combustible products can be made by collecting the particles or passing the smoke through a saline solution. Detection of some individual ROS/RNS components can be measured directly, such as superoxide (luminol, dihydroethidium)

TABLE 3-3 Summary of Preclinical Studies for the Evaluation of Toxicity from Cigarette Smoke Products

Type of Assays	Summary of Results	Assay References
Cell Based Assays		
Oxidative/Nitrosative Stress	CSE has been shown to induce ROS/RNS in cell culture assays. CSE increases superoxide and nitric oxide generation in cells and also increases oxidative modifications of macromolecules (8-hydroxy-deoxyguanosine, 4-hydroxynonenal, 3-nitrotyrosine), which can be detected by antibody-based methods. Additionally, the antioxidant activity (i.e., glutathione), which can be measured by colorimetric reactions, is reduced after CSE exposure, resulting in enhanced expression of antioxidant genes, which can be assessed by quantitative PCR.	(Bertram et al., 2009; Bond et al., 1989; Malhotra et al., 2008; Peluffo et al., 2009; Sussan et al., 2009)
Inflammation	Inflammation is assessed in cell-based models via expression of cytokines (measured by ELISA and Western blot) and proinflammatory signaling cascade proteins (e.g., NF-kB, MAPK).	
Epithelial cells	CSE has been shown to induce expression of proinflammatory cytokines IL-8 and MCP-1.	(Starrett and Blake, 2011)
Smooth muscle cells	CSE has been shown to induce secretion of IL-8, eotaxin, and VEGF-A in smooth muscle cells.	(Baarsma et al., 2011)
Inflammatory cells	Dendritic cells and monocytes secrete cytokines (measured by ELISA and Western blot) and up-regulate NF-kB (measured by DNA-binding activity assay).	(Mortaz et al., 2009; Zhou et al., 2011)
Mucus Production	CSE results in increased mucin synthesis (antibody-based) and glycoprotein secretion in tracheal epithelial cells.	(Lin et al., 1989; Wu et al., 1991)
Biphasic culture	Mucus-secreting epithelial cells are maintained in an air-liquid interface to mimic *in vivo* conditions.	(Whitcutt et al., 1988)
Endothelial Activation	CSE leads to enhanced expression of cell adhesion proteins, such as ICAM, VCAM, and E-selectin, and coagulating factors such as thrombomodulin and von Willebrand's factor, which are measured by ELISA.	(Chen et al., 2004, 2009; Furie et al., 2000; Guarino et al., 2011)

continued

TABLE 3-3 Continued

Type of Assays	Summary of Results	Assay References
Monocyte adhesion	Activated endothelial cells adhere to monocytes, which can be assessed by coculture experiments in which monocytes are added to a monolayer of endothelial cells.	(Reilly et al., 2004)
Animal Models		
Inflammation and Emphysema		
Whole-body exposure	This CS exposure system subjects the entire animal to CS for a duration of several hours per day. In this model, CS induces oxidative stress, inflammation, apoptosis, and airspace enlargement.	(Clauss et al., 2011; Ma et al., 2005; Sussan et al., 2009; Yoshida et al., 2010)
Nose-only exposure	This exposure model requires the placement of the rodent's nose in a small chamber for relatively short periods of time. This exposure results in a relatively potent exposure that is shorter in duration than the whole-body exposure. This CS exposure model results in oxidative stress, inflammation, apoptosis, and airspace enlargement comparable to the whole-body exposure.	(Churg et al., 2009; Hautamaki et al., 1997)
Innate Immune Function		
Bacterial infection	Exposure to CS results in innate immune dysfunction, resulting in increased bacterial exacerbations. CS increases inflammation and decreases bacterial clearance and killing in response to *P. aeruginosa*, *H. influenzae*, *S. aureus*, and others.	(Harvey et al., 2011; Huvenne et al., 2011)
Viral infection	CS exposure heightens the inflammatory responses in lungs to influenza H1N1 and rhinoviruses. Additionally, CS enhances viral-induced emphysema caused by the viral PAMP poly(I:C).	(Bauer et al., 2010; Kang et al., 2008; Mallia et al., 2011)

NOTE: CS = cigarette smoke; CSE = cigarette smoke extract; H1N1= Influenza A virus; ICAM = inter-cellular adhesion molecule; PAMP = pathogen-associated molecular patterns; poly(I:C) = Polyinosinic:polycytidylic acid; VCAM = vascular cell adhesion protein; VEGF-A = vascular endothelial growth factor A.

or nitric oxide (2,3-diaminonapthalene) (Bertram et al., 2009; Peluffo et al., 2009). Additionally, many of the oxidative modifications to macromolecules can also be detected, including oxidatively modified DNA (8-hydroxy-deoxyguanosine) (Bond et al., 1989), lipids (malodialdehyde, 4-hydroxynonenal), and proteins (3-nitrotyrosine). Mammalian cells contain large concentrations of the antioxidant glutathione, which scavenges ROS/RNS, resulting in oxidation of glutathione. The ratio of reduced/oxidized glutathione can be quantified to determine the antioxidant capacity of the cells (Sussan et al., 2009). Cells that are undergoing oxidative stress have a decline in their pool of reduced glutathione and an increase in oxidized glutathione. Furthermore, cells respond to oxidative stress by up-regulating a large number of stress response antioxidant and phase II detoxification genes that are aimed at removing the stress and restoring homeostatic glutathione levels. The expression or activity of many of these proteins, including NRF2, SOD1, NQO1, and HMOX-1, can be quantified by quantitative polymerase chain reaction (PCR), Western blots, or commercially available activity assays (Malhotra et al., 2008).

Evaluation of Inflammation *In vitro* measures of inflammation are primarily based on production of cytokines and chemokines by epithelial, smooth muscle, and inflammatory cells (Baarsma et al., 2011; Mortaz et al., 2009; Starrett and Blake, 2011). Individual cytokines can be measured at both the protein and messenger RNA (mRNA) level. Enzyme-linked immunosorbent assays (ELISA) and Western blot assays can be used to measure protein levels, while quantitative PCR can be used to measure mRNA levels. Additionally, nuclear factor kappa B (NF-kB) represents a major pro-inflammatory transcription factor, and its activity often correlates with inflammation (Zhou et al., 2011). Thus, transcriptional activity of NF-kB can be quantified via a DNA-binding assay. Other signaling cascades, such as mitogen-activated protein kinase (MAPK) signaling, can also result in proinflammatory responses, and activation of these pathways can be detected via specific antibodies that detect the phosphorylated forms of key effector proteins (Cheng et al., 2009).

Evaluation for Mucus Production (Biphasic Culture) Mucus is secreted by airway epithelial cells. A recent advance in culturing airway epithelial cells in vitro is the development of a biphasic culture system in which epithelial cells are maintained in an air-liquid medium interface (Whitcutt et al., 1988). This culture system reflects the *in vivo* situation and allows further cell differentiation. Quantifying airway mucin synthesis in culture often relies on the characteristics of several biochemical properties of mucin, such as amino acid and carbohydrate compositions, molecular size and enzymatic characterization, and the presence of O-glycosidic bonds in the isolated molecules (Kim et al., 1985; Wu et al., 1985, 1991). However, these characteristics cannot be used practically in the routine quantification of mucin synthesis and mucous cell population in culture. Several monoclonal antibodies that are useful in the identification of mucous cell population and the quantification of mucin synthesis have been developed (Basbaum et al., 1986; Lin et al., 1989; St. George et al., 1985).

Evaluation for Endothelial Activation Recent studies suggested the involvement of endothelial cells in the pathogenesis of cigarette smoke-induced diseases like emphysema, chronic obstructive pulmonary disease (COPD), and cancer. Extracts of smokeless tobacco also induce proinflammatory changes in cultured human vascular endothelial cells. Activation of endothelial cells following exposure with cigarette smoke extract can be assessed by measuring several different biochemical markers (Chen et al., 2004, 2009; Furie et al., 2000; Guarino et al., 2011). For example, expression of adhesion molecules, such as intercellular adhesion molecule-1 (ICAM-1), E-selectin, and vascular cell adhesion molecule 1 (VCAM-1) can be assessed by Western blot or ELISA. Other markers of endothelial activation, including von Willebrand's factor and thrombomodulin can also be assessed by ELISA. Activated endothelial cells also express cytokines, such as interleukin 8 (IL-8) and monocyte chemotactic protein-1 (MCP-1), and can be measured as described above. Expression of these activation markers result in enhanced binding of endothelial cells to leukocytes, which can be observed in coculture experiments where monocytes are added to a monolayer of endothelial cells. In these experiments, adhesion is determined through quantification of bound monocytes.

Animal Models

Experiments exposing animals to tobacco smoke have been conducted in hamsters, rats, mice, dogs, rabbits, nonhuman primates, and ferrets. While it is informative to observe the effects of tobacco products in live animal models, it is not possible to mimic human use patterns of combusted products in laboratory animals. This necessarily introduces some artificiality to the experiments, and limits meaningful extrapolation of the findings from animal models to human effects.

Non-cancer Disease Rodent Models for Combusted Products Combusted tobacco products present a risk for pulmonary inflammation and COPD that needs to be evaluated in preclinical models. Multiple animal models of emphysema exist, although the only true inhalation model is the cigarette smoke model of emphysema (Harvey et al., 2011; Rangasamy et al., 2004). There are a variety of commercially available exposure systems, which consist primarily of either whole-body exposure systems or nose-only exposure systems. Whole-body exposure systems are advantageous in their ability to more carefully regulate the concentration of smoke in the exposure chamber over a period of hours. On the other hand, nose-only exposures typically expose individual mice directly to the smoke from one or a small number of cigarettes, resulting in a potent, although relatively short, exposure. Comparisons of the two methods demonstrate increased levels of carboxyhemoglobin in the rodents exposed via the nose-only method compared to whole-body exposure (Mauderly et al., 1989). Both methods are widely used, and emphysema has been demonstrated after 6 months in whole-body (Clauss et al., 2011; Ma et al., 2005; Sussan et al., 2009; Yoshida et al., 2010) and nose-only exposure systems (Churg et al., 2009; Hautamaki et al., 1997). Both exposure models result in increased oxidative stress, inflammation, and apoptosis in the lungs, and also result in alveolar destruction and airspace enlargement. These responses are all hallmarks of emphysema. However, chronic bronchitis cannot be replicated in rodents. Thus, the combustible products can be assessed for oxidative stress, inflammation, apoptosis, and emphysema in lungs of rodent models.

Chronic exposure to combusted products also causes defects in pulmonary innate immune response that increases bacterial and viral exacerbations in COPD and other diseases (Anzueto et al., 2007; Brusselle et al., 2011). Exposure to chronic cigarette smoke causes immune dysfunction in mice leading to bacterial exacerbations (Harvey et al., 2011). A 1-month cigarette smoke exposure and staphylococcus enterotoxin-induced exacerbation mouse model has also been established that shows heightened T-cell and B-cell responses (Huvenne et al., 2011).

In addition, elastase-induced emphysema has been used as a model to determine the effects of bacterial colonization and emphysematous lesion formation and inflammation in hamsters (Wang et al., 2010). Cigarette smoke exposure heightens inflammatory responses in lungs of mice infected with H1N1 (Bauer et al., 2010). Studies have also established rhinovirus infections as a mediator of viral exacerbations in COPD patients (Mallia et al., 2011). Enhanced secretion of chemokines and proteases were seen in each model. Thus, assessment of inhalable tobacco products should be evaluated for their synergistic action on enhancing the inflammatory response to virus infection or viral PAMP (poly[I:C]) in the lungs of rodent models.

Cancer Disease Rodent Models for Combusted Tobacco Products Some studies have shown that inhaled tobacco smoke can induce tumors and cancers in animal models, but the data are inconsistent. Studies in hamsters have produced convincing evidence that exposure to cigarette smoke induced an increased incidence of larynx alterations and cancers. In these experiments, the severity of the alterations correlated to dose and duration, while control hamsters did not develop any alterations (Dontenwill et al., 1973). Studies in mice and rats have produced less consistent results, but two relatively recent studies demonstrated significant incidences of respiratory tract tumors. In a study by Mauderly et al. (2004), rats exposed to cigarette smoke had a convincing, although moderate, increase in tumors of the lung and nasal mucosa. In a study by Hutt et al. (2005) with mice, exposure to cigarette smoke significantly increased incidence of lung adenoma (28.2 percent in treated, 6.7 percent in control), adenocarcinoma (20.3 percent versus 2.8 percent), total benign pulmonary neoplasms (30.9 percent versus 6.7 percent), and other changes. Both the Mauderly and Hutt studies were characterized by lengthy exposures to high concentrations of cigarette smoke. It would be important to replicate these results and to determine whether either of these protocols could become a standard model for cigarette smoke evaluation (Hutt et al., 2005; Mauderly et al., 2004).

The A/J mouse is highly susceptible to lung tumor induction and has been widely used as a screening test system in carcinogenicity evaluations. K-*ras* oncogene activation is associated with enhanced risk for lung tumor susceptibility, illustrated by presentation of pulmonary adenoma. In one replicated exposure protocol, benign lung tumors are reproducibly induced in this strain by a mixture of 89 percent cigarette sidestream smoke and 11 percent mainstream smoke, using an exposure period of 5 months followed by a 4-month recovery period. The response was due to the gas phase of cigarette smoke, and can be used to investigate the effect of secondhand smoke on lung tumorigenesis (Witschi, 2004). Whole-body exposure to diluted cigarette mainstream smoke for 5 months followed by a 4-month postinhalation period gave a concentration dependent tumorigenic response, mainly as pulmonary adenomas in A/J mice as well as in Swiss SWR/J mice. Using this protocol, Stinn et al. (2010) demonstrated that the particulate phase presented the major tumorigenic potency. Further exploration of these models for routine evaluation of combusted products would be desirable.

Experiments exposing animals to fractions of tobacco smoke and its condensate have been conducted to evaluate the carcinogenicity of tobacco smoke constituents. Mouse skin testing of smoke condensate and its subfractions has consistently demonstrated induction of both benign and malignant tumors. Mouse skin testing is particularly sensitive to polycyclic aromatic hydrocarbons, tumor promoters, and cocarcinogens, and should be part of any battery of evaluative assays. Skin application studies have also been conducted in rats, Syrian hamsters, and rabbits (IARC, 2004). Table 3-4 summarizes selected studies of carcino-

TABLE 3-4 Selected Studies of Carcinogenicity in Response to Exposure to Cigarette-Smoke Condensate in Mouse, Rat, and Rabbit

Strain	Sex	No. of Treated Animals/ Group	Type of Exposure	Exposure Dosage and Duration	Tumor Incidence	Reference
Mouse						
CAF1	M + F, 1:1	44–112	Skin painting (dorsal) of CSC	CSC/acetone solution (40 mg CSC/ application), × 3/wk, lifetime	36/81 (skin epidermoid carcinoma), 0/30 (acetone controls)	(Wynder et al., 1953)
ICR Swiss	F	5,200	Skin painting (dorsal) of CSC	CSC/acetone solution (150 mg or 300 mg CSC/week), × 6/wk, 78 wks	482/5,200 (skin carcinoma), 3/800 (acetone controls)[a]	(Gargus et al., 1976)
ICR Swiss	F	4,900	Skin painting (dorsal) of CSC	CSC/acetone solution (25 mg or 50 mg CSC/application), × 6/wk, 78 wks	1,157/4,900 (skin carcinoma), 0/800 (acetone controls)	(Gori, 1976)
ICR/Ha Swiss	F	100	Topical application with CSC to oral mucosa (lips and oral area)	CSC/acetone (26 mg CSC/ application), × 5/wk, 15 months	52/81 (lung tumors[b] ($P < .0001$), 20/89[b] (acetone controls)	(DiPaolo and Levin, 1965)
Initiation Study						
ICR/Ha Swiss	F	30	Skin painting (dorsal) with CSC active fraction with or without subsequent painting of the skin with croton oil	CSC active fraction/ acetone (2.5 mg of 0.6% SC/ application), 10 × on alternate days croton oil (2.5%), × 3/ wk, up to 15 months, 10 days after last CSC active fraction application	After 12 and 15 months: 4/30 (skin carcinoma), 0/65 (croton oil controls)	(Hoffmann and Wynder, 1971)

continued

TABLE 3-4 Continued

Strain	Sex	No. of Treated Animals/ Group	Type of Exposure	Exposure Dosage and Duration	Tumor Incidence	Reference
Promotion Study						
Swiss	F	30–50	Skin painting (dorsal) of CSC with or without initiation by DMBA application	DMBA (75 μg); CSC/acetone (75 mg CSC/application, start: 1 wk after DMBA application), × 2–3/wk, 12 months—animals observed 3 months later	DMBA: 2/30 (skin carcinoma) (7%) 2 × CSC: 1/40 (skin carcinoma) (3%) DMBA + 2 × CSC: 8/30 (skin carcinoma) (27%) 3 × CSC: 11/50 (skin carcinoma) (22%) DMBA + 3 × CSC: 11/30 (skin carcinoma) (37%)	(Wynder and Hoffmann, 1961)
Other Tobacco						
Swiss albino	M	15	Oral gavage of Indian bidi smoke condensate	1 mg bidi smoke condensate/0.1 mg DMSO, 5 days/wk, 55 wks, termination 90 wks	4 hepatic hemangiomas, 1 stomach papilloma and carcinoma, and 1 esophageal carcinoma/15 mice; 0/15 (untreated or DMSO-treated controls)	(Pakhale et al., 1988)
Rat						
Osborne Mendel	F	NG	Intrapulmonary administration of CSC pellet	CSC/beeswax:tricaprylin (24 mg CSC/injection), up to 107 wks after implantation	14/40[c] (lung squamous-cell carcinoma), 0/63[c] (beeswax:tricaprylin controls)	(Stanton et al., 1972)

OM/NCR	F	120[d]	Intrapulmonary administration of CSC pellet	CSC/beeswax:tricaprylin (5, 10, 20 or 67 mg CSC/injection), 120 wks after implantation	4%, 10%, 20%, and 42% pulmonary carcinoma prevalence; 0% carcinoma prevalence for three control groups of about 190 rats each	(Dagle et al., 1978)
Rabbit						
Albino New Zealand	M + F	38	Skin painting of CSC (both ears)	SC/acetone solution (100 mg CSC/application/ear), × 5/wk, lifetime (4–6 yrs)	4/38 (2 skin carcinoma + 1 skin liposarcoma + 1 skin fibrosarcoma), 0/7 (acetone controls)	(Graham et al., 1957)

NOTE: CSC = cigarette-smoke condensate; NG = not given; wk = week; yrs = years.
[a] Skin papillomas.
[b] Mostly adenomas.
[c] Incidence in animals that died 43–107 weeks after injection.
[d] 4 × 10 rats/group terminated before 120 weeks.
SOURCE: Adopted from the International Agency for Research on Cancer (IARC). 2004. *IARC Monographs on the Evaluation of Carcinogenic Risks to Humans: Volume 83. Tobacco smoke and involuntary smoking*. Lyon, France: IARC.

genicity in response to different methods of tobacco smoke condensate administration.

Human

Detection of mutagens in the urine of smokers has been shown to be an effective and reliable method of quantifying human exposure to mutagens created by combusted tobacco (Kriebel et al., 1985; Putzrath et al., 1981; Yamasaki and Ames, 1977). These methods involve concentrating organic compounds from urine and evaluating the mutagenicity of the resulting mixture with the Ames test. Aromatic amines and heterocyclic aromatic amines have particularly high activities in these assays, so the results obtained from studies of smokers' urine may mainly reflect the concentrations of these compounds. Studies have shown that urinary mutagenicity increases with the number of cigarettes smoked (Kuenemann-Migeot et al., 1996; Tuomisto et al., 1986), and that mutagenicity of urine from individuals who used products that heat rather than burn tobacco is similar to that of nonsmokers (DeBethizy et al., 1990; Doolittle et al., 1989; Smith et al., 1996).

Cytogenetic damage, including micronuclei (Bonassi et al., 2003), sister chromatid exchange, and other chromosomal aberrations can also be detected in the cells of smokers. Sister chromatid exchange in peripheral lymphocytes of smokers has been shown to be consistently higher in smokers than in nonsmokers (Rowland and Harding, 1999; Sarto et al., 1985).

Summary of Preclinical Studies

Although preclinical assays for toxicity and carcinogenicity can provide relevant and meaningful data about tobacco products, these assays are limited in their usefulness in this regulatory context. For the purposes of evaluating MRTPs, scientific evidence should be able to support the inference that a particular MRTP will reduce the rates of tobacco-related disease compared to another conventional product.

Preclinical assays alone are fundamentally incapable of supporting such a claim. The majority of the technologies used to test *in vitro* toxicology were not generated for testing tobacco products and their toxicity (Johnson et al., 2009). These methods "are not reliably quantitative to allow valid comparisons of substantially different tobacco products with differing yields of complex chemical mixtures" and "provide data that cannot reliably be extrapolated to infer human cancer risk" (Johnson et al., 2009). As such, evidence produced by these methods cannot by itself support the inference that an MRTP will produce less harm than another product.

Nevertheless, preclinical assays of toxicity still play an integral role in the evaluation of MRTPs. These toxicology methods are primarily intended to be used as screening methods to identify potential human carcinogens. These assays are essential in identifying particularly risky or toxic products that should not be tested in humans and for identifying products that have reasonable potential for success and should therefore proceed to clinical evaluation. The role of these tests is to ensure that products that proceed to clinical evaluation in people are not unnecessarily risky and have a reasonable potential to ultimately reduce harm. No one assay can do this alone, because each assay is limited in its scope. A complete battery of preclinical assays should be required prior to committing a product to clinical evaluation. At a minimum, the battery should include assays with consistent and reproducible results and that reach across a wide spectrum of mechanisms and types of toxicity, such as: (1) *in vitro* toxicity and genetic toxicology tests; (2) appropriate animal studies; and (3) urinary mutagenicity and sister chromatid exchange in smokers. The proper role of these assays is as gatekeepers to long-term studies in humans, which include studies of not only health effects in individuals but also population effects and behavioral effects.

Going forward, it should be anticipated that new assays that specifically focus on tobacco products, that are intended to produce evidence upon which reliable comparisons can be made between products, and for which inferences about human effects can be reliably made, will be developed and should be added to the evaluation process. Over time, the assays discussed in this section may become outdated as technology advances and develops. In the future, it is possible and indeed likely that new assays will be developed that may specifically focus on tobacco products. These assays could be designed to produce evidence intended for comparisons between products, or evidence intended for inference about human effects. These assays should be added to the evaluation process.

CLINICAL STUDIES

Clinical Trial Methods

The use of appropriately designed clinical trials will be important to establish whether use of the MRTP reduces exposure to toxicants or induces positive changes in surrogate markers as claimed by the manufacturer. Although people who have never used tobacco products cannot be randomized to begin using tobacco products (including MRTPs) in the longer term, there may be advantages from certain trial designs involving substitution of conventional tobacco products with the MRTPs. This design has similarities with a clinical trial evaluation of a smoking cessation intervention. This topic has recently been reviewed (Hatsukami et al., 2009), and so the committee will not reiterate this material here. Short- and intermediate-term clinical trials—where the research participants use the product regularly throughout the day rather than in the confines of a laboratory setting—are thought to provide a better approximation to real-world use. This is particularly true in regard to the question of an MRTP's ability to be a substitute for cigarettes. Typically, there are "forced switching" studies, where the participant ceases using traditional cigarettes and uses the MRTP for a fixed period of time. Use patterns of the MRTP can be prescribed (controlled use) or can be left to the participant (*ad libitum*). Such studies can be conducted in the field (i.e., research participant brings the MRTP home) or in a residential setting (i.e., research participant is confined for the duration of the study). Residential settings offer the advantage of stricter control over exogenous factors that could affect biomarkers of exposure or risk (e.g., diet, environmental exposures, etc.) and facilitate compliance with product use. However, these are necessarily contrived and so represent a best-case scenario for product use. Nonresidential studies are more difficult to control and compliance is more difficult to assure, but they are more accurate representations of user behavior. Studies in this idiom have consisted of 12–120 participants, typically containing 10–20 participants per experimental arm. Intermediate-term trials have typically been conducted in the field, but designs have been more varied, ranging from relatively tight prescription of product use to more observational designs. Intermediate-term trials have an advantage of stabilization of use of the MRTP with time.

As one can readily appreciate, demonstrating that an MRTP can achieve measurable changes on clinical endpoints may require large, long-term trials. These designs are sometimes questionable from a perspective of feasibility, are undoubtedly costly, and can only provide useful data after years of investigation. In most studies described in the literature, biomarkers of exposure (e.g., NNAL, cotinine, 1-HOP) and/or risk (8-epi-prostaglandin $F_{2\alpha}$, forced expiratory volume in one second [FEV1], CRP) have been assessed as the main outcomes. The ability to demonstrate in a randomized trial the significant reduction of a range of biomarkers of exposure and/or risk, in the absence of significant elevation of others, will be critical to the consideration of an MRTP application.

There may be other situations where randomized trials can be employed for the evaluation of specific health effects of MRTPs. While the pathogenesis of the primary tobacco-related chronic diseases (e.g., various cancers, heart disease, and stroke) is thought to take place over many years, there are a number of conditions where MRTP effects could

BOX 3-2
Some Examples of Short-Term Health Outcomes for Which MRTPs Might Be Evaluated

1. Short-term vascular phenomena, such as intermittent claudication or Raynaud's disease, which may be responsive over a short term Ankle-Brachial index
2. Mitigation of tobacco-related skin conditions, such as psoriasis or hyperhidrosis
3. Alterations in surgical wound healing, which are known to be tobacco sensitive
4. Variation in the progression and impact of periodontal disease, which is sensitive to tobacco use
5. Alteration in the progression or regression of precancerous mucosal lesions in the oral cavity, where frequent evaluation is feasible
6. Time required for a fracture to heal, also related to tobacco exposure
7. Alteration in the rates of tobacco-related outcomes of pregnancy associated with MRTP use, including fetal death, premature labor and delivery, and low birth weight infants, could be assessed in a relatively short period of time
8. Lung function, pulmonary function testing
9. Blood pressure

be evaluated over a relatively short (< 2 years) time frame. An emphasis on shorter-term clinical outcomes might be one important way to achieve relevant information about the potential health impact of an MRTP. Although not an exhaustive list, Box 3-2 presents a list of examples of health outcomes that MRTPs might be evaluated for relative to smoking and smoking cessation.

In clinical trial design, the use of at least one control arm is crucial. Previous trials have employed various control groups, including arms involving those that continued smoking, those that undertook smoking cessation, and those that switched to medicinal nicotine. Use of a continued smoking arm is necessary to compare exposure and risk reduction while using novel products with levels associated with traditional product use. A cessation arm (where participants may quit with or without pharmaceutical aids) provides researchers with a comparison of the MRTP with the greatest possible exposure or risk reduction. Broadly speaking, a desirable outcome for MRTPs would be a pattern of exposure and risk biomarkers closer to the cessation level than the smoking level. Standard analytical techniques, such as the intention-to-treat principle, would generally be applied to this fundamental design. It is important to recognize that no single randomized controlled trial can address all of the health effects caused by tobacco use. Replication of clinical trial results is an almost universal requirement in the regulation of drugs. Although replication frequently is interpreted as the replication of results using an identical protocol design, replication requirements can be met by the confirmatory evidence standard. In fact, from a psychometric point of view, stronger conclusions are possible if congruent results are obtained using different measures and methods.

Participant selection and recruitment are important considerations for the generalizability of clinical trial findings. Typically, pregnant and breastfeeding women, children, and those with unstable physical or mental illness have been excluded from MRTP studies. Typically, minimum daily cigarette consumption values are specified for smokers (often > 10 cigarettes per day), and concurrent use of other forms of nicotine or prior experience with the MRTP is proscribed. Research participants have typically been recruited through community advertising (e.g., flyers, newspapers) seeking smokers willing to test new and potentially less risky products.

The Role of Clinical Trials in the Evaluation of MRTPs for Health Effects

Overall, despite the limitations of clinical trials for product evaluation, the committee recognizes the critical role for clinical trials in evaluating the effects of MRTPs on human health. The committee suggests that clinical trial designs consider the following key points, adapted from recommendations provided by Hatsukami et al. (2009):

- Trial designs where biomarkers are used as primary or secondary endpoints should be informed by the half-life of the biomarker(s) examined and the time needed to stabilize use behavior of the MRTP. Determining the stabilization of product use behavior may require a longitudinal trial.
- Clinical trial designs should use both a controlled use approach and an *ad libitum* approach in complementary studies.
- Any clinical trial should include at least two control conditions—usual brand use and cessation—to allow examination of the relative effects of the MRTP on biomarkers of exposure or risk.
- Short-term residential and nonresidential studies and intermediate-term clinical studies have different strengths and limitations, and proper evaluation of MRTP effects may require several or even all of these study designs. Use of these different study designs will assist for cross-validation.
- Participants in trials should be drawn from a broad cross-section of the population, considering sex, race or ethnicity, smoking or tobacco use history, degree of dependence, stage of change, socioeconomic status, and genetic makeup (e.g., rate of nicotine metabolism).

Observational Methods

Observational epidemiologic studies play a critical and central role in the evaluation of MRTPs. Although they will rarely, if ever, have the compelling scientific credibility of experimental designs, these methods form the basis for most evaluation studies of regulated products in the community. This is true particularly in the postlicensure/postcertification period, but also during the initial regulatory evaluation.

Given the great diversity of health consequences of tobacco use (see Table 1-1 in Chapter 1), determining the contrasting potential effects of MRTPs on disease outcomes and population health is a difficult matter. Long, intensive, and robust studies of actual health outcomes would be required to fully evaluate the net effects of MRTPs relative to conventional tobacco products.

An exhaustive, multidisciplinary approach to plan and execute epidemiologic studies to evaluate the relative impact of various MRTPs on health status and outcomes—behavioral, biochemical, genetic, and pathophysiological—are all necessary at some level. In some cases, full answers may not be possible; however, in many cases, rigorously designed studies are likely to be extremely useful in making important policy decisions.

This section provides an overview of the types of epidemiologic and related studies that can address the issues noted above. It is divided into four sections relevant to regulatory and related policy decisions:

1. Considerations on studying disease-exposure associations
2. General design issues for epidemiologic and related studies
3. Evaluating outcomes for various conditions, including the selection of research conditions and the contingencies for each disease category
4. Types of feasible study designs

Preliminary Considerations in Studying the Disease Outcomes Associated with MRTPs

There May Be Many Potential Types of MRTPs and Many Patterns of Usage

One of the critical general issues in exploring the health impact of MRTPs is the multiplicity of products that may become available and the potential variation in their characteristics. Product type and the purported mechanisms by which it is expected to reduce disease risk by necessity inform the type of epidemiologic studies that can be effective in evaluating its health effects. If there are many products with potentially diverse pharmacological and biological effects, evaluating each separately could be a great logistical challenge. In observational studies, it is axiomatic that the product(s) involved should be unambiguously identified so the effects of MRTP exposure can be disentangled from other tobacco products.

Based on available information, it may be necessary to combine various products into a manageable number of analytical categories in order to conduct statistically robust studies. The construction of these categories should, however, be scientifically credible. A related issue is that over time, individuals using MRTPs may switch products at irregular intervals, use them at varying rates, use them interchangeably, or even use them simultaneously with conventional tobacco products, making it very difficult to credibly document use patterns that can be related to health outcomes in observational studies. A similar issue arises if many products are not widely used in the general population; in this case, there may be insufficient population exposure to confidently assess particular health outcomes. It is likely that only products with substantial and long-term general market sales in the general population will be suitable for epidemiologic assessment of MRTP-related disease occurrence, that is, largely for postcertification activities. It is also possible that MRTPs that have been consumed in the community over a long period may have changed in content and exposure yields, thus complicating exposure assessment.

The Diversity of Diseases and Conditions Caused by Conventional Tobacco Products

Tobacco use, particularly cigarette smoking, causes a large number of diseases and conditions (HHS, 2004a, 2006). Diseases and conditions caused by active cigarette smoking are summarized in Table 1-1. Thus, an important conceptual issue is which conditions should be evaluated for alteration when evaluating MRTPs. It is obvious that not all tobacco-related conditions can be assessed, and policy decisions on evaluative strategies need to be made. Also, it is possible that different MRTPs will have different effects on different disease processes. For example, there is no necessary *a priori* reason to believe that an MRTP that reduces the risk of atherosclerotic disease may yield the same effects on risk of various cancers, bone fracture, premature delivery, or Alzheimer's disease. The multiplicity of potentially available health outcomes requires careful consideration when selecting epidemiologic study designs. Epidemiologic assessments will be much more efficient if targeted to specific diseases and conditions based on hypotheses grounded in previous literature reviews of the product—disease associations, the known chemical constituents of the MRTP, the constituents to which product users are exposed, and other suggestions from professionals or the public. It is conceivable or even likely that different types of study designs may be needed for different disease outcomes. For example, a study evaluating the effects of an

MRTP on lung cancer may be structured differently from one assessing atherosclerotic outcomes. In fact, it is entirely possible that an MRTP may decrease the risk of some conditions while increasing the risk of others, even those that are not necessarily caused by tobacco smoking. The design of studies to assess offsetting risk can be extremely complex, and policy decisions will have to be made as to how this issue should be regarded. All of this reinforces the need to mandate epidemiologic precertification studies that are directed by the best exposure and toxicological data available.

General Design Issues for Epidemiologic and Related Studies of MRTP-Disease Associations

The Importance of Determining "Acceptable" Effect Size Differences Among Products

As population studies are developed for evaluating the health impact of MRTPs, there may be value in establishing in advance the policy for interpreting various study effect sizes as the differences between outcomes of MRTPs versus conventional tobacco products emerge. That is, how much of a decrease in disease rates is important to individuals trying to change their smoking habits, and what differences should lead to certain regulatory decisions? And how much difference should occur before a product can be called an MRTP? In general, such policies should be determined aside from statistical significance, although the latter is important. For example, if an MRTP, *ceteris paribus*, yielded a hypothetical 2 percent reduction in lung or bladder cancer rates over a defined time period, would that be a suitable basis for regulatory decisions or compelling enough for a smoker to change products? And what if the effect is different over a longer time period? Such regulatory decisions may be more problematic given that the impact of the MRTP on other conditions may not be well understood. Considering acceptable "effect sizes" early on may help define the sample sizes and other design features of proposed studies.

The structure of studies that contrast risk of disease among MRTPs and conventional tobacco products is of paramount interest, and, speculatively, many potential MRTPs with substantial reduction in toxic exposures may show reductions in disease risks. This is likely given the high toxic exposures that occur due to use of conventional cigarettes. However, studies that contrast disease risks conferred by competing MRTPs may be more challenging because exposures are lower and confounding factors may become more important. This problem should be considered in structuring such studies.

Strategies to Increase the Efficiency of Study Designs in Exploring MRTP-Associated Disease Risks

In conventional cohort studies, as discussed below in more detail, health outcomes among those persons using MRTPs are prospectively compared to those using conventional tobacco products, and it may take many years for answers to appear. This is because the incubation period of many smoking-related conditions may extend to decades, and there may be no basis for understanding how long it may take to show differences among those using MRTPs, even with lesser cumulative exposures to certain tobacco constituents. This is especially true because many persons in these study cohorts would be former smokers, and disease pathogenesis is already under way. However, some strategies exist that may be explored to enable acquisition of earlier answers:

1. *Some diseases emerge earlier than others* after tobacco initiation, or decrease more rapidly when conventional tobacco products are withdrawn, and in these situations it may be possible to acquire earlier answers regarding MRTP health effects. An important example is coronary heart disease, where withdrawal of cigarette smoking is associated with a clear reduction in disease risk within a few years of smoking cessation. Another important example could be evaluating the relative effects of MRTPs and conventional tobacco products on pregnancy outcomes. While all pregnant women should be strongly discouraged from all tobacco use, those who cannot or will not quit may be approached to use alternative products, and answers to problems such as fetal loss or premature delivery may be available relatively quickly. Such study designs will require substantial consideration and thorough ethical review.

2. An additional approach is to focus on various *population groups that are at particularly high risk of disease outcomes* of interest. One obvious approach is to enrich disease outcome studies with individuals possessing high-risk factor levels (other than tobacco use) for those conditions. This may allow smaller sample sizes and possibly shorter study intervals. For example, rates of cardiovascular disease outcomes would be increased by enrolling those with elevated blood pressure and cholesterol levels, or familial hypercholesterolemia and diabetics. Selected occupational groups where smoking levels are high in addition to job-related exposures may be at special risk of lung tumors, such as in uranium or asbestos miners, and textile workers. Smokers who have had one tumor that is "cured" are at greater risk of a second tumor (e.g., those with head and neck cancers) and may be important research participants. All of these groups may be suitable for clinical trials or observational studies of MRTP health effects.

3. A related general approach to possibly accelerate informative studies on the role of MRTPs in altering risks and rates for important diseases and conditions is to *focus on population groups with higher prevalence rates of conventional tobacco product use.* Examples of groups with higher cigarette smoking rates include certain minority groups, persons with lower socioeconomic status, sexual minorities, individuals with psychiatric conditions and substance abuse other than tobacco, and disabled individuals. Focusing on such populations may lead to efficiencies in study recruitment, and because of higher rates of smoking, such populations should be given special consideration for emphasis in reducing smoking-related morbidity and mortality. As above, these high-risk populations may be important candidates for trials or observational studies.

4. The use of composite health outcomes could also increase the efficiency of MRTP evaluation studies. Conventional cohort studies (see below) can yield data on all disease outcomes for which information is sought. However, understandably, these studies examine single disease outcomes separately, following specified hypotheses and exploring biologically and toxicologically plausible causal pathways. But because these studies yield many tobacco-related outcomes of importance, such as major diseases and causes of death associated with cigarette smoking, there may be scientifically credible value in *pre-specifying and exploring composite outcomes,* possibly increasing their efficiency and decreasing their duration. One of the most obvious strategies would be to create an outcome consisting of any of these major diseases, whichever comes first. In fact, this is an approach used in some cohort studies and clinical trials. The issue has been perhaps best evaluated with respect to varying causes of death, where smoking leads to any number of important illnesses, most precluding the occurrence of the others, a phenomenon called "competing mortality." Because it is an important goal for MRTPs to prevent and alleviate suffering from a variety of diseases, composite outcomes may more accurately reflect general health outcomes, in the same way that self-reported health status and disability-adjusted life years summarize health status across individual disease states. It might even be worthwhile to weight disease outcomes in terms of clinical importance or likely relation to product use (e.g., with lung cancer receiving a higher weighting than chronic bronchitis). The use of a composite category in no way precludes evaluating individual disease outcomes.

Consideration of Confounding Factors in Epidemiologic Studies of Tobacco, MRTPs, and Altered Disease Risk

Almost all epidemiologic studies can be subverted if confounding factors—factors associated with both the likelihood of exposure and disease outcomes—are not taken into account. Often, these are the very risk factors that explain why some persons are at greater risk for various diseases, such as hypertension, hypercholesterolemia and diabetes, and atherosclerotic diseases. One would not want to falsely attribute an altered disease risk associated with an MRTP when the contrast groups actually differ in other risk factors that can explain the observed differences.

Another example of a confounding factor is the common situation where epidemiologic studies contrast continuing cigarette smokers with those who change to MRTPs. The latter group may be less addicted to tobacco and nicotine and thus may have quantitative tobacco exposure differences that need to be considered in assessing disease risk. Determining patterns of MRTP use and levels of exposure will be very important in assessing product-disease associations. Comparative studies of these groups should attempt to adjust for these exposure differences among the contrast groups.

Also, as suggested above, an important example of a confounding factor associated with tobacco addiction is the fact that cigarette smoking is associated with increased prevalence rates of a variety of psychiatric illnesses, including various substance use and abuse syndromes including alcohol and illicit drugs. Thus, for maximum analytical specificity in evaluating MRTPs, it would generally be important to try to acquire a history of psychiatric illnesses and related substance abuse activities that may in themselves lead to adverse health outcomes. If the illnesses and substance use rates are lower among those able to switch to MRTPs, this could confound study findings.

Other situations exist where confounding factors may be important when considering studies that contrast disease outcomes of MRTPs versus conventional tobacco products:

a. Cigarette smokers often try to stop smoking, as documented in this report, but it may be important to understand some of the motivations. Some smokers stop smoking or change tobacco products because of overt incident diseases or the self-perception of abnormal symptoms or related clinical problems. It is important to obtain a history of these events when conducting epidemiologic studies; otherwise, product use may appear to be associated with increased disease risk when in fact the product was initiated because of the advent of clinical problems.

b. Another related issue that occurs with such studies is that smokers often use aids to assist in smoking cessation, such as nicotine-containing products or other medications. It is documented that many of these smoking cessation aids have their own set of adverse clinical events (Singh et al., 2011), and to the extent possible their use should be carefully monitored and not be confused with MRTP-associated effects.

c. Comorbid conditions in smokers and MRTP users are also potential confounders that will need attention. It is not surprising that current and former smokers may have higher rates of various medical conditions than nonsmoking populations. The presence of such conditions and their treatments prior to study onset can confound the evaluation and interpretation of MRTP or conventional tobacco disease outcomes and should be scrupulously documented in all epidemiologic and related studies of product contrasts. It should be emphasized that assessment of disease treatments is also extremely important. Extensive treatments of important diseases may be indicative of more severe disease processes, and they may be associated with higher rates of secondary complications, such as from percutaneous coronary stents or adjuvant chemotherapy.

d. In the genetics/genomics era, gene variants have been discovered that may affect the pharmacological and pathogenetic effects of both conventional cigarettes and MRTPs, as well as various disease outcomes (NCI, 2009). In a sense, for the purposes of product evaluation, genes may become confounders of product-outcome assessments, because they may relate both to product use behavior and to the clinical outcomes. The relevant genetic literature should be monitored so genetic studies can be made if a gene variant becomes an important part of causal pathways. The committee recognizes that people may be increasingly likely to have genome scans or other genetic tests, and availability of such information should be monitored.

e. For the past several years in the United States, cigarette smoking rates have been higher among persons with lower socioeconomic status (CDC, 2011a); that is, those with lower educational attainment, those with lower personal and family income, and "blue-collar" workers are more likely to encumber higher rates of adverse occupational or environmental exposures. Epidemiologic studies that compare conventional cigarette smokers with nonsmokers or MRTP users thus need to scrupulously adjust for socioeconomic differences among these groups in order to avoid confounding by this potent factor, which is related to both rates of tobacco use and adverse health outcomes.

f. Because of social and regulatory pressure on smoking behaviors, cigarette smokers tend to congregate with each other, or find themselves together in designated smoking venues. Thus, in epidemiologic studies of MRTPs that include biomarkers or more concrete health outcomes, the role of secondhand smoke exposure can be an important determinant. This problem should lead to routine data collection of secondhand smoke exposure as part of observational study methodology.

g. In this era of rapid changes in tobacco-related public health policies, legislation (e.g., increased tobacco taxes), and health information, it is possible that changes in secular events could significantly influence such outcomes as tobacco use and cessation, likelihood of adoption of MRTP use, and engagement in other health-relevant behaviors (exercise, use of statin drugs). This calls for careful evaluation of not only such secular events but also the possible consequences of such events, so that these can be used as covariates or time-varying covariates, depending on the nature of the research design.

h. In certain observational studies, ascertainment and detection bias may be an issue. For example, ex-smokers switching to an MRTP might be under more surveillance than other populations, or higher-risk subjects may undergo additional diagnostic tests or screening, which may skew the results. Consideration of detection and ascertainment bias is particularly important in the design and evaluation of longer-term observational studies.

Benchmarking the Health Effects of MRTPs

A generally useful but sometimes tacit presumption in evaluation studies of MRTPs is that conventional tobacco product use is the health benchmark against which MRTPs are evaluated. However, this could be difficult to execute in the common situation where a credible lifetime history of cigarette smoking is difficult to obtain. Researchers may also benchmark MRTPs with each other and with the health outcomes of nonsmokers, and this may also be of value in making policy decisions. An explicit approach to benchmarking health outcome levels is extremely useful and could encompass a range of tobacco products or MRTPs. These should be declared in advance of proposed health studies.

Evaluating Health and Disease Outcomes in the Study of MRTPs

There are several potential types of response variables (outcomes) to MRTPs and tobacco products in observational studies and other clinical and population research. The advantages and limitations of using biomarkers and surrogate endpoints were discussed earlier in this chapter. It should be noted that with respect to reflecting true disease outcomes, biomarkers have been controversial. In general, because there have been many documented instances where pharmacological alteration of biomarker levels has not led to disease progress in the predicted direction, biomarkers have received limited credibility as disease endpoints (Hatsukami et al., 2006; Hecht et al., 2010). In general, they are not acceptable alternatives to true disease endpoints.

Furthermore, for many years there has been substantial concern about adopting surrogate endpoints as the sole measure of therapeutic efficacy in clinical trials, particularly because there are very important counterexamples in the history of drug regulation where surrogate endpoint control did not lead to disease prevention or amelioration; such intermediate endpoints included blood pressure control, antiarrhythmic treatments, and cholesterol-lowering agents.

General Epidemiologic and Related Study Designs for Assessing Altered Disease Risk or Mitigation Associated with MRTP Use

In general, except for short-term pharmacological or toxicological studies and some behavioral interventions, disease risks associated with MRTPs will be assessed with observational studies, although there is certainly room for clinical trial methodology, as noted previously, because of their growing importance to translational science. A panoply of observational research designs is available, and only a few of the most basic and central will be discussed here.

Cohort Studies in MRTP Assessment

Cohort studies are obvious candidates for the evaluation of MRTPs, and over the years they have been an important instrument of tobacco product evaluation (FDA, 2011b). In this type of study design, persons with various product use habits are followed into the future to assess variation in clinical outcomes. These studies have several important strengths:

- Biochemical tobacco and MRTP exposure assessments can be made at baseline, offering "unbiased" exposure assessment before health outcomes occur.
- There is less of a problem with retrospective recall of product use, because this information is summarized at the start of the study and followed prospectively.
- Changing product use habits can be monitored concurrently as the study progresses.
- Outcomes are documented as they occur, and verification becomes more efficient.
- A wide variety of outcomes can be evaluated in the same study, particularly those that are more common.

Indeed, cohort studies allow assessment of overall health status and outcomes. However, there are also prominent or at least potential limitations to this design:

- Important and severe chronic illnesses may be uncommon and take many years to occur, even in a population of cigarette smokers, and thus the studies may require many years, large sample sizes, and substantial resources to complete.

- If exposure to MRTPs is limited in the community, these studies may be underpowered and inefficient.
- MRTP or conventional tobacco product use habits may change over time, and thus determining and analyzing differential exposures may be complex.

Efficiencies could be obtained in part by enrolling only persons who use certain tobacco products or MRTPs in a cohort study. Depending on whether the contrast group for MRTP evaluation is nonsmokers or never smokers, one variation of a general cohort study approach is to create "inception cohorts" of those beginning MRTP use for the first time. This is similar to a "new-user" cohort when evaluating drug use outcomes. However, the logistics of this type of study and maintaining the cohort for many years would always be challenging.

One additional, possibly more efficient approach would be a *retrospective cohort study*, where the data have already been collected. This might occur in the situation, for example, where the product (conventional tobacco and MRTP) purchasing behavior of a large group of persons has been previously recorded, and the population had been monitored for relevant health outcomes. However, this would not apply to MRTPs never on the market, and there would be a problem of ascertaining important confounding variables to conduct a credible analysis. It is not likely that many such retrospective cohorts would be available, and in such situations, important confounding factors may not have been collected.

Finally, an additional strategy to increase efficiency of prospective cohort studies is to include additional questions and measures of biomarkers to concurrent studies.

The Important Role of Case-Control Studies

Another important instrument of observational epidemiology is the case-control study, where persons with a particular health outcome or disease are the cases and a healthy control group is used to contrast the history of exposure to whatever exposure is being evaluated. Case-control studies are commonly used because of their efficiency in assembling study participants, including the circumstance where the outcomes are not common in general populations. Further, this method has been used widely to evaluate the impact of preventive interventions (Weiss, 1994). It is possible to contrast product exposures among those with varying levels of biomarkers or disease outcomes—intermediate or clinically overt—with "controls" who have no evidence of disease and have normal disease-related biomarker levels.

As in the situation of cohort studies, however, case-control studies encumber many important methodological issues that require attention, in addition to the confounding problem noted above. These include

- cases and controls should have the same source population to allow more credible contrast;
- exposures to MRTPs should have had sufficient time to occur and be generally available so usage can be evaluated;
- diseases (cases) can only be assessed one at a time; no overall health impact is usually possible; and
- exposures can only be assessed retrospectively, which can be a problem because of lapses in recall or memory (the so-called recall bias).

As in all case-control studies, the accuracy of retrospective recall of exposure can decrease scientific credibility and usefulness. Nonetheless, for evaluating MRTPs that have had community usage, the case-control study will remain an important tool.

Crossover Designs

When the outcomes are short term and/or recurrent, particularly when using biomarkers or intermediate endpoints, an observational crossover (or "case-crossover") design becomes feasible and informative. In its most simple form, research participants serve as their own control, and the outcome of interest is assessed during each of the exposures of interest. Then, for example, the effects of cigarette smoking can be compared with exposure to an MRTP. This "self-control" approach eliminates many potential "within-person" confounders, but it assumes that the effect of the first exposure does not carry over when the switch to the other exposure occurs. Crossover designs could be used to evaluate participants who switch from one MRTP to another, or who switch from an MRTP back to conventional cigarettes or other tobacco products.

Applying the Methodology of Comparative Effectiveness Research (CER)

The methodology of CER is basically designed to more critically inform health care and policy decisions by comparing health outcomes associated with different clinical interventions (usually therapies) for a particular disease or other clinical situations. Although CER methods include clinical trials, most approaches have been developed for observational study application, particularly in the analysis of clinical cohorts.

CER can sharpen or extend observational methodology that could provide additional approaches for comparing smokers and nonusers of tobacco with those using certain MRTPs. Methods such as propensity scoring (the likelihood of switching options, e.g., MRTP or conventional products) and instrumental variable analysis (to adjust for unmeasured confounders) are routinely used in non-CER research, but offer additional techniques for exploring causation. CER also offers techniques to review and synthesize the medical literature and identify important gaps, promote new analytical tools, and translate research findings to diverse stakeholders (IOM, 2009). Although CER does not claim to provide the same level of causal inference that might be derived from a randomized experimental design, its promise is to provide more credible answers when only observational data are available.

Summary of Observational Studies

There is no overriding conflict between observational and experimental methods and designs; rather, the contributions of both study designs are complementary and will be necessary for thorough evaluation of MRTPs. It is important to note that even when randomized clinical trials for health and behavioral outcomes are feasible and performed, subgroup analyses of these data are essentially observational in nature. Each general approach to scientific inquiry is really a large suite of study designs, to be chosen and exploited in the combinations that yield the best possible answers to the health and safety questions of interest. Consideration of study designs in general depend on them having suitable feasibility in execution, scientific credibility, responsiveness to informing policy decisions, and efficient use of available resources. The scientific and regulatory reality is that most of the population outcome studies can only be satisfied with the best observational studies possible. There are several reasons for the centrality of observational methods:

a. There are substantial ethical limitations on the application of MRTPs or contrasting conventional tobacco products in planned intervention studies, although some situations do allow for such interventions. A discussion on the ethical considerations of tobacco research is found in Chapter 2.

b. Research participants in randomized trials can rarely be expected to adhere to a particular intervention or product for long periods of time, as is true of drug or other intervention trials, so that summarization of usage patterns will require detailed and complex observational techniques. This is crucial to measuring personal exposure to MRTPs and conventional tobacco products, so the exposure "dose" can be assessed as accurately as possible and related to health or behavioral outcomes of interest.

c. Behavioral patterns of MRTP use and myriad health outcomes that may be anticipated yield a level of complexity that can often not be captured in experimental designs, no matter how desirable. This complexity limits the nature and execution of experimental designs.

d. Data on MRTP marketing and distribution are in essence observational. These data play an increasingly important role in the evaluation of population exposure to various MRTPs or conventional tobacco products, and they help set boundaries to better understand potential rates of potential adverse events.

e. MRTP manufacturers and marketers will change product formulations, designs, and advertising modes and presentations in order to maximize sales. Such activities will likely work to subvert experimental studies where adherence to a particular MRTP is a fundamental part of the study design.

f. Community surveillance for adverse effects of MRTPs is for the most part an observational activity, including such sources as (a) citizen reports to health agencies; (b) individual case reports or case series reported by health professionals, which are uncontrolled clinical observations; (c) ad hoc institutional or multicenter disease registers, which in general do not have a geographic-based catchment area; or (d) monitoring of electronic medical records for health events not otherwise anticipated.

g. Most adverse health and behavioral effects of MRTPs will be detected and validated in the longer term, to the extent possible, using observational methods such as cohort, case-control, or other related study designs.

h. As with most other consumer products released in the community, MRTPs are subject to personal misuse and abuse, accidental contamination, conscious adulteration, faulty manufacturing, the release of imitation (and thus unregulated) products, and long-term unanticipated alterations in product content and potential health outcomes. The development of protocols for inspection and other problem-control protocols for these problems, as is true for other consumer products in general, are essentially observational in nature and rarely subject to experimental designs.

THE USE OF MODELING IN ESTIMATING HEALTH EFFECTS OF MRTPS

While evaluating the empirical evidence emerging from studies on the health effects of MRTPs, researchers and regulators should anticipate how an intended exposure reduction affects disease risks. For this, models based on scientific data, rather than on speculation, can provide relevant insight. Mathematical modeling for estimating health effects of tobacco products is one method to improve the quantification of exposure response data from product development. Modeling can generate data on complex issues of product and constituent interaction and can provide insight for trials in specific subpopulations.

Risk assessment models, developed to represent the mechanistic pathways leading to clinical endpoints, can be used to study disease endpoints. There is a history of using models to understand the health impacts of tobacco use. For example, following the emergence of evidence linking smoking to lung cancer in the 1950s, Levin proposed a model linking smoking to lung cancer; this method is still in use today (Levin, 1953).

A model linking a reduced carcinogen exposure to a reduced risk of cancer should include a causality assessment, which details how the targeted carcinogen affects an individual's health and risk for cancer. The model should also include knowledge of the dose-response relationship for the carcinogen, as well as an individual susceptibility assessment. Additionally, it should include an understanding of the targeted carcinogen in context of the other carcinogens present in the product (IOM, 2001). Exposure can be measured using a validated biomarker, rather than by individual constituents present in the tobacco product or its smoke.

While designing a model, researchers should take into account the potential limitations of its inputs. For example, a dose-response curve may change for individuals with different histories of tobacco use. An integrated mathematical model for tobacco harm reduction should consider dose-response relationships for multiple disease outcomes. That is, a model with a dose-response relationship for only a single disease outcome will limit the relevance of the data, as tobacco product use leads to multiple health outcomes.

Failures in modeling design can lead to unsuccessful future studies and other product safety issues (FDA, 2009). It is likely that discussion of quantitative tobacco product development methods between the FDA and product sponsors will improve these results.

REFERENCES

Adam, T., S. Mitschke, T. Streibel, R. R. Baker, and R. Zimmermann. 2006. Quantitative puff-by-puff-resolved characterization of selected toxic compounds in cigarette mainstream smoke. *Chemical Research in Toxicology* 19(4):511-520.

Andrei, G. 2006. Three-dimensional culture models for human viral diseases and antiviral drug development. *Antiviral Research* 71(2-3):96-107.

Andrei, G., S. Duraffour, J. Van den Oord, and R. Snoeck. 2010. Epithelial raft cultures for investigations of virus growth, pathogenesis and efficacy of antiviral agents. *Antiviral Research* 85(3):431-449.

Anzueto, A., S. Sethi, and F. J. Martinez. 2007. Exacerbations of chronic obstructive pulmonary disease. *Proceedings of the American Thoracic Society* 4(7):554-564.

Ashrafi, S., A. Das, R. Worowongvasu, B. Mehdinejad, and J. Waterhouse. 1992. A light, transmission and scanning electron microscope study of snuff-treated hamster cheek pouch epithelium. *Scanning Microscopy* 6(1):183.

Baarsma, H. A., H. Meurs, A. J. Halayko, M. H. Menzen, M. Schmidt, H. A. Kerstjens, and R. Gosens. 2011. Glycogen synthase kinase-3 regulates cigarette smoke extract- and IL-1β-induced cytokine secretion by airway smooth muscle. *American Journal of Physiology—Lung Cellular and Molecular Physiology* 300(6):L910-L919.

Bagchi, D., M. Bagchi, E. A. Hassoun, and S. J. Stohs. 1996. Cadmium-induced excretion of urinary lipid metabolites, DNA damage, glutathione depletion, and hepatic lipid peroxidation in Sprague-Dawley rats. *Biological Trace Element Research* 52(2):143-154.

Bagchi, M., J. Balmoori, D. Bagchi, S. D. Ray, C. Kuszynski, and S. J. Stohs. 1999. Smokeless tobacco, oxidative stress, apoptosis, and antioxidants in human oral keratinocytes. *Free Radical Biology and Medicine* 26(7-8):992-1000.

Bagchi, M., C. A. Kuszynski, J. Balmoori, S. S. Joshi, S. J. Stohs, and D. Bagchi. 2001. Protective effects of antioxidants against smokeless tobacco-induced oxidative stress and modulation of *Bcl-2* and *p53* genes in human oral keratinocytes. *Free Radical Research* 35(2):181-194.

Bagchi, M., J. Balmoori, D. Bagchi, S. J. Stohs, J. Chakrabarti, and D. K. Das. 2002. Role of reactive oxygen species in the development of cytotoxicity with various forms of chewing tobacco and pan masala. *Toxicology* 179(3):247-255.

Banerjee, A., V. Gopalakrishnan, and J. Vishwanatha. 2007. Inhibition of nitric oxide-induced apoptosis by nicotine in oral epithelial cells. *Molecular and Cellular Biochemistry* 305(1):113-121.

Basbaum, C. B., A. Chow, B. A. Macher, W. E. Finkbeiner, D. Veissiere, and L. S. Forsberg. 1986. Tracheal carbohydrate antigens identified by monoclonal antibodies. *Archives of Biochemistry and Biophysics* 249(2):363-373.

Batariova, A., V. Spevackova, B. Benes, M. Cejchanova, J. Smid, and M. Cerna. 2006. Blood and urine levels of Pb, Cd and Hg in the general population of the Czech Republic and proposed reference values. *International Journal of Hygiene and Environmental Health* 209(4):359-366.

Bauer, C. M., C. C. Zavitz, F. M. Botelho, K. N. Lambert, E. G. Brown, K. L. Mossman, J. D. Taylor, and M. R. Stampfli. 2010. Treating viral exacerbations of chronic obstructive pulmonary disease: Insights from a mouse model of cigarette smoke and H1N1 influenza infection. *PLoS One* 5(10):e13251.

Bernzweig, E., J. B. Payne, R. A. Reinhardt, J. K. Dyer, and K. D. Patil. 1998. Nicotine and smokeless tobacco effects on gingival and peripheral blood mononuclear cells. *Journal of Clinical Periodontology* 25(3):246-252.

Bertram, K. M., C. J. Baglole, R. P. Phipps, and R. T. Libby. 2009. Molecular regulation of cigarette smoke induced-oxidative stress in human retinal pigment epithelial cells: Implications for age-related macular degeneration. *American Journal of Physiology—Cell Physiology* 297(5):C1200-1210.

Boffetta, P., S. Clark, M. Shen, R. Gislefoss, R. Peto, and A. Andersen. 2006. Serum cotinine level as predictor of lung cancer risk. *Cancer Epidemiology, Biomarkers & Prevention* 15(6):1184-1188.

Bonassi, S., M. Neri, C. Lando, M. Ceppi, Y. Lin, W. P. Chang, N. Holland, M. Kirsch-Volders, E. Zeiger, and M. Fenech. 2003. Effect of smoking habit on the frequency of micronuclei in human lymphocytes: Results from the human micronucleus project. *Mutation Research/Reviews in Mutation Research* 543(2):155-166.

Bond, J. A., B. T. Chen, W. C. Griffith, and J. L. Mauderly. 1989. Cigarette smoke induces DNA adducts in lungs of rats after inhalation. *Experimental Pathology* 37(1-4):190-193.

Borgerding, M., and H. Klus. 2005. Analysis of complex mixtures—cigarette smoke. *Experimental and Toxicologic Pathology* 57(Suppl. 1):43-73.

Brunnemann, K. D., J. Qi, and D. Hoffmann. 2002. Chemical profile of two types of oral snuff tobacco. *Food and Chemical Toxicology* 40(11):1699-1703.

Brusselle, G. G., G. F. Joos, and K. R. Bracke. 2011. New insights into the immunology of chronic obstructive pulmonary disease. *Lancet* 378(9795):1015-1026.

Burns, D. M., E. Dybing, N. Gray, S. Hecht, C. Anderson, T. Sanner, R. O'Connor, M. Djordjevic, C. Dresler, P. Hainaut, M. Jarvis, A. Opperhuizen, and K. Straif. 2008. Mandated lowering of toxicants in cigarette smoke: A description of the World Health Organization TobReg proposal. *Tobacco Control* 17(2):132-141.

Carmella, S. G., M. Chen, S. Han, A. Briggs, J. Jensen, D. K. Hatsukami, and S. S. Hecht. 2009. Effects of smoking cessation on eight urinary tobacco carcinogen and toxicant biomarkers. *Chemical Research in Toxicology* 22(4):734-741.

CAST II Investigators. 1992. Effect of the antiarrhythmic agent moricizine on survival after myocardial infarction. *NEJM* 327(4):227-233.

CDC (Centers for Disease Control and Prevention). 1999. Tobacco use—United States, 1990–1999. *Morbidity and Mortality Weekly Report* 48(43):986-993.

CDC. 2010. Vital signs: Current cigarette smoking among adults aged ≥ 18 years—United States, 2009. *Morbidity and Mortality Weekly Report* 59(35):1135-1140.

CDC. 2011a. Cigarette smoking—United States, 1965–2008. *Morbidity and Mortality Weekly Report* 60(1):109-113.

CDC. 2011b. Ten great public health achievements—United States, 2001–2010. *Morbidity and Mortality Weekly Report* 60(19):619-623.

Chen, C., and J. F. Pankow. 2009. Gas/particle partitioning of two acid-base active compounds in mainstream tobacco smoke: Nicotine and ammonia. *Journal of Agricultural and Food Chemistry* 57(7):2678-2690.

Chen, H. W., M. L. Chien, Y. H. Chaung, C. K. Lii, and T. S. Wang. 2004. Extracts from cigarette smoke induce DNA damage and cell adhesion molecule expression through different pathways. *Chemico-Biological Interactions* 150(3):233-241.

Chen, H. W., C. K. Lii, H. J. Ku, and T. S. Wang. 2009. Cigarette smoke extract induces expression of cell adhesion molecules in HUVEC via actin filament reorganization. *Environmental and Molecular Mutagenesis* 50(2):96-104.

Chen, P., and S. Moldüveanu. 2003. Mainstream smoke chemical analyses for 2R4F Kentucky reference cigarette. *Beiträge zur Tabak forschung International* 20(7):448-458.

Cheng, S. E., S. F. Luo, M. J. Jou, C. C. Lin, Y. R. Kou, I. T. Lee, H. L. Hsieh, and C. M. Yang. 2009. Cigarette smoke extract induces cytosolic phospholipase A2 expression via NADPH oxidase, MAPKs, AP-1, and NF-kappaB in human tracheal smooth muscle cells. *Free Radical Biology and Medicine* 46(7):948-960.

Chung, F. L., L. Zhang, J. E. Ocando, and R. G. Nath. 1999. Role of 1, N-2-propanodeoxyguanosine adducts as endogenous DNA lesions in rodents and humans. *Exocyclic DNA Adducts in Mutagenesis and Carcinogenesis*(150):45-54.

Church, T. R., K. E. Anderson, N. E. Caporaso, M. S. Geisser, C. T. Le, Y. Zhang, A. R. Benoit, S. G. Carmella, and S. S. Hecht. 2009. A prospectively measured serum biomarker for a tobacco-specific carcinogen and lung cancer in smokers. *Cancer Epidemiology, Biomarkers & Prevention* 18(1):260-266.

Church, T. R., K. E. Anderson, C. Le, Y. Zhang, D. M. Kampa, A. R. Benoit, A. R. Yoder, S. G. Carmella, and S. S. Hecht. 2010. Temporal stability of urinary and plasma biomarkers of tobacco smoke exposure among cigarette smokers. *Biomarkers* 15:345-352.

Churg, A., S. Zhou, X. Wang, R. Wang, and J. L. Wright. 2009. The role of interleukin-1β in murine cigarette smoke-induced emphysema and small airway remodeling. *American Journal of Respiratory Cell and Molecular Biology* 40(4):482-490.

Clauss, M., R. Voswinckel, G. Rajashekhar, N. L. Sigua, H. Fehrenbach, N. I. Rush, K. S. Schweitzer, A. O. Yildirim, K. Kamocki, A. J. Fisher, Y. Gu, B. Safadi, S. Nikam, W. C. Hubbard, R. M. Tuder, H. L. Twigg, 3rd, R. G. Presson, S. Sethi, and I. Petrache. 2011. Lung endothelial monocyte-activating protein 2 is a mediator of cigarette smoke-induced emphysema in mice. *Journal of Clinical Investigation* 121(6):2470-2479.

Copp, J., G. Manning, and T. Hunter. 2009. TORC-specific phosphorylation of mammalian target of rapamycin (mTOR): Phospho-Ser2481 is a marker for intact mTOR signaling complex 2. *Cancer Research* 69(5):1821-1827.

Correa, E., P. A. Joshi, A. Castonguay, and H. M. Schuller. 1990. The tobacco-specific nitrosamine 4-(methylnitrosamino)-1-(3-pyridyl)-1-butanone is an active transplacental carcinogen in Syrian golden hamsters. *Cancer Research* 50(11):3435-3438.

Counts, M. E., F. S. Hsu, S. W. Laffoon, R. W. Dwyer, and R. H. Cox. 2004. Mainstream smoke constituent yields and predicting relationships from a worldwide market sample of cigarette brands: ISO smoking conditions. *Regulatory Toxicology and Pharmacology* 39(2):111-134.

Dagle, G., K. McDonald, L. Smith, and D. Stevens Jr. 1978. Pulmonary carcinogenesis in rats given implants of cigarette smoke condensate in beeswax pellets. *Journal of the National Cancer Institute* 61(3):905.

DeBethizy, J., M. Borgerding, D. Doolittle, J. Robinson, K. McManus, C. Rahn, R. Davis, and G. Burger. 1990. Chemical and biological studies of a cigarette that heats rather than burns tobacco. *The Journal of Clinical Pharmacology* 30(8):755.

Ding, Y. S., X. J. Yan, R. B. Jain, E. Lopp, A. Tavakoli, G. M. Polzin, S. B. Stanfill, D. L. Ashley, and C. H. Watson. 2006. Determination of 14 polycyclic aromatic hydrocarbons in mainstream smoke from U.S. Brand and non-U.S. Brand cigarettes. *Environmental Science and Technology* 40(4):1133-1138.

Ding, Y. S., D. L. Ashley, and C. H. Watson. 2007. Determination of 10 carcinogenic polycyclic aromatic hydrocarbons in mainstream cigarette smoke. *Journal of Agricultural and Food Chemistry* 55(15):5966-5973.

Ding, Y. S., B. C. Blount, L. Valentin-Blasini, H. S. Applewhite, Y. Xia, C. H. Watson, and D. L. Ashley. 2009. Simultaneous determination of six mercapturic acid metabolites of volatile organic compounds in human urine. *Chemical Research in Toxicology* 22(6):1018-1025.

DiPaolo, J. A. 1962. Effect of tobacco diets on rodents. *Nature* 195:1316.

DiPaolo, J. A., and M. L. Levin. 1965. Tumor incidence in mice after oral painting with cigarette smoke condensate. *Journal of the National Cancer Institute* 34:595.

Djordjevic, M. V., J. Fan, L. P. Bush, K. D. Brunnemann, and D. Hoffann. 1993. Effects of storage conditions on levels of tobacco-specific N-nitrosamines and N-nitrosamino acids in U.S. moist snuff. *Journal of Agricultural and Food Chemistry* 41(10):1790-1794.

Dontenwill, W., H. Chevalier, H. Harke, U. Lafrenz, G. Reckzeh, and B. Schneider. 1973. Investigations on the effects of chronic cigarette-smoke inhalation in Syrian golden hamsters. *Journal of the National Cancer Institute* 51(6):1781.

Doolittle, D., C. Rahn, G. Burger, R. Davis, J. deBethizy, G. Howard, C. Lee, S. McKarns, E. Riccio, and J. Robinson. 1989. Human urine mutagenicity study comparing cigarettes which burn or only heat tobacco. *Mutation Research/Genetic Toxicology* 223(2):221-232.

Dunham, L. J., C. S. Muir, and J. E. Hamner, 3rd. 1966. Epithelial atypia in hamster cheek pouches treated repeatedly with calcium hydroxide. *British Journal of Cancer* 20(3):588-593.

Dunham, L. J., K. C. Snell, and H. L. Stewart. 1975. Argyrophilic carcinoids in two Syrian hamsters (mesocricetus auratus). *Journal of the National Cancer Institute* 54(2):507-513.

FDA (Food and Drug Administration). 2009. *Guidance for industry: End-of-phase 2a meetings.* http://www.fda.gov/downloads/Drugs/GuidanceComplianceRegulatoryInformation/Guidances/ucm079690.pdf (accessed November 28, 2011).

FDA. 2011a. Harmful and potentially harmful constituents in tobacco products and tobacco smoke. *Federal Register* 76(156).

FDA. 2011b. *FDA and NIH announce joint study on tobacco use and risk perceptions*. http://www.fda.gov/NewsEvents/Newsroom/PressAnnouncements/ucm274626.htm (accessed November 28, 2011).

FDA. 2011c. *Draft initial list of harmful/potentially harmful constituents in tobacco smoke or smokeless tobacco products*. http://www.fda.gov/downloads/AdvisoryCommittees/CommitteesMeetingMaterials/TobaccoProductsScientificAdvisoryCommittee/UCM221804.pdf (accessed October 26, 2011).

Federal Trade Commission. 2007. *Smokeless tobacco reports for the years 2002-2005*. Washington, DC: Federal Trade Commission.

Feng, S., H. J. Roethig, Q. Liang, R. Kinser, Y. Jin, G. Scherer, M. Urban, J. Engl, and K. Riedel. 2006a. Evaluation of urinary 1-hydroxypyrene, S-phenylmercapturic acid, trans,trans -muconic acid, 3-methyladenine, 3-ethyladenine, 8-hydroxy-2'-deoxyguanosine and thioethers as biomarkers of exposure to cigarette smoke. *Biomarkers* 11(1):28-52.

Feng, Z., W. Hu, Y. Hu, and M. S. Tang. 2006b. Acrolein is a major cigarette-related lung cancer agent. Preferential binding at p53 mutational hotspots and inhibition of DNA repair. *Proceedings of the National Academy of Sciences* 103:15404-15409.

Fisher, M. A., G. W. Taylor, and K. R. Tilashalski. 2005. Smokeless tobacco and severe active periodontal disease, NHANES III. *Journal of Dental Research* 84(8):705-710.

Fleming, T. R., and D. L. DeMets. 1996. Surrogate end points in clinical trials: Are we being misled? *Annals of Internal Medicine* 125(7):605-613.

Forster, K., R. Preuss, B. Rossbach, T. Bruning, J. Angerer, and P. Simon. 2008. 3-hydroxybenzo[a]pyrene in the urine of workers with occupational exposure to polycyclic aromatic hydrocarbons in different industries. *Occupational and Environmental Medicine* 65(4):224-229.

Frost-Pineda, K., Q. Liang, J. Liu, L. Rimmer, Y. Jin, S. Feng, S. Kapur, P. Mendes, H. Roethig, and M. Sarkar. 2011. Biomarkers of potential harm among adult smokers and nonsmokers in the total exposure study. *Nicotine & Tobacco Research* 13(3):182-193.

Furie, M. B., J. A. Raffanello, E. I. Gergel, T. J. Lisinski, and L. D. Horb. 2000. Extracts of smokeless tobacco induce pro-inflammatory changes in cultured human vascular endothelial cells. *Immunopharmacology* 47(1):13-23.

Gargus, J. L., M. B. H. Powers, R.T., and J. R. Everly. 1976. Mouse dermal bioassays of cigarette smoke condensates. In *Report no. 1. Toward less hazardous cigarettes. The first set of experimental cigarettes*. Vol. DHEW Publ. No. (NIH) 76-905, edited by G. B. Gori. Washington, DC: Department of Health, Education, and Welfare, Public Health Service, National Institutes of Health, National Cancer Institute. Pp. 85–94.

Gijare, P. S., K. V. Rao, and S. V. Bhide. 1989. Effects of tobacco-specific nitrosamines and snuff extract on cell proliferation and activities of ornithine decarboxylase and aryl hydrocarbon hydroxylase in mouse tongue primary epithelial cell cultures. *Journal of Cancer Research and Clinical Oncology* 115(6):558-563.

Goniewicz, M. L., M. D. Eisner, E. Lazcano-Ponce, W. Zielinska-Danch, B. Koszowski, A. Sobczak, C. Havel, P. Jacob, and N. L. Benowitz. 2011. Comparison of urine cotinine and the tobacco-specific nitrosamine metabolite 4-(methylnitrosamino)-1-(3-pyridyl)-1-butanol (NNAL) and their ratio to discriminate active from passive smoking. *Nicotine & Tobacco Research* 13(3):202-208.

Gori, G. B. 1976. Report on the second set of experimental cigarettes. In *Report no. 2. Toward less hazardous cigarettes. The second set of experimental cigarettes*. Vol. DHEW Publ. No. (NIH) 76-1111, edited by G. B. Gori. Washington, DC: Department of Health, Education, and Welfare, Public Health Service, National Institutes of Health, National Cancer Institute. Pp. 4-15.

Graham, E. A., A. B. Croninger, and E. L. Wynder. 1957. Experimental production of carcinoma with cigarette tar. *Cancer Research* 17(11):1058-1066.
Grasso, P., and A. H. Mann. 1998. Smokeless tobacco and oral cancer: An assessment of evidence derived from laboratory animals. *Food and Chemical Toxicology* 36(11):1015-1029.
Gregg, E., C. Hill, M. Hollywood, M. Kearney, K. McAdam, D. McLaughlin, S. Purkis, and M. Williams. 2004. The UK smoke constituents testing study. Summary of results and comparison with other studies. *Beiträge zur Tabakforschung International* 21:117-138.
Gregory, R. L., and L. E. Gfell. 1996. Effect of nicotine on secretory component synthesis by secretory epithelial cells. *Clinical and Diagnostic Laboratory Immunology* 3(5):578-583.
Guarino, F., G. Cantarella, M. Caruso, C. Russo, S. Mancuso, G. Arcidiacono, R. R. Cacciola, R. Bernardini, and R. Polosa. 2011. Endothelial activation and injury by cigarette smoke exposure. *Journal of Biological Regulators and Homeostatic Agents* 25(2):259-268.
Hakura, A., H. Shimada, M. Nakajima, H. Sui, S. Kitamoto, S. Suzuki, and T. Satoh. 2005. Salmonella/human S9 mutagenicity test: A collaborative study with 58 compounds. *Mutagenesis* 20(3):217-228.
Hammond, D., and R. J. O'Connor. 2008. Constituents in tobacco and smoke emissions from Canadian cigarettes. *Tobacco Control* 17(Suppl. 1):i24-i31.
Harvey, C. J., R. K. Thimmulappa, S. Sethi, X. Kong, L. Yarmus, R. H. Brown, D. Feller-Kopman, R. Wise, and S. Biswal. 2011. Targeting Nrf2 signaling improves bacterial clearance by alveolar macrophages in patients with COPD and in a mouse model. *Science Translational Medicine* 3(78):78ra32.
Hasséus, B., M. Wallström, B.-G. Österdahl, J.-M. Hirsch, and M. Jontell. 1997. Immunotoxic effects of smokeless tobacco on the accessory cell function of rat oral epithelium. *European Journal of Oral Sciences* 105(1):45-51.
Hatsukami, D. K., C. Lemmonds, Y. Zhang, S. E. Murphy, C. Le, S. G. Carmella, and S. S. Hecht. 2004. Evaluation of carcinogen exposure in people who used "reduced exposure" tobacco products. *Journal of the National Cancer Institute* 96(11):844-852.
Hatsukami, D. K., N. L. Benowitz, S. I. Rennard, C. Oncken, and S. S. Hecht. 2006. Biomarkers to assess the utility of potential reduced exposure tobacco products. *Nicotine & Tobacco Research* 8(2):169-191.
Hatsukami, D. K., R. M. Feuer, J. O. Ebbert, I. Stepanov, and S. S. Hecht. 2007a. Changing smokeless tobacco products: New tobacco delivery systems. *American Journal of Preventive Medicine* 33:S368-S378.
Hatsukami, D. K., A. M. Joseph, M. LeSage, J. Jensen, S. E. Murphy, P. Pentel, M. Kotlyar, E. Borgida, C. Le, and S. S. Hecht. 2007b. The science base for reducing tobacco harm. *Nicotine & Tobacco Research* 9:S537-S553.
Hatsukami, D. K., K. Hanson, A. Briggs, M. Parascandola, J. M. Genkinger, R. O'Connor, and P. G. Shields. 2009. Clinical trials methods for evaluation of potential reduced exposure products. *Cancer Epidemiology, Biomarkers & Prevention* 18(12):3143-3195.
Hatsukami, D. K., M. Kotlyar, L. A. Hertsgaard, Y. Zhang, S. G. Carmella, J. A. Jensen, P. G. Shields, S. E. Murphy, I. Stepanov, and S. S. Hecht. 2010. Reduced nicotine content cigarettes: Effects on toxicant exposure, dependence and cessation. *Addiction* 105:343-355.
Hautamaki, R. D., D. K. Kobayashi, R. M. Senior, and S. D. Shapiro. 1997. Requirement for macrophage elastase for cigarette smoke-induced emphysema in mice. *Science* 277(5334):2002-2004.
Hecht, S. S. 1999. Tobacco smoke carcinogens and lung cancer. *Journal of the National Cancer Institute* 91:1194-1210.
Hecht, S. S. 2002. Human urinary carcinogen metabolites: Biomarkers for investigating tobacco and cancer. *Carcinogenesis* 23:907-922.
Hecht, S. S. 2003a. Tobacco carcinogens, their biomarkers, and tobacco-induced cancer. *Nature Reviews. Cancer* 3:733-744.

Hecht, S. S. 2003b. Carcinogen derived biomarkers: Applications in studies of human exposure to secondhand tobacco smoke. *Tobacco Control* 13(Suppl. 1):i48-i56.

Hecht, S. S. 2008. Progress and challenges in selected areas of tobacco carcinogenesis. *Chemical Research in Toxicology* 21:160-171.

Hecht, S. S. 2010. Tobacco smoke carcinogens and lung cancer. In *Chemical carcinogenesis*, edited by T. M. Penning. New York: Springer.

Hecht, S. S., A. Rivenson, J. Braley, J. DiBello, J. D. Adams, and D. Hoffmann. 1986. Induction of oral cavity tumors in F344 rats by tobacco-specific nitrosamines and snuff. *Cancer Research* 46(8):4162-4166.

Hecht, S. S., M. Chen, A. Yoder, J. Jensen, D. Hatsukami, C. Le, and S. G. Carmella. 2005. Longitudinal study of urinary phenanthrene metabolite ratios: Effect of smoking on the diol epoxide pathway. *Cancer Epidemiology, Biomarkers & Prevention* 14(12):2969-2974.

Hecht, S. S., J. M. Yuan, and D. Hatsukami. 2010. Applying tobacco carcinogen and toxicant biomarkers in product regulation and cancer prevention. *Chemical Research in Toxicology* 23(6):1001-1008.

Herrold, K. M., and L. J. Dunham. 1962. Induction of carcinoma and papilloma of the Syrian hamster by intratracheal instillation of benzo[a]-pyrene. *Journal of the National Cancer Institute* 28:467-491.

HHS (U.S. Department of Health and Human Services). 2004a. *The health consequences of smoking: A report of the Surgeon General*. Rockville, MD: U.S. Department of Health and Human Services, Public Health Service, Centers for Disease Control and Prevention, Center for Health Promotion and Education, Office on Smoking and Health.

HHS. 2004b. *Report on carcinogens, 11th edition*. Research Triangle Park, NC: HHS, National Toxicology Program.

HHS. 2006. *The health consequences of involuntary exposure to tobacco smoke: A report of the Surgeon General.* Washington, DC: U.S. Department of Health and Human Services, Centers for Disease Control and Prevention, National Center for Chronic Disease Prevention and Health Promotion, Office on Smoking and Health.

HHS. 2010. *How tobacco smoke causes disease: The biology and behavioral basis for smoking-attributable disease: A report of the Surgeon General.* Atlanta, GA: U.S. Department of Health and Human Services, Centers for Disease Control and Prevention, National Center for Chronic Disease Prevention and Health Promotion, Office on Smoking and Health.

Hirsch, J. M., and S. L. Johansson. 1983. Effect of long-term application of snuff on the oral mucosa: An experimental study in the rat. *Journal of Oral Pathology* 12(3):187-198.

Hirsch, J. M., and H. Thilander. 1981. Snuff-induced lesions of the oral mucosa—an experimental model in the rat. *Journal of Oral Pathology* 10(5):342-353.

Hirsch, J. M., S. L. Johansson, and A. Vahlne. 1984. Effect of snuff and herpes simplex virus-1 on rat oral mucosa: Possible associations with the development of squamous cell carcinoma. *Journal of Oral Pathology* 13(1):52-62.

Hoffmann, D., and J. D. Adams. 1981. Carcinogenic tobacco-specific N-nitrosamines in snuff and in the saliva of snuff dippers. *Cancer Research* 41(11 Pt 1):4305-4308.

Hoffmann, D., and E. L. Wynder. 1971. A study of tobacco carcinogenesis. XI. Tumor initiators, tumor accelerators, and tumor promoting activity of condensate fractions. *Cancer* 27(4):848-864.

Hoffmann, D., A. Castonguay, A. Rivenson, and S. S. Hecht. 1981. Comparative carcinogenicity and metabolism of 4-(methylnitrosamino)-1-(3-pyridyl)-1-butanone and N′-nitrosonornicotine in Syrian golden hamsters. *Cancer Research* 41(6):2386-2393.

Hoffmann, D., A. Rivenson, and S. S. Hecht. 1992. Carcinogenesis of smokeless tobacco. In *NCI smoking and tobacco control monograph*. Vol. 92-3461. Bethesda, MD: National Cancer Institute, National Institutes of Health.

Hoffmann, K., K. Becker, C. Friedrich, D. Helm, C. Krause, and B. Seifert. 2000. The German environmental survey 1990/1992 (GerES II): Cadmium in blood, urine and hair of adults and children. *Journal of Exposure Analysis and Environmental Epidemiology* 10:126-135.

Homburger, F. 1971. Mechanical irritation, polycyclic hydrocarbons, and snuff. Effects on facial skin, cheek pouch, and oral mucosa in Syrian hamsters. *Archives of Pathology* 91(5):411-417.

Hukkanen, J., P. Jacob, III, and N. L. Benowitz. 2005. Metabolism and disposition kinetics of nicotine. *Pharmacological Reviews* 57(1):79-115.

Hutt, J. A., B. R. Vuillemenot, E. B. Barr, M. J. Grimes, F. F. Hahn, C. H. Hobbs, T. H. March, A. P. Gigliotti, S. K. Seilkop, and G. L. Finch. 2005. Life-span inhalation exposure to mainstream cigarette smoke induces lung cancer in B6C3F1 mice through genetic and epigenetic pathways. *Carcinogenesis* 26(11):1999.

Huvenne, W., E. A. Lanckacker, O. Krysko, K. R. Bracke, T. Demoor, P. W. Hellings, G. G. Brusselle, G. F. Joos, C. Bachert, and T. Maes. 2011. Exacerbation of cigarette smoke-induced pulmonary inflammation by Staphylococcus aureus enterotoxin B in mice. *Respiratory Research* 12:69.

IARC (International Agency for Research on Cancer). 1985. Tobacco habits other than smoking; betel-quid & areca-nut chewing; and some related nitrosamines. In *IARC monographs on the evaluation of carcinogenic risks to humans*. Vol. 37. Lyon, France: IARC.

IARC. 1987. Overall evaluations of carcinogenicity: An updating of IARC monographs volumes 1 to 42. In *IARC monographs on the evaluation of the carcinogenic risk of chemicals to humans, suppl. 7*. Lyon, France: IARC. Pp. 91-122.

IARC. 1995. Acrolein. In *IARC monographs on the evaluation of carcinogenic risks to humans*. Vol. 63. Lyon, France: IARC. Pp. 337-391.

IARC. 1999a. Beryllium, cadmium, mercury, and exposures in the glass manufacturing industry. In *IARC monographs on the evaluation of carcinogenic risks to humans*. Vol. 58. Lyon, France: IARC. Pp. 119-237.

IARC. 1999b. Re-evaluation of some organic chemicals, hydrazine and hydrogen peroxide (part one). In *IARC monographs on the evaluation of carcinogenic risks to humans*. Vol. 71. Lyon, France: IARC. Pp. 43-108.

IARC. 2004. *IARC monographs on the evaluation of carcinogenic risks to humans: Volume 83. Tobacco smoke and involuntary smoking*. Lyon, France: IARC.

IARC. 2006. Formaldehyde, 2-butoxyethanol and 1- tert -butoxypropan-2-ol. In *IARC monographs on the evaluation of carcinogenic risks to humans, vol. 88*. Lyon, France: IARC. Pp. 37-325.

IARC. 2007. *Smokeless tobacco and tobacco-specific nitrosamines*. Lyon, France: IARC.

IARC. 2008. 1,3-butadiene, ethylene oxide and vinyl halides (vinyl fluoride, vinyl chloride and vinyl bromide). In *IARC monographs on the evaluation of carcinogenic risks to humans, vol. 97*. Lyon, France: IARC. Pp. 45-309.

IOM (Institute of Medicine). 2001. *Clearing the smoke: Assessing the science base for tobacco harm reduction*. Washington, DC: National Academy Press.

IOM. 2009. *Initial national priorities for comparative effectiveness research*. Washington, DC: The National Academies Press.

IOM. 2010. *Evaluation of biomarkers and surrogate endpoints in chronic disease*. Washington, DC: The National Academies Press.

Ismail, A. I., B. A. Burt, and S. A. Eklund. 1983. Epidemiologic patterns of smoking and periodontal disease in the United States. *Journal of the American Dental Association* 106(5):617-621.

Jacob, P., III, M. Wilson, and N. L. Benowitz. 2007. Determination of phenolic metabolites of polycyclic aromatic hydrocarbons in human urine as their pentafluorobenzyl ether derivatives using liquid chromatography-tandem mass spectrometry. *Analytical Chemistry* 79(2):587-598.

Jaju, R., R. Patel, S. Bakshi, A. Trivedi, B. Dave, and S. Adhvaryu. 1992. Chromosome damaging effects of pan masala. *Cancer Letters* 65(3):221-226.

Jansson, T., L. Romert, J. Magnusson, and D. Jenssen. 1991. Genotoxicity testing of extracts of a Swedish moist oral snuff. *Mutation Research/Genetic Toxicology* 261(2):101-115.

Johansson, S. L., J. M. Hirsch, P. A. Larsson, J. Saidi, and B. G. Österdahl. 1989. Snuff-induced carcinogenesis: Effect of snuff in rats initiated with 4-nitroquinoline N-oxide. *Cancer Research* 49(11):3063.

Johansson, S. L., J. M. Hirsch, P. A. Larsson, J. Saidi, and B. G. Österdahl. 1991a. Lack of promoting ability of snuff in rats initiated with 4-nitroquinoline-N-oxide. *IARC Scientific Publications* (105):507-509.

Johansson, S. L., J. Saidi, B. G. Österdahl, and R. A. Smith. 1991b. Promoting effect of snuff in rats initiated by 4-nitroquinoline-N-oxide or 7, 12-dimethylbenz (a) anthracene. *Cancer Research* 51(16):4388.

Johnson, M. D., J. Schilz, M. V. Djordjevic, J. R. Rice, and P. G. Shields. 2009. Evaluation of in vitro assays for assessing the toxicity of cigarette smoke and smokeless tobacco. *Cancer Epidemiology, Biomarkers & Prevention* 18(12):3263-3304.

Jorquera, R., A. Castonguay, and H. M. Schuller. 1992. Placental transfer of 4-(methylnitrosamino)-1-(3-pyridyl)-1-butanone instilled intratracheally in Syrian golden hamsters. *Cancer Research* 52(12):3273-3280.

Kang, M. J., C. G. Lee, J. Y. Lee, C. S. Dela Cruz, Z. J. Chen, R. Enelow, and J. A. Elias. 2008. Cigarette smoke selectively enhances viral pamp- and virus-induced pulmonary innate immune and remodeling responses in mice. *Journal of Clinical Investigation* 118(8):2771-2784.

Kavvadias, D., G. Scherer, F. Cheung, G. Errington, J. Shepperd, and M. McEwan. 2009a. Determination of tobacco-specific N-nitrosamines in urine of smokers and non-smokers. *Biomarkers* 14(8):547-553.

Kavvadias, D., G. Scherer, M. Urban, F. Cheung, G. Errington, J. Shepperd, and M. McEwan. 2009b. Simultaneous determination of four tobacco-specific N-nitrosamines (TSNA) in human urine. *Journal of Chromatography B* 877(11-12):1185-1192.

Kim, K. C., J. I. Rearick, P. Nettesheim, and A. M. Jetten. 1985. Biochemical characterization of mucous glycoproteins synthesized and secreted by hamster tracheal epithelial cells in primary culture. *Journal of Biological Chemistry* 260(7):4021-4027.

Klaunig, J. E. 2008. Acrylamide carcinogenicity. *Journal of Agricultural and Food Chemistry* 56(15):5984-5988.

Kotlyar, M., L. A. Hertsgaard, B. R. Lindgren, J. A. Jensen, S. G. Carmella, I. Stepanov, S. E. Murphy, S. S. Hecht, and D. K. Hatsukami. 2011. Effect of oral snus and medicinal nicotine in smokers on toxicant exposure and withdrawal symptoms: A feasibility study. *Cancer Epidemiology, Biomarkers & Prevention* 20(1):91-100.

Kriebel, D., J. Henry, J. Gold, A. Bronsdon, and B. Commoner. 1985. The mutagenicity of cigarette smokers' urine. *Journal of Environmental Pathology, Toxicology and Oncology* 6(2):157.

Kuenemann-Migeot, C., F. Callais, I. Momas, and B. Festy. 1996. Urinary promutagens of smokers: Comparison of concentration methods and relation to cigarette consumption. *Mutation Research/Genetic Toxicology* 368(2):141-147.

Larsson, P. A., S. L. Johansson, A. Vahlne, and J. M. Hirsch. 1989. Snuff tumorigenesis: Effects of long-term snuff administration after initiation with 4-nitroquinoline-N-oxide and herpes simplex virus type 1. *Journal of Oral Pathology and Medicine* 18(4):187-192.

Leslie, E. M., G. Ghibellini, K. Nezasa, and K. L. Brouwer. 2007. Biotransformation and transport of the tobacco-specific carcinogen 4-(methylnitrosamino)-1-(3-pyridyl)-1-butanone (NNK) in bile duct-cannulated wild-type and Mrp2/Abcc2-deficient (TR^-) wistar rats. *Carcinogenesis* 28(12):2650-2656.

Levin, M. 1953. The occurrence of lung cancer in man. *Acta-Unio Internationalis Contra Cancrum* 9(3):531.

Lin, H., D. M. Carlson, J. A. St. George, C. G. Plopper, and R. Wu. 1989. An ELISA method for the quantitation of tracheal mucins from human and nonhuman primates. *American Journal of Respiratory Cell and Molecular Biology* 1(1):41-48.

Lindemann, R. A., and N. H. Park. 1988. Inhibition of human lymphokine-activated killer activity by smokeless tobacco (snuff) extract. *Archives of Oral Biology* 33(5):317-321.

Liu, J., Q. Liang, K. Frost-Pineda, R. Muhammad-Kah, L. Rimmer, H. Roethig, P. Mendes, and M. Sarkar. 2011. Relationship between biomarkers of cigarette smoke exposure and biomarkers of inflammation, oxidative stress, and platelet activation in adult cigarette smokers. *Cancer Epidemiology, Biomarkers & Prevention* 20(8):1760-1769.

Lowe, F. J., E. O. Gregg, and M. McEwan. 2009. Evaluation of biomarkers of exposure and potential harm in smokers, former smokers and never-smokers. *Clinical Chemistry and Laboratory Medicine* 47(3):311-320.

Ma, B., M. J. Kang, C. G. Lee, S. Chapoval, W. Liu, Q. Chen, A. J. Coyle, J. M. Lora, D. Picarella, R. J. Homer, and J. A. Elias. 2005. Role of CCR5 in IFN-γ-induced and cigarette smoke-induced emphysema. *Journal of Clinical Investigation* 115(12):3460-3472.

Malhotra, D., R. Thimmulappa, A. Navas-Acien, A. Sandford, M. Elliott, A. Singh, L. Chen, X. Zhuang, J. Hogg, P. Pare, R. M. Tuder, and S. Biswal. 2008. Decline in NRF2-regulated antioxidants in chronic obstructive pulmonary disease lungs due to loss of its positive regulator, DJ-1. *American Journal of Respiratory and Critical Care Medicine* 178(6):592-604.

Mallia, P., S. D. Message, V. Gielen, M. Contoli, K. Gray, T. Kebadze, J. Aniscenko, V. Laza-Stanca, M. R. Edwards, L. Slater, A. Papi, L. A. Stanciu, O. M. Kon, M. Johnson, and S. L. Johnston. 2011. Experimental rhinovirus infection as a human model of chronic obstructive pulmonary disease exacerbation. *American Journal of Respiratory and Critical Care Medicine* 183(6):734-742.

Mangipudy, R. S., and J. K. Vishwanatha. 1999. Role of nitric oxide in the induction of apoptosis by smokeless tobacco extract. *Molecular and Cellular Biochemistry* 200(1-2):51-57.

Mariner, D. C., M. Ashley, C. J. Shepperd, G. Mullard, and M. Dixon. 2010. Mouth level smoke exposure using analysis of filters from smoked cigarettes: A study of eight countries. *Regulatory Toxicology and Pharmacology*.

Mauderly, J. L., W. E. Bechtold, J. A. Bond, A. L. Brooks, B. T. Chen, R. G. Cuddihy, J. R. Harkema, R. F. Henderson, N. F. Johnson, K. Rithidech, et al. 1989. Comparison of 3 methods of exposing rats to cigarette smoke. *Experimental Pathology* 37(1-4):194-197.

Mauderly, J. L., A. P. Gigliotti, E. B. Barr, W. E. Bechtold, S. A. Belinsky, F. F. Hahn, C. A. Hobbs, T. H. March, S. K. Seilkop, and G. L. Finch. 2004. Chronic inhalation exposure to mainstream cigarette smoke increases lung and nasal tumor incidence in rats. *Toxicological Sciences* 81(2):280.

McElroy, J. A., M. M. Shafer, A. Trentham-Dietz, J. M. Hampton, and P. A. Newcomb. 2007. Urinary cadmium levels and tobacco smoke exposure in women age 20-69 years in the United States. *Journal of Toxicology and Environmental Health. Part A* 70(20):1779-1782.

Melikian, A. A., M. V. Djordjevic, S. Chen, J. Richie, and S. D. Stellman. 2007. Effect of delivered dosage of cigarette smoke toxins on the levels of urinary biomarkers of exposure. *Cancer Epidemiology, Biomarkers & Prevention* 16(7):1408-1415.

Mendes, P., S. Kapur, J. Wang, S. Feng, and H. Roethig. 2008. A randomized, controlled exposure study in adult smokers of full flavor Marlboro cigarettes switching to Marlboro lights or Marlboro ultra lights cigarettes. *Regulatory Toxicology and Pharmacology* 51(3):295-305.

Merne, M., K. Heikinheimo, I. Saloniemi, and S. Syrjanen. 2004. Effects of snuff extract on epithelial growth and differentiation in vitro. *Oral Oncology* 40(1):6-12.

Morin, A., C. J. Shepperd, A. C. Eldridge, N. Poirier, and R. Voisine. 2011. Estimation and correlation of cigarette smoke exposure in Canadian smokers as determined by filter analysis and biomarkers of exposure. *Regulatory Toxicology and Pharmacology* 61(3, Supplement):S3-S12.

Mortaz, E., A. D. Kraneveld, J. J. Smit, M. Kool, B. N. Lambrecht, S. L. Kunkel, N. W. Lukacs, F. P. Nijkamp, and G. Folkerts. 2009. Effect of cigarette smoke extract on dendritic cells and their impact on T-cell proliferation. *PLoS One* 4(3):e4946.

Muns, G., J. K. Vishwanatha, and I. Rubinstein. 1994. Effects of smokeless tobacco on chemically transformed hamster oral keratinocytes: Role of angiotensin I-converting enzyme. *Carcinogenesis* 15(7):1325-1327.

Naufal, Z. S., K. M. Marano, S. J. Kathman, and C. L. Wilson. 2011. Differential exposure biomarker levels among cigarette smokers and smokeless tobacco consumers in the National Health and Nutrition Examination Survey 1999-2008. *Biomarkers* 16(3):222-235.

NCI (National Cancer Institute). 1998. *Cigars: Health effects and trends, Smoking and tobacco control monograph no. 9*. Bethesda, MD: U.S. Department of Health and Human Services, National Institutes of Health. NIH Pub. No. 98-4645.

NCI. 2001. *Risks associated with smoking cigarettes with low machine-measured yields of tar and nicotine, Smoking and tobacco control monograph no. 13*. Bethesda, MD: U.S. Department of Health and Human Services, National Institutes of Health, National Cancer Institute. NIH Pub. No. 99-4645.

NCI. 2009. *Phenotypes and endophenotypes: Foundations for genetic studies of nicotine use and dependence*. Bethesda, MD: U.S. Department of Health and Human Services, National Institutes of Health, National Cancer Institute.

Nelson, D. E., P. Mowery, S. Tomar, S. Marcus, G. Giovino, and L. Zhao. 2006. Trends in smokeless tobacco use among adults and adolescents in the United States. *American Journal of Public Health* 96(5):897-905.

Niphadkar, M. P., A. N. Bagwe, and R. A. Bhisey. 1996. Mutagenic potential of Indian tobacco products. *Mutagenesis* 11(2):151.

Pakhale, S., S. Sarkar, K. Jayant, and S. Bhide. 1988. Carcinogenicity of Indian bidi and cigarette smoke condensate in Swiss albino mice. *Journal of Cancer Research and Clinical Oncology* 114(6):647-649.

Palladino, G., J. D. Adams, K. D. Brunnemann, N. J. Haley, and D. Hoffmann. 1986. Snuff-dipping in college students: A clinical profile. *Military Medicine* 151(6):342-346.

Pankow, J. F., A. D. Tavakoli, W. Luo, and L. M. Isabelle. 2003. Percent free base nicotine in the tobacco smoke particulate matter of selected commercial and reference cigarettes. *Chemical Research in Toxicology* 16(8):1014-1018.

Papageorge, M. B., E. Cataldo, and E. G. Jahngen. 1996. The effect of N-nitrosonornicotine on the buccal mucosa of Syrian hamsters. *Journal of Oral and Maxillofacial Surgery* 54(2):187-190.

Park, N. H., J. P. Sapp, and E. G. Herbosa. 1986. Oral cancer induced in hamsters with herpes simplex infection and simulated snuff dipping. *Oral Surgery, Oral Medicine, Oral Pathology* 62(2):164-168.

Paschal, D. C., V. Burt, S. P. Caudill, E. W. Gunter, J. L. Pirkle, E. J. Sampson, D. T. Miller, and R. J. Jackson. 2000. Exposure of the U.S. population aged 6 years and older to cadmium: 1988-1994. *Archives of Environmental Contamination and Toxicology* 38:377-383.

Patel, R. K., R. J. Jaju, S. R. Bakshi, A. H. Trivedi, B. J. Dave, and S. G. Adhvaryu. 1994. Pan masala—a genotoxic menace. *Mutation Research/Genetic Toxicology* 320(3):245-249.

Pauly, J. L., R. J. O'Connor, G. M. Paszkiewicz, K. M. Cummings, M. V. Djordjevic, and P. G. Shields. 2009. Cigarette filter-based assays as proxies for toxicant exposure and smoking behavior—a literature review. *Cancer Epidemiology, Biomarkers & Prevention* 18(12):3321-3333.

Peacock, E. E., Jr., and B. W. Brawley. 1959. An evaluation of snuff and tobacco in the production of mouth cancer. *Plastic and Reconstructive Surgery and the Transplantation Bulletin* 23(6):628-635.

Peacock, E. E., Jr., B. G. Greenberg, and B. W. Brawley. 1960. The effect of snuff and tobacco on the production of oral carcinoma: An experimental and epidemiological study. *Annals of Surgery* 151:542-550.

Peluffo, G., P. Calcerrada, L. Piacenza, N. Pizzano, and R. Radi. 2009. Superoxide-mediated inactivation of nitric oxide and peroxynitrite formation by tobacco smoke in vascular endothelium: Studies in cultured cells and smokers. *American Journal of Physiology—Heart and Circulatory Physiology* 296(6):H1781-H1792.

Phillips, D. H. 2002. Smoking-related DNA and protein adducts in human tissues. *Carcinogenesis* 23(12):1979-2004.

Prentice, R. L. 1989. Surrogate endpoints in clinical trials: Definition and operational criteria. *Statistics in Medicine* 8(4):431-440.

Putzrath, R. M., D. Langley, and E. Eisenstadt. 1981. Analysis of mutagenic activity in cigarette smokers' urine by high performance liquid chromatography. *Mutation Research/Environmental Mutagenesis and Related Subjects* 85(3):97-108.

Rainey, C. L., P. A. Conder, and J. V. Goodpaster. 2011. Chemical characterization of dissolvable tobacco products promoted to reduce harm. *Journal of Agricultural and Food Chemistry* 59(6):2745-2751.

Rangasamy, T., C. Y. Cho, R. K. Thimmulappa, L. Zhen, S. S. Srisuma, T. W. Kensler, M. Yamamoto, I. Petrache, R. M. Tuder, and S. Biswal. 2004. Genetic ablation of Nrf2 enhances susceptibility to cigarette smoke-induced emphysema in mice. *Journal of Clinical Investigation* 114(9):1248-1259.

Reilly, K. B., S. Srinivasan, M. E. Hatley, M. K. Patricia, J. Lannigan, D. T. Bolick, G. Vandenhoff, H. Pei, R. Natarajan, J. L. Nadler, and C. C. Hedrick. 2004. 12/15-lipoxygenase activity mediates inflammatory monocyte/endothelial interactions and atherosclerosis in vivo. *Journal of Biological Chemistry* 279(10):9440-9450.

Richter, P., and F. W. Spierto. 2003. Surveillance of smokeless tobacco nicotine, pH, moisture, and unprotonated nicotine content. *Nicotine & Tobacco Research* 5(6):885-889.

Richter, P., K. Hodge, S. Stanfill, L. Zhang, and C. Watson. 2008. Surveillance of moist snuff: Total nicotine, moisture, pH, un-ionized nicotine, and tobacco-specific nitrosamines. *Nicotine & Tobacco Research* 10(11):1645-1652.

Rickert, W. S., P. J. Joza, M. Sharifi, J. Wu, and J. H. Lauterbach. 2008. Reductions in the tobacco specific nitrosamine (TSNA) content of tobaccos taken from commercial Canadian cigarettes and corresponding reductions in TSNA deliveries in mainstream smoke from such cigarettes. *Regulatory Toxicology and Pharmacology* 51(3):306-310.

Rodgman, A., and T. Perfetti. 2009. *The chemical components of tobacco and tobacco smoke*. Boca Raton, FL: CRC Press.

Roemer, E., R. Stabbert, K. Rustemeier, D. J. Veltel, T. J. Meisgen, W. Reininghaus, R. A. Carchman, C. L. Gaworski, and K. F. Podraza. 2004. Chemical composition, cytotoxicity and mutagenicity of smoke from U.S. commercial and reference cigarettes smoked under two sets of machine smoking conditions. *Toxicology* 195(1):31-52.

Roethig, H. J., B. K. Zedler, R. D. Kinser, S. Feng, B. L. Nelson, and Q. Liang. 2007. Short-term clinical exposure evaluation of a second-generation electrically heated cigarette smoking system. *Journal of Clinical Pharmacology* 47(4):518-530.

Roethig, H. J., S. Munjal, S. Feng, Q. Liang, M. Sarkar, R. A. Walk, and P. E. Mendes. 2009. Population estimates for biomarkers of exposure to cigarette smoke in adult U.S. cigarette smokers. *Nicotine & Tobacco Research* 11(10):1216-1225.

Rohatgi, N., J. Kaur, A. Srivastava, and R. Ralhan. 2005. Smokeless tobacco (khaini) extracts modulate gene expression in epithelial cell culture from an oral hyperplasia. *Oral Oncology* 41(8):806-820.

Rowland, R., and K. Harding. 1999. Increased sister chromatid exchange in the peripheral blood lymphocytes of young women who smoke cigarettes. *Hereditas* 131(2):143-146.

Rundle, A., and H. Ahsan. 2008. Molecular epidemiological studies that can be nested within cohorts. In *Molecular epidemiology of chronic diseases*, edited by C. Wild, P. Vineis, and S. Garte. West Sussex, England: J.W. Wiley. Pp. 23-37.

SAMHSA (Substance Abuse and Mental Health Services Administration). 2007. *Results from the 2006 national survey on drug use and health: National findings.* DHHS Publication No. SMA 07-4293. Rockville, MD: Substance Abuse and Mental Health Services Administration.

SAMHSA. 2011. *Results from the 2010 national survey on drug use and health: National findings.* NSDUH Series H-41, HHS Publication No. (SMA) 11-4658. Rockville, MD: Substance Abuse and Mental Health Services Administration.

Sarkar, M., S. Kapur, K. Frost-Pineda, S. Feng, J. Wang, Q. Liang, and H. Roethig. 2008. Evaluation of biomarkers of exposure to selected cigarette smoke constituents in adult smokers switched to carbon-filtered cigarettes in short-term and long-term clinical studies. *Nicotine & Tobacco Research* 10(12):1761-1772.

Sarto, F., M. Faccioli, I. Cominato, and A. Levis. 1985. Aging and smoking increase the frequency of sister-chromatid exchanges (SCE) in man. *Mutation Research Letters* 144(3):183-187.

Scherer, G. 2005. Biomonitoring of inhaled complex mixtures—ambient air, diesel exhaust and cigarette smoke. *Experimental and Toxicologic Pathology* 57(Suppl. 1):75-110.

Scherer, G. 2006. Carboxyhemoglobin and thiocyanate as biomarkers of exposure to carbon monoxide and hydrogen cyanide in tobacco smoke. *Experimental and Toxicologic Pathology* 58(2-3):101-124.

Scherer, G., M. Urban, J. Engl, H. W. Hagedorn, and K. Riedel. 2006. Influence of smoking charcoal filter tipped cigarettes on various biomarkers of exposure. *Inhalation Toxicology* 18(10):821-829.

Scherer, G., J. Engl, M. Urban, G. Gilch, D. Janket, and K. Riedel. 2007a. Relationship between machine-derived smoke yields and biomarkers in cigarette smokers in Germany. *Regulatory Toxicology and Pharmacology* 47(2):171-183.

Scherer, G., M. Urban, H. W. Hagedorn, S. Feng, R. D. Kinser, M. Sarkar, Q. Liang, and H. J. Roethig. 2007b. Determination of two mercapturic acids related to crotonaldehyde in human urine: Influence of smoking. *Human & Experimental Toxicology* 26(1):37-47.

Scherer, G., M. Urban, H. W. Hagedorn, R. Serafin, S. Feng, S. Kapur, R. Muhammad, Y. Jin, M. Sarkar, and H. J. Roethig. 2010. Determination of methyl-, 2-hydroxyethyl- and 2-cyanoethylmercapturic acids as biomarkers of exposure to alkylating agents in cigarette smoke. *Journal of Chromatography B* 878(27):2520-2528.

Schuller, H. M., R. Jorquera, A. Reichert, and A. Castonguay. 1993. Transplacental induction of pancreas tumors in hamsters by ethanol and the tobacco-specific nitrosamine 4-(methylnitrosamino)-1-(3-pyridyl)-1-butanone. *Cancer Research* 53(11):2498-2501.

Schuller, H. M., R. Jorquera, X. Lu, A. Riechert, and A. Castonguay. 1994. Transplacental carcinogenicity of low doses of 4-(methylnitrosamino)-1-(3-pyridyl)-1-butanone administered subcutaneously or intratracheally to hamsters. *Journal of Cancer Research and Clinical Oncology* 120(4):200-203.

Schwartz, J., and X. Gu. 2002. Chapter 8: Hamster oral cancer model. In *Tumor models in cancer research*. Vol. 10, edited by B. A. Teicher. Totowa, NJ: Humana Press, Inc.

Schwartz, J. L., K. D. Brunnemann, A. J. Adami, S. Panda, S. C. Gordon, D. Hoffmann, and G. R. Adami. 2010. Brand specific responses to smokeless tobacco in a rat lip canal model. *Journal of Oral Pathology and Medicine* 39(6):453-459.

Secretan, B., K. Straif, R. Baan, Y. Grosse, F. El Ghissassi, V. Bouvard, L. Benbrahim-Tallaa, N. Guha, C. Freeman, L. Galichet, and V. Cogliano. 2009. A review of human carcinogens—Part E: Tobacco, areca nut, alcohol, coal smoke, and salted fish. *Lancet Oncology* 10(11):1033-1034.

Shirname-More, L. 1991. Smokeless tobacco extracts mutate human cells. *Carcinogenesis* 12(5):927-930.

Shklar, G., K. Niukian, M. Hassan, and E. G. Herbosa. 1985. Effects of smokeless tobacco and snuff on oral mucosa of experimental animals. *Journal of Oral and Maxillofacial Surgery* 43(2):80-86.

Singh, R., and H. S. Nalwa. 2011. Medical applications of nanoparticles in biological imaging, cell labeling, antimicrobial agents, and anticancer nanodrugs. *Journal of Biomedical Nanotechnology* 7(4):489-503.

Singh, S., Y. K. Loke, J. G. Spangler, and C. D. Furberg. 2011. Risk of serious adverse cardiovascular events associated with varenicline: A systematic review and meta-analysis. *Canadian Medical Association Journal* 183(12):1359-1366.

Smith, C., S. McKarns, R. Davis, S. Livingston, B. Bombick, J. Avalos, W. Morgan, and D. Doolittle. 1996. Human urine mutagenicity study comparing cigarettes which burn or primarily heat tobacco. *Mutation Research/Environmental Mutagenesis and Related Subjects* 361(1):1-9.

Song, S. Q., and D. L. Ashley. 1999. Supercritical fluid extraction and gas chromatography mass spectrometry for the analysis of tobacco-specific nitrosamines in cigarettes. *Analytical Chemistry* 71(7):1303-1308.

St. George, J. A., D. L. Cranz, S. C. Zicker, J. R. Etchison, D. L. Dungworth, and C. G. Plopper. 1985. An immunohistochemical characterization of rhesus monkey respiratory secretions using monoclonal antibodies. *American Review of Respiratory Disease* 132(3):556-563.

Stamm, S. C., B. Z. Zhong, W. Z. Whong, and T. Ong. 1994. Mutagenicity of coal-dust and smokeless-tobacco extracts in *salmonella typhimurium* strains with differing levels of O-acetyltransferase activities. *Mutation Research* 321(4):253-264.

Stanton, M., E. Miller, C. Wrench, and R. Blackwell. 1972. Experimental induction of epidermoid carcinoma in the lungs of rats by cigarette smoke condensate. *Journal of the National Cancer Institute* 49(3):867.

Starrett, W., and D. J. Blake. 2011. Sulforaphane inhibits de novo synthesis of IL-8 and MCP-1 in human epithelial cells generated by cigarette smoke extract. *Journal of Immunotoxicology* 8(2):150-158.

Stenstrom, B., C. M. Zhao, A. B. Rogers, H. O. Nilsson, E. Sturegard, S. Lundgren, J. G. Fox, T. C. Wang, T. M. Wadstrom, and D. Chen. 2007. Swedish moist snuff accelerates gastric cancer development in *Helicobacter pylori*-infected wild-type and gastrin transgenic mice. *Carcinogenesis* 28(9):2041-2046.

Stepanov, I., and S. S. Hecht. 2005. Tobacco-specific nitrosamines and their N-glucuronides in the urine of smokers and smokeless tobacco users. *Cancer Epidemiology, Biomarkers & Prevention* 14:885-891.

Stepanov, I., J. Jensen, D. Hatsukami, and S. S. Hecht. 2008. New and traditional smokeless tobacco: Comparison of toxicant and carcinogen levels. *Nicotine & Tobacco Research* 10:1773-1782.

Stepanov, I., S. G. Carmella, A. Briggs, L. Hertsgaard, B. Lindgren, D. K. Hatsukami, and S. S. Hecht. 2009. Presence of the carcinogen N'-nitrosonornicotine in the urine of some users of oral nicotine replacement therapy products. *Cancer Research* 69:8236-8240.

Stepanov, I., P. W. Villalta, A. Knezevich, J. Jensen, D. Hatsukami, and S. S. Hecht. 2010. Analysis of 23 polycyclic aromatic hydrocarbons in smokeless tobacco by gas chromatography-mass spectrometry. *Chemical Research in Toxicology* 23:66-73.

Stinn, W., J. H. E. Arts, A. Buettner, E. Duistermaat, K. Janssens, C. F. Kuper, and H. J. Haussmann. 2010. Murine lung tumor response after exposure to cigarette mainstream smoke or its particulate and gas/vapor phase fractions. *Toxicology* 275(1-3):10-20.

Straif, K., R. Baan, Y. Grosse, B. Secretan, F. El Ghissassi, and V. Cogliano. 2005. Carcinogenicity of polycyclic aromatic hydrocarbons. *Lancet Oncology* 6(12):931-932.

Sussan, T. E., T. Rangasamy, D. J. Blake, D. Malhotra, H. El-Haddad, D. Bedja, M. S. Yates, P. Kombairaju, M. Yamamoto, K. T. Liby, M. B. Sporn, K. L. Gabrielson, H. C. Champion, R. M. Tuder, T. W. Kensler, and S. Biswal. 2009. Targeting Nrf2 with the triterpenoid CDDO- imidazolide attenuates cigarette smoke-induced emphysema and cardiac dysfunction in mice. *Proceedings of the National Academy of Sciences* 106(1):250-255.

Suwan-ampai, P., A. Navas-Acien, P. T. Strickland, and J. Agnew. 2009. Involuntary tobacco smoke exposure and urinary levels of polycyclic aromatic hydrocarbons in the United States, 1999 to 2002. *Cancer Epidemiology, Biomarkers & Prevention* 18(3):884-893.

Thompson, C. A., and P. C. Burcham. 2008. Genome-wide transcriptional responses to acrolein. *Chemical Research in Toxicology* 21:2245-2256.

Trivedi, A. H., B. J. Dave, and S. G. Adhvaryu. 1993. Genotoxic effects of tobacco extract on Chinese hamster ovary cells. *Cancer Letters* 70(1-2):107-112.

Tuomisto, J., S. Kolonen, M. Sorsa, and P. Einistö. 1986. No difference between urinary mutagenicity in smokers of low-tar and medium-tar cigarettes: A double-blind crossover study. *Archives of Toxicology. Supplement* 9:115.

Wang, Y., E. Rotem, F. Andriani, and J. A. Garlick. 2001. Smokeless tobacco extracts modulate keratinocyte and fibroblast growth in organotypic culture. *Journal of Dental Research* 80(9):1862-1866.

Wang, D., Y. Wang, and Y. N. Liu. 2010. Experimental pulmonary infection and colonization of *Haemophilus influenzae* in emphysematous hamsters. *Pulmonary Pharmacology and Therapeutics* 23(4):292-299.

Weed, D. L. 2000. Epidemiologic evidence and causal inference. *Hematology/Oncology Clinics of North America* 14(4):797-807.

Weiss, N. S. 1994. Application of the case-control method in the evaluation of screening. *Epidemiologic Reviews* 16(1):102-108.

Whincup, P. H., J. A. Gilg, J. R. Emberson, M. J. Jarvis, C. Feyerabend, A. Bryant, M. Walker, and D. G. Cook. 2004. Passive smoking and risk of coronary heart disease and stroke: Prospective study with cotinine measurement. *BMJ* 329(7459):200-205.

Whitcutt, M. J., K. B. Adler, and R. Wu. 1988. A biphasic chamber system for maintaining polarity of differentiation of cultured respiratory tract epithelial cells. *In Vitro Cellular and Developmental Biology* 24(5):420-428.

Whong, W. Z., R. G. Ames, and T. M. Ong. 1984. Mutagenicity of tobacco snuff: Possible health implications for coal miners. *Journal of Toxicology and Environmental Health* 14(4):491-496.

Whong, W. Z., J. D. Stewart, and T. Ong. 1985. Formation of bacterial mutagens from the reaction of chewing tobacco with nitrite. *Mutation Research/Genetic Toxicology* 158(3):105-110.

Witschi, H. 2004. A/J mouse as a model for lung tumorigenesis caused by tobacco smoke: Strengths and weaknesses. *Experimental Lung Research* 31(1):3-18.

Worawongvasu, R., S. Ashrafi, A. Das, J. Waterhouse, and H. Medak. 1991. A light and scanning electron microscopic study of snuff induced early changes in hamster cheek pouch mucosa. *Biomedical Research (India)* 2:240-250.

Wu, R., E. Nolan, and C. Turner. 1985. Expression of tracheal differentiated functions in serum-free hormone-supplemented medium. *Journal of Cellular Physiology* 125(2):167-181.

Wu, R., C. G. Plopper, and P. W. Cheng. 1991. Mucin-like glycoprotein secreted by cultured hamster tracheal epithelial cells. Biochemical and immunological characterization. *Biochemical Journal* 277(3):713-718.

Wu, Z., P. Upadhyaya, S. G. Carmella, S. S. Hecht, and C. L. Zimmerman. 2002. Disposition of 4-(methylnitrosamino)-1-(3-pyridyl)-1-butanone (NNK) and 4-(methylnitrosamino)-1-(3-pyridyl)-1-butanol (NNAL) in bile duct-cannulated rats: Stereoselective metabolism and tissue distribution. *Carcinogenesis* 23(1):171-179.

Wynder, E. L., and D. Hoffmann. 1961. A study of tobacco carcinogenesis. VIII. The role of the acidic fractions as promoters. *Cancer* 14(6):1306-1315.

Wynder, E. L., E. A. Graham, and A. B. Croninger. 1953. Experimental production of carcinoma with cigarette tar. *Cancer Research* 13(12):855-864.

Yamasaki, E., and B. N. Ames. 1977. Concentration of mutagens from urine by absorption with the nonpolar resin XAD-2: Cigarette smokers have mutagenic urine. *Proceedings of the National Academy of Sciences* 74(8):3555.

Yildiz, D., Y. S. Liu, N. Ercal, and D. W. Armstrong. 1999. Comparison of pure nicotine- and smokeless tobacco extract-induced toxicities and oxidative stress. *Archives of Environmental Contamination and Toxicology* 37(4):434-439.

Yoshida, T., I. Mett, A. K. Bhunia, J. Bowman, M. Perez, L. Zhang, A. Gandjeva, L. Zhen, U. Chukwueke, T. Mao, A. Richter, E. Brown, H. Ashush, N. Notkin, A. Gelfand, R. K. Thimmulappa, T. Rangasamy, T. Sussan, G. Cosgrove, M. Mouded, S. D. Shapiro, I. Petrache, S. Biswal, E. Feinstein, and R. M. Tuder. 2010. Rtp801, a suppressor of mTOR signaling, is an essential mediator of cigarette smoke-induced pulmonary injury and emphysema. *Nature Medicine* 16(7):767-773.

Yuan, J. M., W. P. Koh, S. E. Murphy, Y. Fan, R. Wang, S. G. Carmella, S. Han, K. Wickham, Y. T. Gao, M. C. Yu, and S. S. Hecht. 2009. Urinary levels of tobacco-specific nitrosamine metabolites in relation to lung cancer development in two prospective cohorts of cigarette smokers. *Cancer Research* 69:2990-2995.

Zedler, B. K., R. Kinser, J. Oey, B. Nelson, H. J. Roethig, R. A. Walk, P. Kuhl, K. Rustemeier, G. Schepers, K. Von Holt, and A. R. Tricker. 2006. Biomarkers of exposure and potential harm in adult smokers of 3-7 mg tar yield (Federal Trade Commission) cigarettes and in adult non-smokers. *Biomarkers* 11(3):201-220.

Zhong, Y., S. G. Carmella, J. B. Hochalter, S. Balbo, and S. S. Hecht. 2010. Analysis of *r* -7, *t* -8,9, *c*-10-tetrahydroxy-7,8,9,10-tetrahydrobenzo[*a*]pyrene in human urine: A biomarker for directly assessing carcinogenic polycyclic aromatic hydrocarbon exposure plus metabolic activation. *Chemical Research in Toxicology* 24:73-80.

Zhou, J., E. A. Eksioglu, N. R. Fortenbery, X. Chen, H. Wang, P. K. Epling-Burnette, J. Y. Djeu, and S. Wei. 2011. Bone marrow mononuclear cells up-regulate toll-like receptor expression and produce inflammatory mediators in response to cigarette smoke extract. *PLoS One* 6(6):e21173.

4

Methods for Investigating Addictive Potential

EVALUATION OF REINFORCEMENT AND ADDICTIVE POTENTIAL

As specified by the Family Smoking Prevention and Tobacco Control Act of 2009 (FSPTCA),[1] the evaluation of a modified risk tobacco product (MRTP) with regard to the public health standard concerns, in part, an evaluation of the product with regard to its tendency to promote the following:

- Initiation and continuation of its regular use
- Switching to its use and cessation of the consumption of more harmful tobacco products (e.g., aid in cessation of use of conventional cigarettes)
- Dual use (use of the MRTP concurrent with continued use of an existing harmful form of tobacco use such as smoking conventional cigarettes)
- Relapse back to more harmful tobacco use (e.g., resume smoking conventional cigarettes after an extended period of abstinence)

All of these outcomes can be logically related to the reinforcing value of the MRTP (how rewarding it is).

[1] *Family Smoking Prevention and Tobacco Control Act of 2009*, Public Law 111-31, 123 Stat. 1776 (June 22, 2009).

The chief reason for testing reinforcement value in the laboratory setting is that measures yielded by such testing show a good correspondence to a product's addiction potential in real-world use (Haney and Spealman, 2008). Specifically, drugs that have a positive subjective evaluation and are self-administered in laboratory tasks are ones that tend to be used and abused recreationally in real-world use (Comer et al., 2008; Haney, 2009).

The reinforcement value of an agent (e.g., a specific drug, such as nicotine) or a product (i.e., a drug[s] provided via a particular delivery system, such as smokeless tobacco or cigarettes) can be gauged through animal research; however, in the present situation, animal research on reinforcement value does not appear optimal. First, animal research is especially warranted when the product poses significant immediate health risks. However, to the extent that an MRTP has been adequately screened in preclinical work, it seems that the MRTP could be safely used in laboratory assessments of reinforcement value or self-administration (where toxic effects of possible prolonged dual use would not pertain). Second, because of the difficulty in modeling certain kinds of delivery systems with particular tobacco products (e.g., snus), human research may present the most externally valid research option. Third, human research methods afford an array of research paradigms that should yield meaningful assessment of MRTP reinforcement potential. Finally, human research requires less extrapolation because of a lack of interspecies differences, which can be substantial in terms of nicotine reinforcement (Rogers et al., 2009).

Key Considerations for Reinforcement and Self-Administration Studies

Almost by definition, an addictive agent must support self-administration. Moreover, there is a long history of research that shows a rough correspondence between the reinforcement capacity of an agent in the laboratory setting and its abuse potential in real-world contexts (Comer et al., 2008; Haney, 2009). *Reinforcement* is generally defined as the capacity of an agent to sustain self-administration. Therefore, one meaningful step in assessing the ability of an MRTP to support self-administration in real-world contexts is to determine whether it supports self-administration in laboratory or controlled settings.

Evaluating reinforcement is complicated by several factors, one of which is a continuum of reinforcement potency. Therefore, methods must capture the reinforcement potential of a product relative to other products or agents to provide meaningful comparisons. In theory, a desirable MRTP should be somewhat more reinforcing than nicotine replacement therapies (NRTs), but perhaps less reinforcing than conventional cigarettes (at least among current smokers who have demonstrated considerable susceptibility to cigarette reinforcement). The relative value of products will be affected by the dose of product tested. Doses may reflect what is considered a meaningful dose in terms of real-world use, they may be based upon brief *ad libitum* use, or they may be established via dose banding methods. Ideally, an MRTP would be sufficiently reinforcing so as to attract smokers away from conventional cigarettes but not encourage the widespread dependent use of the product by individuals who were previously nonusers or who would have quit smoking. NRTs represent a meaningful lower bound of reinforcement magnitude because they tend not to support addictive or dependent use (Shiffman et al., 2008a). Further, there appear to be interactions between specific products and individuals, such that individuals differ in terms of the hierarchy of reinforcement potential across products (Perkins, 2009). The determinants of such individual differences in product-relative reinforcement are unknown but no doubt reflect multiple influences such as prior experience (because reinforcement changes with exposure), genetic factors, and social influences. Thus, the level of reinforcement value may lie more in the type of research participant than in the type of product.

Reinforcement and Self-Administration Methods

Likelihood of initiation, as well as maintenance or persistence of use, can be studied across multiple types of studies ranging from laboratory studies, to randomized controlled trials (RCTs), to population-based cohort studies. Different methodological principles and standards apply to each type of study. As in all research, research methods are determined in part by the question(s) being addressed. In the case of the evaluation of an MRTP, the core questions in this area involve the extent to which the product will attract and support heavy self-administration and abusive use. Several relevant experimental contexts can be used in the effort to determine the self-administration and use or abuse potential of an MRTP:

1. Subjective evaluation of the product both initially, and with repeated exposure or use in laboratory contexts relative to appropriate comparison products
2. Acute self-administration in laboratory contexts (only reflecting use within laboratory sessions), relative to appropriate comparison products
3. Use in extended residence facilities
4. Natural environment contexts where long-term use can be studied in real-world contexts, via
    a. long-term use in RCTs,
    b. cross-sectional survey studies, and
    c. longitudinal cohort studies.

Additionally, methodological approaches must be tailored to each research context. Unless otherwise specified, these considerations apply to both acute laboratory and residential stay experiments.

Size and Nature of the Sample

Recruited participants must permit appropriate inferences regarding the populations and questions to be addressed. No standard sample size can be specified confidently for the studies described in this section. Each study must be powered consistent with the study questions posed and the comparison products used. Some guidance on power might be gleaned from studies in which high- and low-preference products are evaluated (e.g., conventional cigarettes and NRT products [Johnson et al., 2004; Perkins et al., 2004a, 2009]). Clearly the *nature* of the sample will differ with regard to the particular research question posed.

Relative Reinforcement Value in Regular Smokers One question of key importance is the extent to which an MRTP is reinforcing among *current heavy smokers*. This would be relevant to the extent to which the product would be used heavily enough by smokers to serve as a cessation aid or a long-term substitute with regard to smoking conventional cigarettes. A very high reinforcement value in smokers of conventional cigarettes would suggest the product could serve as a cessation aid or long-term substitute to conventional cigarettes and could also present a meaningful risk of initiation of use among nonsmokers or ex-smokers. The use of a population of current smokers has the advantage of ensuring that the tested population is sensitive to nicotine reward (Carter and Griffiths, 2009). If current smokers are used, the researcher should ensure that the research participants have no strong desire to quit, so the findings relate to smoking behavior in regular smokers and not quitting behavior (Perkins et al., 1997). About 45 percent of current smokers attempt to quit each year (CDC, 2009), and including such smokers in the sample might not only produce greater within-cell error, but also distort outcomes systematically. Such smokers, for instance, might be especially willing to self-administer a perceived safer alternative to smoking conventional cigarettes and more likely to try to avoid smoking conventional cigarettes. Thus, their self-administration data might not validly reflect the actual reinforcement value of the product.

Relative Reinforcement Value in Nonsmokers, Ex-Smokers, and Adolescents Testing an MRTP among nonsmokers would provide some evidence of attractiveness and reinforcement potential in people who are essentially nicotine naïve.[2] If multiple sessions are used, the research could yield some evidence on how much drug experience might be needed to show an increase in reward value. To increase the likelihood that the tested population comprises at-risk individuals, some selection factors could be used such as high levels of impulsivity, extreme delay discounting (Bickel et al., 2010), use of other abused drugs, or risk haplotypes for tobacco dependence (Weiss et al., 2008). Also, because there may be a relation between age and reaction to nicotine and vulnerability to dependence (Weiss et al., 2008), it may be important to use relatively young individuals in such research. Adolescents might be optimal, but research methods and oversight would have to be appropriate for their participation. Adolescents who have experimented with smoking might constitute a particularly high-risk population with high public health significance. Finally, the use of ex-smokers would suggest the potential reinforcement value of MRTP use in this population, which has demonstrated sensitivity to nicotine reinforcement. Of course, inclusion of ex-smokers would require a careful assessment of the risks and benefits of participation.

In addition, because reinforcement from nicotine or tobacco can vary with gender, age, tobacco experience, and other factors, the researchers should ensure that such dimensions are appropriately represented or controlled (e.g., used for blocking or as exclusion criteria) in the sample to the extent that it is compatible with the question addressed.

Characterization of the Sample

A comprehensive characterization of the sample is important because it defines the population to which the conclusions may be most directly related. It also permits tests of the interaction of person factors with MRTP liking or use—factors that appear to modulate product reinforcing value. Variables that may be important to measure, based on prior research on tobacco reinforcement, are gender, age, ethnicity, educational and socioeconomic status, tobacco and nicotine use history (including peak tobacco use levels, prior quitting history, age of initial use, and use histories of different tobacco and nicotine products), expectations about the effects of the products to be tested, tobacco or nicotine dependence, blood or breath levels of tobacco or nicotine exposure, health and mental health status and history, and use of psychoactive products including psychiatric medications. These variables are important because they have been related to nicotine dependence, tobacco self-administration, and ability to control tobacco use.

---

[2] Non-tobacco users are defined as those who have never smoked more than 10 cigarettes and who have never used any other form of tobacco.

In terms of tobacco dependence assessment, the Fagerstrom Test for Nicotine Dependence or one of the new multifactorial dependence assessments (the Nicotine Dependence Syndrome Scale [Shiffman and Sayette, 2005; Shiffman et al., 2004] or the Wisconsin Inventory of Smoking Dependence Motives [Smith et al., 2010]) appear to provide more accurate appraisal of dependence than do the *Diagnostic and Statistical Manual of Mental Disorders* criteria (Hughes et al., 2011). In addition, researchers should ensure that the dependence instrument used is one that is appropriate to the population in question. For instance, there is concern that some dependence instruments may not be appropriate for young or light smokers, so researchers should use an instrument validated with such populations (Colby et al., 2000).

Standardization of Pre-Session Experiences

Investigators should ensure that research participants have similar experiences prior to experimental sessions. Standard durations of abstinence from, or controlled use of, nicotine, caffeine, and other psychoactive agents or products before sessions is needed so subjects enter sessions at similar motivational states. Deprivation tends to significantly increase motivation to use tobacco and its self-administration (Fant et al., 1995; Perkins et al., 1994a; Zinser et al., 1999). Studies designed to test maximal motivation would impose a period of deprivation, such as overnight deprivation, which could be tested with a carbon monoxide (CO) test in the case of deprivation of combustible products. Another approach would be to impose a modest but standard level of deprivation (e.g., 1–2 hours) to model a motivational state that would typically occur throughout the day. The most comprehensive approach to assessing self-administration would be to test products across a variety of deprivation levels. Deprivation prior to clinical studies may add complications in data interpretation. An alternative, although more costly and time consuming, is observation of *ad libitum* self-administration so that the response measured reflects real use.

It is probably not a concern if subjects take their normal prescription medication, including psychiatric medication, on the days of sessions or measurement. This is because the main outcome data will be relative preference for, or use of, the tested products, and this presumably will not be differentially affected by chronic use of psychiatric medications.

It is important that subjects have similar expectations about the experiment and what it entails (e.g., the nature of the tested products) unless manipulation of expectations is an explicit element of the study design (because expectations can significantly affect response to a tobacco product [Perkins et al., 2010]). One possible strategy is to provide subjects with considerable superfluous information, which may reduce disparities in expectations (Griffiths et al., 2003). Finally, to the extent that measures are complex (e.g., with certain types of cognitive performance tasks) it is important that practice effects be reduced by pre-session task familiarization.

Reinforcement and Self-Administration Measures

Biochemical Measures

Biochemical measures of tobacco or nicotine exposure are important because they reflect prior self-administration intensity or tolerance, and therefore they should serve as useful covariates for laboratory-based self-administration. The appropriate measure could be CO level for cigarette smokers, or nicotine or cotinine levels (from blood, saliva, or urine) in other types of nicotine or tobacco users (those using smokeless tobacco or NRT). In particular, acute blood nicotine absorption profiles in response to both single and repeated use of products is a relevant component in assessing the addictive potential of MRTPs. Cotinine might be preferred to CO and nicotine because of its longer half-life. This could be extremely useful if long-term abstinence is imposed prior to experimental sessions or if subjects have engaged in only infrequent use of a nicotine or tobacco product. Also, if a noncombustible MRTP is studied, CO levels during or after the experiment will not provide measures of effective dosing. Therefore, to obtain a true baseline for such later measures, either nicotine or cotinine should be measured at baseline. In deciding between assessing cotinine versus nicotine, if the intent is to study effective self-dosing acutely (over minutes or 1–3 hours), then nicotine is the measure of choice, while cotinine would be the measure of choice if the effects of dosing over an extended time period (many hours or days) are targeted. The best predictor of plasma cotinine may be measurement of urine cotinine corrected for creatinine concentration (Benowitz et al., 2009). Finally, the investigator might wish to measure both nicotine and 3-hydroxy-cotinine in order to estimate nicotine metabolism (Schnoll et al., 2009). However, cotinine may be a poor choice for dual-use studies because it can reflect nicotine from multiple sources.

Selection of a biochemical assay depends upon the particular experiment, the questions posed, and the nature of the product. If relatively sensitive determination of nicotine receipt is sought, then it would be necessary to measure venous or arterial nicotine levels (typically via a venous catheter) and to obtain multiple measures over time to determine boost peak (peak baseline level) and area under the curve (see Benowitz [2006] for calculation).

Imaging methods such as positron emission tomography or functional magnetic resonance imaging could be used to further characterize the addiction potential of MRTPs. There is increasing evidence that particular neurotransmitter systems and associated brain regions are critically involved in the motivational processing of nicotine cues and nicotine anticipation: e.g., the dorsal striatum, nucleus accumbens, and anterior cingulate cortex (Gloria et al., 2009; McClernon et al., 2009). Therefore, amongst experienced MRTP users, MRTP cues or anticipation of MRTP delivery would be expected to activate such brain regions. However, at present there is little evidence that such measures possess the sensitivity to yield accurate rank-orderings of the addictive potential of different products or delivery systems.

Nature of the Comparison Stimuli

The selection of products or stimuli to be compared should be determined by the goals of the experiment and the need to obtain a sufficient number of comparators to permit an informative interpretive context. However, as discussed elsewhere, it seems that use of both conventional cigarettes (when smokers or ex-smokers are used as subjects) and NRT would be informative, because these represent products with very high versus modest reinforcement value. The study by Kotlyar et al. (2007) reveals how MRTPs can be meaningfully compared with NRTs on the basis of subjective evaluation and effective nicotine delivery.

It may be important to compare the product with nonpharmacologic stimuli as a means of providing a generally meaningful anchor point for the comparison of the pharmacologic products (including the MRTP). For instance, nicotine or tobacco products might be compared with pictorial stimuli (e.g., the International Affective Picture System), attractive music, compounds that stimulate taste buds, or money (Perkins et al., 1997). It is especially important to use nonpharmacologic stimuli as comparison stimuli (e.g., money) when using nonsmokers as subjects because it would be important to compare the MRTP with a stimulus of meaningful reinforcing value.

If the study is using current smokers as subjects, it would be informative to use the subject's own or preferred brand of cigarettes, as this could represent an optimally reinforcing product against which to compare the MRTP. However, another strategy would be to use cigarettes with a range of known nicotine contents, which would provide a range of reinforcement value against which the MRTP could be compared.

Operant Self-Administration

One standard method of evaluating reinforcement value is to use an operant self-administration paradigm in which some sort of instrumental response (key presses for instance) is executed to "earn" doses of the product. How hard an individual is willing to work for a dose is related to the addictive potential of the product. For example, the subject might be given the opportunity to earn either puffs of a conventional cigarette, inhalations from a nicotine inhaler, or doses from an MRTP. Such operant paradigms permit collection of many different sorts of measures, such as: (1) response rates including peak response rates for each type of product; (2) relative response rates on concurrent schedules (Perkins et al., 1997); and (3) demand elasticity for each type of product (the extent to which responding is affected by increasing the response requirement or dose). The last index may be especially useful because it permits meaningful interproduct (or interstimulus) comparisons on the basis of demand curves (Johnson and Bickel, 2006), in essence permitting more direct inferences regarding reinforcement magnitude.

Timing and Exposure Parameters

Experiments aimed at characterizing reinforcement value could present MRTPs and other products in diverse ways. The mode of presentation should be dictated by the experimental paradigm used, as well as the research question. In acute dose-effect comparison studies conducted in laboratories settings, presentation of discrete doses of products or stimuli should be counterbalanced, controlling for amount and order of delivery. In self-administration studies or behavioral economic studies, the researcher could use progressive ratio schedules in separate sessions for each product or concurrent schedules (e.g., comparing each product with monetary payment) or could test products individually across different response requirements to construct demand curves. In either acute dose-effect studies or self-administration studies, relatively standard doses with cigarettes can be achieved either with puff duration signals or with devices that control puff volume mechanically (Perkins et al., 1997). Timing signals might be the best way to manage dose parameters with products such as smokeless tobacco or NRT (Shiffman et al., 2003).

There are many things to consider in setting up and interpreting such experiments. One concern is how much experience or exposure to permit in the experiment. There is certainly evidence that preference or reinforcing value changes over exposure. This could occur because of tolerance to aversive effects, sensitization, familiarity (learning how to self-dose), development of dependence, and so forth. Thus, the researcher must structure the study so the person's experience prior to the study and the exposure during the study are designed to match the experimental goals. An important principle, however, is that the best estimate of the ultimate reinforcement potential of an MRTP may be obtained after fairly extensive use.

Another concern is the interdose interval and amount of exposure (dose) to the products. Different delivery systems may deliver different doses of nicotine and doses with different pharmacodynamics. The investigator must consider whether standard dosing or exposure parameters do not "unlevel" the playing field for the various products (e.g., creating toxic effects or different levels of withdrawal for one product versus another). Investigators may also want to mimic extreme use, because some users may overuse the product. Interdose intervals should be determined based on the anticipated pharmacodynamics of the tested products.

Because the ordering of stimuli or products might affect the response (such as when an earlier product might satiate the subject, thereby reducing his or her motivation to self-administer additional nicotine), it is especially important to counterbalance stimulus presentations in acute dose-effect comparison studies so order effects are not inextricably confounded with stimulus effects. In essence, great care must be taken to ensure that exposures to products relatively late in the exposure sequence are meaningful. To the extent that earlier exposures result in high nicotine levels, reduced withdrawal, or priming effects, the subject's motivational state is altered and therefore the subject's responses are not similarly meaningful across the sequence. One strategy that could be used to address this is to have subjects "earn" dosings during a session but not consume them until after the session (Perkins et al., 1997). This may not be appropriate where delay would distort the motivational value of exposure. There is evidence that immediate versus delayed access to addictive agents or products makes a substantial difference in motivational and evaluative response (Gloria et al., 2009; Sayette and Hufford, 1994).

Another concern with timing of the experimental sessions is to ensure the anticipated end of the experimental session does not bias subjects' responses. For instance, if one of the measures of product evaluation is instrumental to secure a dose of the product or amount of money needed to purchase a dose of the product from the subject, these measures could be distorted if the subject knows that he or she will shortly be released from the session and have ready access to nicotine or tobacco. Therefore, a postsession waiting period (which might range from 30 to 90 minutes) is often imposed so the only prospect of imminent tobacco receipt is that which will occur in the session (Perkins et al., 1999).

Additionally, with some procedures such as instrumental self-administration (behavioral economic strategies) or with unusual controlled dosing procedures, it may be desirable to allow the subjects some practice with the procedure so learning or familiarization effects are not confounded with changes in reinforcement value that develop with drug use experience.

In most self-administration experiments it would probably be important to determine the efficiency of self-administration, meaning the relation between self-administration and effective drug delivery of doses consumed (measured by biochemical indices of product receipt, such as CO and nicotine). This would allow one to distinguish gross self-administration behaviors from effective drug delivery. This distinction, for instance, might be relevant to questions about whether compensation occurs because of use of an MRTP. For instance, use of an MRTP might decrease the number of conventional cigarettes that a person smokes. However, this does not necessarily mean the person is actually exposed to less smoke or takes in less nicotine (Benowitz et al. [2006] provides a compensation determination formula for cigarettes with known machine determined yields). Multiple measures are available to assess self-administration behavior so as to capture effective delivery more accurately (Rose et al., 2003; Strasser et al., 2007). This could be done by the use of especially sensitive assessments of self-administration. One example of this is the use of smoking topography measures that permit assessment of puff duration, inhalation force, and so on via force or flow transducers (Strasser et al., 2009). Video cameras and monitors have also been used to assess puff number and duration (Benowitz et al., 2006).

Finally, one could indirectly infer the effective dose by repeatedly measuring physiological responses that are acutely sensitive to nicotine dose and rise-time effects (e.g., nicotine-induced tachycardia or skin temperature effects [Benowitz et al., 2006; Perkins et al., 1994b]) and deriving peak and area under the curve indices.

Reinforcement and Self-Administration Study Designs

Acute Dose-Effect Comparison Studies

This approach has been labeled as a standard with regard to human abuse liability drug testing, because of the correspondence between subjective ratings of drug effects and real-world abuse potential (Carter and Griffiths, 2009). This sort of research is faster and more economical to conduct than human self-administration studies. In this research, appropriate subject groups are given discrete agent or product exposures and asked to rate them on validated scales. These are generally placebo-controlled, blinded, within-subject crossover designs. However, the apparent differences among some tobacco products (snus versus conventional cigarettes or e-cigarettes) may compromise the ability to achieve true placebo control or blinding. Each product, though, could have a placebo preparation, which should control for some expectancy effects. Ideally, subjects should be allowed to rate a variety of dose levels or exposures to the products to obtain a more comprehensive index of product effects. In addition, it would be important in at least a subset of studies to test at multiple intervals postexposure to ensure the pharmacodynamics of response are characterized for each product. This is important in part because pharmacodynamics may greatly affect reinforcement value and abuse potential (Dewit et al., 1993; Mumford et al., 1995). While acute dose-effect comparison studies are often conducted on closed or residential wards when using illicit drugs, this seems unnecessary for the type of research discussed because the tested products will not be significantly intoxicating, the product would not be a controlled substance, and biochemical and self-report means can be used to determine intersession use.

Measures for Use in Acute Dose-Effect Paradigms Certainly researchers would collect self-report measures of subjective responses to the MRTP and other products, either in anticipation of receipt of the product (after the subject has some familiarity with it) or following its effects. There are well-characterized scales that permit the assessment of a variety of relevant rating dimensions (e.g., the Duke Cigarette Evaluation Scale and the Duke Sensory Questionnaire [Benowitz et al., 2006; Rose et al., 1999; Westman et al., 1996]; also cf. [Kotlyar et al., 2007]), including physical and affective reactions to the rated products (Benowitz et al., 2006). There is substantial evidence attesting to the validity of such self-report assessments. For instance, similar items have been shown to be sensitive to degree of drug deprivation (Carter and Tiffany, 1999; Sayette et al., 2003; Zinser et al., 1999) and have been shown to be sensitive to the actual nicotine content of cigarettes (Benowitz et al., 2006; Rose et al., 1999). However, they are not consistently strongly related to actual self-administration (Hughes et al., 1996; Perkins et al., 1997), leading to suggestions that self-administration and subjective ratings capture different facets of reinforcement value.

The short form of the Addiction Research Center Inventory is a self-report measure that has been used most extensively to index subjective reactions to nonnicotine drug effects (Jasinski, 1977). This measure contains the Morphine-Benzedrine Group scale, which purportedly measures euphoria (Bigelow, 1991; Foltin and Fischman, 1991; Jasinski, 1977). While this scale appears to reflect subjective evaluations of multiple drugs of abuse, it is unclear at present whether it is ideal for measuring nicotine reinforcement.

Other measures could be incorporated into acute dose-effect comparison studies. For instance, 24-hour retrospective recall of reinforcement would reveal the extent to which postexposure processing alters the memorial representation of incentive properties (Carter and Griffiths, 2009). It is important to use exactly the same questions in those recall tests as used in earlier tests to ensure that change is due to passing of time, not altered assessment formats. Also, the multiple-choice procedure can be used to monetize the worth of additional product doses or exposures at the end of sessions to provide additional data on reinforcement value (Griffiths et al., 1993). A study by Hatsukami et al. (2011) shows how subjective evaluation measures can be paired with tobacco product use measures and product choice measures to enhance the validity of the subjective evaluation measures.

Behavioral Economic Self-Administration Studies

When addictive agents are self-administered in the laboratory context, there is a meaningful relation between laboratory assessed self-administration on the one hand, and clinical evidence of dependence and drug motivation on the other hand (Bickel and Madden, 1999a; Madden and Bickel, 1999; Perkins et al., 2004b; Piasecki et al., 2010). If a contingency is established between the receipt of an agent or product on the one hand, and execution of an instrumental response (e.g., access to MRTP dosing and pressing a lever) on the other hand, then instrumental responses for the agent or product would constitute a key indication of reinforcement potency.

In an acute laboratory setting subjects could work for products across several different contexts: under differing levels of tobacco withdrawal, with different response requirements, and using different instrumental paradigms (progressive ratio schedules for individual products, concurrent schedules for the MRTP versus conventional cigarettes and/or money; and with varying response requirements to generate demand curves). Product exposures could be controlled with smoking topography equipment for cigarettes, while the investigator might have to rely upon duration of use (e.g., duration of oral exposure to smokeless tobacco) and number of self-administrations (e.g., nicotine nasal spray, gum) for noncigarette products (Perkins et al., 2004b; Shiffman et al., 2003). Effective exposure could be indexed by biochemical indices for all products.

Measures Gathered in Behavioral Economic Self-Administration Studies The key measure would certainly be counts of the instrumental response, but it could also include biochemical measures of nicotine or smoke exposure, subjective product evaluations, and withdrawal symptoms. That is, the study could not only assess self-administration under various conditions, but also gather data on perceived reinforcement value and the ability of the product to alleviate withdrawal symptoms (combining the goals of both acute dose-effect studies and behavioral economic studies).

Other self-administration studies could be conducted that do not rely upon instrumental self-administration methods. For instance, there is substantial evidence that tobacco withdrawal plays a major role in spurring relapse back to tobacco use, which may occur because smokers try to escape aversive withdrawal symptoms or because withdrawal enhances the incentive value of smoking cues (Baker et al., 2004a, 2004b; Piasecki et al., 2003). Therefore, researchers might explore the extent to which the MRTP, used either *ad libitum* or under controlled dosing, can ameliorate withdrawal symptoms caused by discontinuation of smoking of conventional cigarettes. Acceptable methods for such studies have been well developed (Shiffman et al., 2003; Welsch et al., 1999). In such research, heavy smokers of conventional cigarettes could be withdrawn from tobacco for an extended period of time and then permitted to use an MRTP. A well-characterized withdrawal scale (Hughes and Hatsukami, 1986; Hughes et al., 1991; Welsch et al., 1999) could be used to measure the extent to which use of the MRTP versus a placebo or other comparison product (e.g., NRT) reduces withdrawal. Such data would be relevant to the notion that an MRTP could substitute for conventional cigarettes and thereby perhaps reduce their use.

In addition to measures of hedonic and evaluative responses, researchers might also gather measures of product impact on other measures such as cognitive performance (attention, memory) and psychomotor performance. Some individuals may use nicotine to enhance their cognitive performance (Heishman et al., 2010; Kleykamp et al., 2011), and such measures could index this source of reinforcement, especially for selected populations such as persons with schizophrenia or attention deficit disorder. Such data would be relevant to the question of whether the MRTP might substitute for conventional tobacco products in such populations.

Finally, while human drug discrimination paradigms can be highly informative in the evaluation of new products (Carter and Griffiths, 2009), they seem less germane to the current questions of interest because the goal is not to compare different types of agents or drugs but instead to compare different nicotine delivery systems.

Analyses Analyses for most of the studies described in this section should be fairly straightforward. For instance, repeated measures of analyses of variance could be used to identify significant main effects associated with the various types of products or stimuli used, and product by repeated measures interactions could be used to determine if products differ in their patterns of change over repeated exposures. Moreover, analyses could be conducted with repeated exposures within sessions crossed with days (or sessions) in order to examine if changes within sessions vary as a function of number of days of exposure or some feature of days (e.g., amount of deprivation preceding a day). Instead of analysis of variance, growth curve modeling (e.g., via hierarchical linear modeling) could be used to estimate intercepts and trajectories and to model days as second-level variables. Appropriate covariates might include gender, starting CO or nicotine level, and dependence. In addition, interaction terms could test whether effects differ significantly as a function of any special subpopulations (e.g., those high versus low in dependence). In all such analyses, the normal analytic considerations pertain such as examining and adjusting scores for distributional deviations, missingness, and autocorrelation.

The analysis of behavioral economic data presents special challenges. For some outcomes such as evaluation of demand elasticity, special formulas are required to model the relation between cost and response (Hursh and Silberberg, 2008; Murphy et al., 2011). Demand elasticity refers to the extent to which work for a substance (e.g., an MRTP or conventional tobacco product) is sensitive to price or work requirements to obtain the substance (e.g., the extent to which self-administration decreases with increased cost). Presumably, the more reinforcing a substance is, the less its self-administration is affected by increased cost. The determination of formal demand curves from self-administration data can be costly in terms of time and resources. Easier to implement strategies are available that may allow for more efficient determination of the relative reinforcement value of different substances: e.g., hypothetical purchase tasks (Murphy et al., 2009). In addition, measures such as peak-response rate and breakpoint are related to the economic measures of maximal output and elasticity of demand and could also be used (Bickel and Madden, 1999b).

General Conclusions

A principal message of the research literature on drug reinforcement value is that no single approach to assessing reinforcement value provides a comprehensive index of value and that using a variety of approaches conveys superior information about relative reinforcing value of pharmacologic agents or products and factors that influence their value.

Therefore, an overarching observation is that a comprehensive assessment of product motivational value includes studies that examine reinforcement value in different relevant populations, with different paradigms, with multiple comparison stimuli and products, and with different types of outcome measures. Specifically, the comprehensive evaluation of the reinforcement value of an MRTP may examine reinforcement value as per the five categories described below.

1. **Subject populations:** Examination of reinforcement value in daily smokers of conventional cigarettes who range in level of tobacco dependence and in beginning smokers (especially young smokers) may be necessary. Other potentially useful populations would be daily smokers interested in cessation, smokeless tobacco users, and nonsmokers.
2. **Experimental paradigm:** Collection of data on subjective evaluations of the MRTP in acute dose-effect comparison studies and in behavioral economic self-administration studies testing over multiple days and extended sessions is necessary. Use of behavioral economic paradigms would permit more informative indices of interstimulus reinforcement value as it could be denominated on the basis of a standard behavioral response. Moreover, some self-administration paradigms may not only examine reactions to, and self-administration of, the MRTP relative to other products, but also examine the ability of the product to quell tobacco withdrawal (especially urges) and to reduce motivation to smoke conventional cigarettes because of preloading with the MRTP. Important in all of these paradigms is the modeling of change over repeated exposure occasions because this could reflect development of increased reinforcement value owing to tolerance to aversive effects or dependence development.
3. **Comparison stimuli and products:** Examination of subjects' reactions to the MRTP relative to conventional cigarettes and acute forms of NRT (nicotine nasal spray, gum, lozenge, or inhaler) may be necessary. It would also be quite informative to conduct evaluations in which preference for each product could be monetized, at least via a multiple-choice procedure (Griffiths et al., 1993).
4. **Outcome measures:** It may be necessary to include measures of self-administration, biochemical indices of effective dosing, and self-report of preference and psychoactive effects. Other measures such as withdrawal severity may be used to explore effects such as withdrawal suppression; it may be efficient to also include putative biomarkers of disease risk (Hatsukami et al., 2006).
5. **Data interpretation:** This may be one of the most challenging aspects of the assessment of liability for adverse effects on the public's health. There is no clear outcome that signals whether the MRTP has the "right" level of reinforcement potential in order to supplant smoking conventional cigarettes but yet not be so reinforcing that its availability poses additional significant threats to the public health. Presumably it will be more reinforcing than NRTs, because NRTs are not sufficiently reinforcing to support even prescribed levels of use (Lam et al., 2005; Liu et al., 2001; Shiffman et al., 2008b; Vogt et al., 2008). But, the MRTP presumably should not be as reinforcing as smoking conventional cigarettes. So, roughly speaking, an MRTP should be intermediate in reinforcement magnitude. Of course, decisions about optimal reinforcement magnitude depend on

other factors such as the product's delivery of toxicants (a product that results in little toxicant exposure would present little risk even if being highly reinforcing) and the results of other research efforts (data from RCTs).

EVALUATION OF PUBLIC HEALTH RISK USING RANDOMIZED CLINICAL TRIAL METHODS

As noted in the Introduction, the evaluation of an MRTP with regard to the public health standard concerns such factors as (1) how heavily it is used, (2) the extent to which its use directly exposes individuals to toxicants, (3) the effect of its use on the consumption of conventional tobacco products, (4) how conjoint use of the MRTP plus conventional tobacco affects health, and (5) how its use affects the initiation of use of conventional tobacco products and relapse back to use of such products (e.g., resumption of smoking conventional cigarettes after an extended period of abstinence).

Some of these issues can be explored via RCTs. In particular, the RCT may be a highly efficient means of examining such related issues as (1) acceptability and use of the MRTP; (2) the ability of the MRTP to increase cessation in users of conventional tobacco products, either by enhancing total cessation or by reducing use of such products; and (3) the likelihood that MRTP availability will lead to dual use. An RCT could also, in theory, produce evidence on such topics as (1) whether and how much individuals use an MRTP after they have used it to help them quit use of conventional tobacco products, (2) changes in perception of the MRTP with its continued use, and (3) the MRTP's ability to suppress tobacco withdrawal symptoms. The last effect would increase the likelihood that the MRTP would serve as an effective cessation aid.

Key Considerations for the Use of Randomized Clinical Trials

An RCT that tests the potential public health impact of an MRTP requires decisions about key issues that will affect the validity and relevance of the resulting data. One issue is the balancing of internal versus external validity. This issue has implications for multiple aspects of the experimental design and methods, such as how heavy the assessment burden should be, whether subjects are asked to pay for the MRTP at some point in the trial, and how to maintain subject involvement in the trial. Therefore, the challenge is to ensure the real-world relevance of the work, while maintaining enough internal validity (experimental control) so strong inferences can be made. Other major decisions concern the nature of the specific comparison products to be tested in the research (see discussion below), whether and how to implement blinding procedures, the nature of the outcomes to be assessed, and the nature of the population to be recruited.

It is important to recognize that no single RCT can address all of the important issues that pertain to the possible public health impact of an MRTP. Therefore, it may be necessary to conduct two or more RCTs in order to address the major questions that exist. For instance, it would seem desirable for one RCT to emphasize internal validity, while another might be designed to emphasize external validity (real-world relevance). Moreover, it may be economical of time and other resources (burden and risk to the individuals who would participate in an RCT) that an RCT be launched only after there is some evidence from laboratory studies that the MRTP (1) has a significantly favorable toxicant profile, (2) is sufficiently reinforcing or nonaversive so as to permit a reasonable level of use by smokers, and (3) is not so reinforcing (or addictive) so as to lead to high levels of use by nonsmoking youth. The suggestion for prior laboratory studies is made despite the fact that clear-cut criteria do not presently exist that would allow definitive determinations with regards to the above issues. A key question for both laboratory studies and RCTs is how data or outcomes can be interpreted so as to have optimal meaning or relevance with regard to public health impact. That is, what patterns of use and effects (e.g., impact on smoking cessation) would suggest a net positive versus harmful effect?

Design and Overarching Methods Considerations

The trial design should reflect the questions targeted. If the major question is how MRTP availability affects future use of conventional tobacco products, the design should contrast a condition where some subjects are randomly assigned to use the MRTP and others a placebo or a comparison product (Robinson et al., 2000). For instance, a meaningful comparison condition would be the provision of acute administration NRT products that have strong patient acceptance and use (e.g., perhaps newer acute NRT products that show relatively high rates of patient acceptance). A highly acceptable and efficacious NRT would be a good benchmark for MRTP evaluation. Such products show modest levels of smoker acceptance and use, tend not to substitute effectively for conventional tobacco use (e.g., smoking) among individuals not making quit attempts (a large portion of the smoker population has *not* switched from conventional tobacco products to NRT as a form of long-term use), and pose little risk of addictive or dependent use. Presumably, if an MRTP has promise to attract individuals away from use of conventional tobacco products, it should be somewhat more reinforcing than NRT, promoting greater sustained use and substituting for conventional tobacco use more effectively than NRT. The value of the use of an NRT as an MRTP comparison product is apparent in a study by Kotlyar et al. (2007). Other criteria could also be forwarded, such as withdrawal from an MRTP should not be as severe as that arising from withdrawal from conventional tobacco products. In addition, NRT makes a meaningful comparison because it is a potential marketplace competitor with the MRTP, meaning that most forms are widely available over the counter. Presumably an MRTP would achieve meaningful use only if it were more appealing than NRT. Thus, NRT would appear to be a more meaningful comparison product than a prescription cessation aid (e.g., varenicline, bupropion) because the latter aids would not be available competitors for chronic use. Although NRT would constitute a meaningful comparison product in an RCT, interpretation of MRTP effects, and estimation of its potential risks and benefits, would be a challenging task (see the "Inference" section below).

Therefore, a reasonable design would be one in which subjects are randomly assigned to either the MRTP or to the comparison product(s) (ideally both a placebo and NRT) with blocking on intention to quit or interest in quitting. Ideally, more than one RCT should be conducted, with trials constituting both efficacy and effectiveness trials. Thus, the former would recruit highly motivated subjects, be double blind, entail fairly heavy assessment with compensation for adherence, and include other features designed to reduce error and encourage high use of the tested products (sustained provision of free products). The effectiveness studies might recruit "all comers," use open label product, use relatively brief nonburdensome assessments, and provide products in a manner that more closely resembles real-world use.

An alternative to such a traditional RCT design would be one in which multiple products were tested in full factorial or fractional factorial design (Collins et al., 2011). This would permit the simultaneous and efficient testing of multiple comparison products and also testing of the interactions among such products. In addition, a crossover design could be used in which participants alternate the tobacco products that they use over standard cycles of use (Hatsukami et al., 2009).

At least some of the RCTs should permit extended use of the MRTP. This is because the impact or acceptability of a product might change with time. For instance, there is evidence that nicotine nasal spray use increases when individuals learn to use it properly (Blondal et al., 1997; Fiore et al., 2008; Sutherland et al., 1992). In addition, over time, other factors might encourage changes in use (e.g., secular events such as tax increases on conventional tobacco products, development of dependence). Also, some patterns of use, such as dual use, might be transitional stages that ultimately convert to more stable use patterns. For instance, there is considerable evidence that chronic conjoint use of an NRT while smoking increases subsequent smoking cessation attempts and success (Carpenter et al., 2004; Chan et al., 2011). Finally, relapse back to smoking occurs at meaningful levels even after a year of cigarette abstinence (Hawkins et al., 2010; Heffner et al., 2010; Hughes et al., 2008); it seems important, therefore, to study MRTP effects up to the point where significant relapse risk has passed. It is possible, in fact, that quitting with the use of an MRTP results in higher than normal relapse because the continued use of the product primes continued or resurgent motivation to resume conventional tobacco use (Shaham et al., 1996, 2003). Any duration recommended for an MRTP RCT would be somewhat arbitrary. But, because the rate of relapse tends to drop to between 2 percent and 4 percent per year after 2 years of abstinence (Krall et al., 2002), a minimum 2-year duration seems advisable. This would suggest that observed cessation rates for conventional tobacco products observed at study end would be fairly stable.

Another important concern is whether the study involves an explicit quit date for those expressing interest in quitting. Setting a quit date for all subjects to make a cessation attempt would probably constitute the most sensitive test of the ability of an MRTP to boost cessation success in a given attempt. If the goal of the RCT is more focused on internal than external validity, and where subjects are motivated to make quit attempts, the investigator could encourage subjects to select a quit date so assessments could be concentrated around this date. This would increase the sensitive measurement of important factors such as quitting-related withdrawal symptoms. However, this design feature would probably not resemble real-world MRTP use, where many people might use an MRTP without intending (at least initially) to make a cessation attempt. Therefore, designs that permit long-term use without formal quit attempts and with individuals not motivated to quit would possess greater external validity. Such designs should certainly be used in at least one or more of the RCTs.

The Consolidated Standards of Reporting Trials (CONSORT) reporting recommendations for clinical trials emphasize the importance of explicitly identifying primary and secondary outcomes on an *a priori* basis. (The CONSORT 2010 checklist is presented in Table 4-1.) Primary outcomes should be few in number and explicit. It seems that a primary outcome should be percentage of smokers of conventional cigarettes (or another conventional tobacco product, depending on the goal of the study) who are abstinent at critical endpoints (e.g., 1 and 2 years post-study initiation). Other outcomes could include percentage of smokers who engage in dual use, amount of smoking of conventional cigarettes by those who engage in dual use, use rates and use prevalence of the MRTP, attitudes and perceptions of the MRTP (in particular, perceptions of rela-

TABLE 4-1 CONSORT 2010 Checklist of Information to Include When Reporting a Randomized Trial*

Section/Topic	Item No.	Checklist Item	Reported on Page No.
Title and abstract			
	1a	Identification as a randomized trial in the title	________
	1b	Structured summary of trial design, methods, results, and conclusions (for specific guidance, see CONSORT for abstracts)	________
Introduction			
Background and objectives	2a	Scientific background and explanation of rationale	________
	2b	Specific objectives or hypotheses	________
Methods			
Trial design	3a	Description of trial design (such as parallel, factorial) including allocation ratio	________
	3b	Important changes to methods after trial commencement (such as eligibility criteria), with reasons	________
Participants	4a	Eligibility criteria for participants	________
	4b	Settings and locations where the data were collected	________
Interventions	5	The interventions for each group with sufficient details to allow replication, including how and when they were actually administered	________
Outcomes	6a	Completely defined prespecified primary and secondary outcome measures, including how and when they were assessed	________
	6b	Any changes to trial outcomes after the trial commenced, with reasons	________

continued

TABLE 4-1 Continued

Section/Topic	Item No.	Checklist Item	Reported on Page No.
Sample size	7a	How sample size was determined	________
	7b	When applicable, explanation of any interim analyses and stopping guidelines	________
Randomization: Sequence generation	8a	Method used to generate the random allocation sequence	________
	8b	Type of randomization; details of any restriction (such as blocking and block size)	________
Allocation Concealment Mechanism	9	Mechanism used to implement the random allocation sequence (such as sequentially numbered containers), describing any steps taken to conceal the sequence until interventions were assigned	________
Implementation	10	Who generated the random allocation sequence, who enrolled participants, and who assigned participants to interventions	________
Blinding	11a	If done, who was blinded after assignment to interventions (for example, participants, care providers, those assessing outcomes) and how	________
	11b	If relevant, description of the similarity of interventions	________
Statistical methods	12a	Statistical methods used to compare groups for primary and secondary outcomes	________
	12b	Methods for additional analyses, such as subgroup analyses and adjusted analyses	________

Results			
Participant flow (a diagram is strongly recommended)	13a	For each group, the numbers of participants who were randomly assigned, received intended treatment, and were analyzed for the primary outcome	________
	13b	For each group, losses and exclusions after randomization, together with reasons	________
Recruitment	14a	Dates defining the periods of recruitment and follow-up	________
	14b	Why the trial ended or was stopped	________
Baseline data	15	A table showing baseline demographic and clinical characteristics for each group	________
Numbers analyzed	16	For each group, number of participants (denominator) included in each analysis and whether the analysis was by original assigned groups	________
Outcomes and estimation	17a	For each primary and secondary outcome, results for each group, and the estimated effect size and its precision (such as 95% confidence interval)	________
	17b	For binary outcomes, presentation of both absolute and relative effect sizes is recommended	________
Ancillary analyses	18	Results of any other analyses performed, including subgroup analyses and adjusted analyses, distinguishing prespecified from exploratory	________
Harms	19	All important harms or unintended effects in each group (for specific guidance see CONSORT for harms)	________

continued

TABLE 4-1 Continued

Section/Topic	Item No.	Checklist Item	Reported on Page No.
Discussion			
Limitations	20	Trial limitations, addressing sources of potential bias, imprecision, and, if relevant, multiplicity of analyses	______
Generalizability	21	Generalizability (external validity, applicability) of the trial findings	______
Interpretation	22	Interpretation consistent with results, balancing benefits and harms, and considering other relevant evidence	______
Other information			
Registration	23	Registration number and name of trial registry	______
Protocol	24	Where the full trial protocol can be accessed, if available	______
Funding	25	Sources of funding and other support (such as supply of drugs), role of funders	______

*CONSORT strongly recommends reading this statement in conjunction with the CONSORT 2010 Explanation and Elaboration for important clarifications on all the items. If relevant, we also recommend reading CONSORT extensions for cluster randomised trials, non-inferiority and equivalence trials, non-pharmacological treatments, herbal interventions, and pragmatic trials.
Additional extensions are forthcoming: for those and for up to date references relevant to this checklist, see www.consort-statement.org.
SOURCE: Schulz, K. F., D. G. Altman, and D. Moher, for the CONSORT Group. CONSORT 2010 Statement: updated guidelines for reporting parallel group randomised trials. BMJ 2010;340:c332. Reprinted by permission of CONSORT Group.

tive health risks, addictiveness, liking of the MRTP, and value in curbing use of conventional tobacco products), motivation and plans to quit smoking among those continuing to do so, self-efficacy estimates of ability to quit with and without the MRTP, severity of the withdrawal syndrome in any quit attempts, incidence of quit attempts, nicotine dependence, and quitting self-efficacy. If the study is a postmarketing study, investigators could also inquire about MRTP use in the subjects' social networks.

Finally, based upon the results of basic research on toxicant exposure or other sources, it might be warranted to include physical health assessments of the participants to determine if MRTP use is associated with changes in toxicants or with other biomarkers of relevant disease processes (e.g., pulmonary function tests).

Randomized Clinical Trial Design and Methods

Nature of the Sample

Perhaps the major question that exists is whether the product will help participants quit use of conventional tobacco products, either resulting in complete abstinence from any tobacco product, including the MRTP, or by the MRTP serving as a long-term substitute. Therefore, chronic smokers of conventional cigarettes (the most harmful conventional tobacco product) are an important target population for an RCT. Today there are more questions than in the past about what constitutes a current smoker because today's smokers are smoking significantly less than in the recent past (CDC, 2005; CDC and National Center for Health Statistics, 2008; Pierce et al., 2011).

Because the results of an RCT should be broadly applicable to today's smokers, the sample should comprise smokers who smoke relatively little (e.g., daily smokers who smoke at least two cigarettes/day) and very heavily (e.g., with no upper limit on daily smoking). The participation of light smokers would be dependent, of course, on the determination that their participation did not pose an unacceptable health risk (e.g., from nicotine toxicity). The participation of very light smokers is warranted for several reasons: (1) they perceive themselves to be at a reduced disease risk and (2) they appear to differ from other smokers in their motives for smoking (Piper et al., 2004). In addition, if the trial is designed to yield data on population-based effects of MRTP availability, then the sample should comprise both those willing and unwilling to quit. The latter population is appropriate because use of the product could encourage smoking reductions or quitting in those not initially wanting to do so, just as NRT encourages quitting in previously unmotivated individuals (Chan et al., 2011; Schuurmans et al., 2004; Stead and Lancaster, 2007). Moreover, those who are not initially interested in quitting smoking might be more likely to engage in long-term MRTP use than would others, or engage in dual use (conventional cigarette smoking plus MRTP use), because they might use the MRTP but have little desire to quit smoking. Either of those outcomes would have public health relevance. Therefore, at least one or more of the RCTs conducted should comprise subjects with a range of intentions or motivations to quit use of conventional tobacco products. Finally, while it might seem difficult to attract smokers into a clinical trial who do not wish to quit, in fact, many such smokers are willing to participate in order to try a new product that might be safer than conventional tobacco or that might allow them to reduce their smoking (Carpenter et al., 2004).

One topic that could be addressed in an RCT is the extent to which the MRTP aids cessation or substitution by young or adolescent smokers. To address this, adolescent or young adult smokers could be recruited into the research either in a main study or a study focused on this topic. As with this and other research, an attempt should be made to recruit a representative sample with regard to gender, ethnicity, and socioeconomic status.

Smokeless tobacco users tend not to respond to NRT medications in the same way as cigarette smokers (Fiore et al., 2008). Therefore, if it is deemed important to study MRTP effects in smokeless tobacco users, then it would be important that the trial is adequately powered so as to permit inferences about smokeless users per se.

Characterization of the Sample

As discussed in the section on reinforcement and self-administration studies, a comprehensive characterization of the sample is important because it defines the population to which the conclusions may be most directly related. It also permits tests of the interaction of person factors with MRTP effects. Variables that should be measured are gender, age, ethnicity, educational and socioeconomic status, history of tobacco and nicotine use (including peak tobacco use levels, prior quitting history, age of initial use, and use histories of different tobacco and nicotine products), expectations about the effects of the products or agents to be tested, motivation to quit using tobacco, self-efficacy regarding ability to quit, tobacco or nicotine dependence, blood or breath levels of tobacco or nicotine exposure, health and mental health status and history, other co-addictions (alcohol, narcotics, etc.), and use of psychoactive products including psychiatric medications. The last factor is important because not only may it signal mental health history, but also because some psychiatric medications are effective smoking cessation agents (e.g., bupropion, nortriptyline) and should be detected for that reason. In addition, measures should also target environmental factors that relate to tobacco cessation success; these include home and work smoking policies and whether the subject lives with a smoker (Bolt et al., 2009). These variables are important because they have been related to nicotine dependence, tobacco self-administration, and ability to control tobacco use.

In terms of tobacco dependence assessment, the Fagerstrom Test for Nicotine Dependence or one of the new multifactorial dependence assessments (the Nicotine Dependence Syndrome Scale [Shiffman and Sayette, 2005; Shiffman et al., 2004] or the Wisconsin Inventory of Smoking Dependence Motives [Smith et al., 2010]) appear to provide more accurate appraisal of dependence than do the *Diagnostic and Statistical Manual of Mental Disorders* criteria (Hughes et al., 2011).

Subject Recruitment and Randomization

Subjects could be recruited via media announcements or via smoker identification methods used at primary care clinics. The former tends to be more appropriate for efficacy studies where highly motivated subjects are targeted, while the latter tends to be more appropriate for effectiveness studies because the primary care recruitment does not focus on "treatment seekers." Care must be taken to ensure that recruitment and screening do not set up expectancies among subjects that bias the findings (e.g., expectations that would not be present in real-world use).

To obtain a large sample, studies might have to be conducted at multiple sites. All sites must be adequately described and methods should be adopted that ensure that recruitment, screening, and research and treatment methods be uniform across sites. Further, poolability analyses should be conducted to determine the consistency of findings across sites.

In terms of sample size, it must be set to detect effects in the primary outcome(s) that would be of public health significance. There is no effect size that has been accepted as having clear public health significance for an outcome such as smoking cessation. One approach would be to test whether an MRTP enhances long-term outcomes to a similar degree as over-the-counter cessation medications, which tend to approximately double 6-month abstinence rates (Fiore et al., 2008). However, research grants in this area are often powered to detect effect sizes in which the experimental intervention increases long-term (e.g., 6-month) cessation rates by at least 10 percent (e.g., 10 percent in controls and 20 percent in experimental subjects). A Cochrane report suggested that a cessation increment of 6 percent could be of public health or clinical significance (Lancaster and Stead, 2005). Because the effects of an MRTP would occur on a population-wide basis with use by many thousands of individuals, it seems prudent to power an RCT to detect relatively small effects. Therefore, consistent with the Cochrane report, it would make sense to power at least one of the RCTs to detect an effect (increment in cessation) of 5–6 percent or greater.

Of course, an RCT permits the collection of information on multiple outcomes, even if many are secondary. If there are especially important secondary outcomes, these too must be considered in setting recruitment goals. For instance, it may be highly efficient to collect data on disease biomarkers or surrogates over the extended use of MRTPs during the trial, although the validity of such biomarkers would need to be considered in such decisions (Hatsukami et al., 2006). Some biomarkers and surrogates may be expected to show changes over the course of a lengthy clinical trial follow-up lasting over a year (e.g., exposure biomarkers or surrogate endpoints like endothelial dysfunction). Biomarkers and surrogates are discussed further in Chapter 3.

The randomization process should follow CONSORT recommendations (Schulz et al., 2010). If multiple sites are used, then randomization should be balanced within sites. Also, the method of randomization should ensure blinding (at least blinding from staff and assessors to the extent possible). Moreover, blocking within each site should be used for factors that might powerfully influence outcomes (e.g., whether or not the research participant has an intention to quit).

Randomized Clinical Trial Measures

Biochemical Measures

As discussed in the section on reinforcement and self-administration measures, biochemical measures of tobacco or nicotine exposure should be collected because they reflect prior self-administration intensity or tolerance and are often related to likelihood of future cessation (al'Absi et al., 2004; Powell et al., 2010). The appropriate measure could be CO level for cigarette smokers, but it could be nicotine or cotinine levels (from blood, saliva, or urine) in other sorts of nicotine or tobacco users (those using smokeless tobacco or NRT). In particular, acute blood nicotine absorption profiles in response to both single and repeated use of products is a meaningful component in assessing the addictive potential of MRTPs. In an RCT where acute effects of self-dosing are not targeted, cotinine may be preferred over nicotine levels, because its longer half-life should provide a more accurate index of chronic consumption. This would be especially important if light smokers are included in the sample. Also, if a noncombustible MRTP is studied, CO levels during or after the experiment will not provide measures of effective dosing. Therefore, to obtain a true baseline for such later measures, either nicotine or cotinine should be measured at baseline. Measurement of urine cotinine corrected for creatinine concentration may be the best predictor of plasma cotinine (Benowitz et al., 2009). Finally, the investigator might wish to measure 3′-hydroxycotinine in order to estimate nicotine metabolism (Schnoll et al., 2009), which might predict heavy product use and the long-term substitution of the MRTP for smoking versus smoking cessation *per se*.

Baseline Assessment

Ideally baseline data should be collected via computerized data acquisition systems that ensure complete data recording and detection of out-of-range values. Baseline measures should be taken of all those variables that are to be used as outcomes, moderators, covariates, or to characterize the sample, such as smoking rate, use of all tobacco products, biochemical measures of heaviness of tobacco use, nicotine dependence, socioeconomic and educational status, withdrawal symptoms, affect, mental health history, physical health status and perceived health status, medication use, aversive events (e.g., due to nicotine toxicity), smoking history (e.g., age of first smoking/daily smoking, longest prior abstinence from a quit attempt, prior use of quitting aids), quitting self-efficacy, perceptions of the MRTP, motivation to quit, home and work smoking policies and restrictions, and alcohol use. Investigators should use psychometrically sound instruments and should routinely report psychometric data for their own sample (e.g., coefficient alpha). Also, to the extent possible investigators should use commonly used instruments to enhance assessment of comparability of the recruited sample with samples used in previous research.

Assessment During the Cessation Trial

The key assessment targets include use of both conventional cigarettes and the MRTP. Such use data can be gathered from a variety of means, such as interactive voice response (IVR) assessments via subjects' cell and landline phones, by mailed questionnaires, or by Internet assessment. If a targeted quit day has been set (e.g., in the context of an efficacy study), then assessments could be concentrated around this time. Otherwise, assessments could occur at intervals of sufficient frequency to permit accurate recall. There is evidence that subjects can complete smoking calendars accurately over 3- to 6-month intervals (Piper et al., 2009), with calendars capturing whether or not smoking occurred on a particular day (i.e., a binary measure of smoking, not a specific amount smoked) over the past 6 months. Assessment of number of cigarettes smoked/day over the past week only would allow for the estimation of current smoking heaviness (and this would also permit point prevalence assessment for the past week). It seems likely that subjects could supply similar information with regard to MRTP use (with estimates of amount of use per day being captured only for recent days [past week]).

Ideally, periodic ecological momentary assessment data (perhaps captured via IVR calls) could be used to assess heaviness of use of both conventional tobacco and the MRTP. These could target use of both products over the past 24 hours and could occur every other week, or even monthly in an extended study, without constituting an undue burden. Recent clinical trials on smoking cessation have used IVR calls with follow-up durations of a year or more (Brendryen and Kraft, 2008; Reid et al., 2007). In an efficacy study such assessment strategies could be maintained for many months, but they might require compensation in order to obtain high completion rates. Such assessments could also track quit attempts, withdrawal symptoms, self-efficacy, and aversive events.

It is sometimes acceptable *not* to collect biochemical confirmation of tobacco use status for follow-up outcome assessment, especially in effectiveness studies where there has been little interpersonal contact between the research staff and subjects (Hughes et al., 2003). However, in any study involving extended and multiple assessment contacts, and where *degree* of product use is important (not just binary measures of use such as targeted in point-prevalence assessments), it would be important to collect biochemical indices of use. Both urine cotinine and CO should be collected, with care taken to collect self-report information on use of any product (e.g., NRT) that could affect levels of biochemical indices of exposure. Therefore, for any RCT it seems highly desirable to schedule in-person visits every 6 months for biological sample collection (Smith et al., 2009). At such in-person visits, researchers could collect additional self-report point-prevalence (past week) data not only on conventional tobacco and MRTP use, but also on such secondary outcomes as MRTP attitudes and liking, motivation to quit, tobacco dependence, changes in important environmental variables (e.g., smokers in the home), and MRTP use in subjects' social networks (if study is postmarketing).

Throughout the trial, the investigators should track all events that need to be reported in CONSORT event diagrams: numbers of individuals contacting the research program and assessed for eligibility, number excluded and reasons for exclusion, number who declined participation during the induction process and when and why they declined, number assigned to each experimental condition, amount of experimental intervention and assessment received, number who formally discontinued participation and reasons for discontinuation, number lost to follow-up (unable to contact), and number analyzed and reasons for any departures from intent-to-treat principles. All data should be reported for the entire sample and with respect to treatment condition for measures collected after random assignment. Of course, CONSORT reporting recommendations will no doubt change over time, and researchers should ensure that their methods reflect the most current standards.

Finally, good experimental design standards demand that aside from the manipulation of the independent variable(s), all procedures in the study, including types and intensity of assessments, be equivalent across all experimental conditions.

Selecting and Delivering the Tobacco Products

Some questions concern the method for product provision, the need for a placebo control, and the need for product blinding. With regard to the method for making the tested products available to subjects, two sets of questions can be distinguished with regard to how an MRTP might affect cessation. One question is, does optimal use of the MRTP help a smoker quit smoking, and if so, how effective is it, and how does it compare in this regard to other widely available cessation aids such as NRTs? This is the sort of question addressed in an efficacy trial, which is designed to gauge intervention effectiveness under near-ideal circumstances. A second question is whether MRTP *availability* per se affects the likelihood of future cessation. This second question is concerned with a real-world effectiveness issue: under conditions of real-world use (or near real-world use), where many individuals may not even use the MRTP or attempt to quit using a conventional tobacco product, how does MRTP availability affect outcomes? If it is deemed important to determine the effectiveness of the product in a formal, structured quit attempt relative to cessation aids such as NRTs, then it should be offered with considerable support for its use. This would entail free product use for the duration of the trial and perhaps training in use, encouragement of use, and perhaps prompting of use. Such a trial would show how effective the MRTP could be in boosting cessation rates (of conventional tobacco products) under ideal conditions. It might even make sense to offer the MRTP in conjunction with adjuvant interventions that are readily available in real-world use, such as quit-line counseling (Miller and Sedivy, 2009; Smith et al., 2009; Tinkelman et al., 2007).

However, it seems that the former question (optimal MRTP efficacy as a cessation aid) is of somewhat less interest than questions that would focus more on real-world use and effects. That is, it seems most relevant to determine if MRTP availability to a group of individuals exerts a net effect on the future use of conventional tobacco products across the population of potential users. The goal of external validity would be served by providing little support for MRTP use (i.e., providing the MRTP at least nominal cost, providing no more use information than would be provided by package instructions). In addition, the MRTP would be offered by itself with no provision of adjuvant therapy or encouragement for its use. However, it might be that this approach would provide even less use support than would occur in real life, where a person's social network for instance, might encourage use and provide information.

A related consideration is that the RCTs will probably be used to address multiple questions (even if only one or two are deemed primary). For instance, not only is it of interest to determine if the MRTP affects future use of conventional tobacco, but also it is important to obtain additional information on the health risks that might attend chronic and unsupervised use, or the extent to which MRTP use affects tobacco withdrawal symptoms. Unless a meaningful portion of the sample uses the MRTP regularly, then no inferences can be made about such topics (Does heavy use increase liking? How does heavy real-world use affect nicotine and toxicant exposure?). Therefore, it seems that a good compromise strategy is to conduct at least one efficacy trial and one effectiveness trial. The effectiveness trial could perhaps start out with free use in the early stages of the trial to ensure some initial trial of the product, and then weaning the subjects off supported use, with their eventual request of the product reducing their supported payments by some meaningful amount.

With regard to the issue of placebo control, it seems as though use of a placebo would be desirable in an efficacy trial but not in the effectiveness trial. The reasons that it would be desirable in the efficacy trial are that (1) there is a history of very strong and persistent responses to placebo tobacco products (e.g., cigarettes that contain no nicotine [Perkins et al., 2008]) and (2) even if the MRTP were compared with another "active" product that contained nicotine (e.g., NRT), this would not control for effects of novelty and "newness" that might accompany the provision of a new or less familiar nicotine delivery system. (If the research occurred in a postmarketing context, then this could affect the need for a placebo control.) If a placebo were used, the research should be double blind. However, subjects would not be blind to the product they were using in an effectiveness study. In any study, to the extent possible, the staff collecting assessment and outcome data should be blinded to treatment assignment or product use. Steps to ensure this and quality assurance measures should be described. Also, if placebo control is used, then data should be collected on subjects' beliefs about the product they were given.

Other intervention procedures should be similar to those used in any well-designed RCT evaluating the use and effectiveness of a cessation aid. Subjects should receive enough product to permit optimal dosing, they should be given instructions for product use that fit the nature of the RCT (efficacy versus effectiveness), they should be given clear information on health risks and how to spot adverse reactions or effects, they should be given a way to communicate about health concerns and get professional advice, and they should have their use of the products tracked in multiple ways (e.g., "pill counts," self-report, ecological momentary assessment self-report, medication recording devices). Finally, it would be important that the subjects not be given clear messages about the possible or targeted effects of the products because this could produce biases in subsequent ratings or behaviors (e.g., disappointment, placebo effects, and so on). Perhaps the subjects could merely be told that the MRTP is being evaluated to determine how much people will use it, how it might affect their use of other forms of tobacco, and their attitudes about it.

There may be instances where cluster assignment of participants may be warranted (e.g., where communities or schools are assigned to various products). This would permit assessment of product effects within larger social units (spread of use within peer groups) and also permit assessment of environmental impacts (community cardiac events or bronchitis incidence).

Long-Term Follow-up Methods

Certain methods have been shown in prior research to boost trial participation and adherence:

1. Clear information early on about the assessment burden
2. Timely payment for assessment information and visits
3. Ability of a subject to reschedule assessments
4. Use of brief, clear questionnaires
5. Use of the same assessor over time to promote the development of a personal relationship
6. Collection of information via multiple contact routes (multiple phone numbers, e-mail addresses, home and work addresses, and collateral informants) to facilitate long-term contact
7. Regular inquiries about the subject possibly moving and likely future addresses
8. Explicit permission for a subject to skip a follow-up contact with the understanding that s/he may resume participation at some future point in time

These methods should be adopted in an effort to reduce attrition and boost ascertainment rates. With such methods it may be possible to track clinical trial participants over several years.

As suggested above, tracking of outcomes should occur via multiple routes: phone calls, mailed questionnaires, Internet questionnaires, and in-person visits. In general, use of multiple data collection routes yields more comprehensive data and higher ascertainment rates. For instance, in-person visits could be made at a periodicity of 6 months to obtain calendar data on smoking and MRTP use (and biochemical samples or physical health tests as needed), but at that periodicity, fine-grained use data (how many cigarettes or MRTP doses were consumed each day) could be obtained only for the past week. Therefore, interval sampling methods using cell phone calls, perhaps on a monthly basis, could provide information on intervening product use and symptoms.

Data Analysis

Important elements of an analytic report include

1. as per CONSORT recommendations, primary and secondary outcomes specified *a priori*;
2. a description of any significant protocol deviations;
3. a complete CONSORT diagram;
4. adherence to intent-to-treat analytic principles and description of exact subject counts included in each analysis;
5. use of experiment-wise error correction, except where primary hypotheses are tested or outcomes important to subject welfare are being evaluated;
6. evaluation of covariates to determine their ability to reduce type II error; and
7. reporting of all adverse events and their relation to MRTP use described.

In addition, the analysis plan should examine relations of MRTP use to outcomes, perhaps with use of formal mediation analytic strategies (MacKinnon et al., 2002; Piper et al., 2008). As with some pharmaceutical products, there may be particular patterns of use that are especially beneficial or harmful; such patterns may be most identifiable through the use of real-time assessments of use patterns, perhaps via electronic monitoring strategies (Cramer et al., 1990; Matsui et al., 1992; Metry, 1999).

At the minimum, it is critical that RCTs analyze the following in order to comprehensively capture the effect of MRTP availability on public health and to support later modeling of such effects:

1. Use of the MRTP
2. Relations of MRTP availability (treatment assignment) and use with measures of use of conventional tobacco products (e.g., cigarette

smoking), with use reflected in both binary and continuous measures (abstinence versus smoking rate data; dual-use rate versus smoking rate data)

3. Relations of MRTP use with occurrence of quit attempts and duration of abstinence achieved in such attempts, and whether MRTP use reduces quit attempts with other sorts of cessation aids (e.g., there may be no net effect on smoking cessation per se, only a shift in type of quitting, as in use of the MRTP versus NRT)

4. Effect of MRTP use on withdrawal and craving during quit attempts and when individuals reduce their use of conventional tobacco products

5. Nicotine dependence with regard to use of both the MRTP as well as conventional tobacco products

6. Changes in perception of conventional tobacco products and of the MRTP as a function of MRTP use over time (e.g., liking, addictiveness, safety)

7. Quitting self-efficacy and quitting intentions in response to use of both conventional tobacco products and the MRTP. Such outcomes should be measured both at discrete endpoints (e.g., abstinence rates at 6-month visits) as well as via ecological momentary assessments that generate data for intensive longitudinal data analysis (e.g., assessment of smoking over time with MRTP use serving as a time varying covariate in growth curve models).

Inferences

Interpretations of the obtained data need to be synthesized in order to attain a comprehensive assessment of the potential public health impact of approving a product as an MRTP. RCTs can yield data on the use of the MRTP over time on the proportion of people who use it and how heavily they use it, the extent to which it produces or sustains nicotine dependence, and the extent to which it reduces use of conventional tobacco products (e.g., smoking) or reduces use of cessation aids. Data from all relevant measures must be integrated, for instance, taking into account not only the size of the effects of the MRTP on important outcomes but also the prevalence of use and safety findings. For instance, if the product is unappealing and infrequently used, then its potential for a positive public health impact is reduced even if it can boost smoking cessation success. Evaluation of the effects of MRTP will be an iterative process, because information gained from postmarket observations may inform or correct assumptions for laboratory and preclinical investigations. In addition, such synthesis may take into account projected costs to the user and society (e.g., via health care impacts). By supplying data on the outcomes noted above (heaviness of use, duration of use, impact on smoking), RCTs should yield evidence that would be useful for modeling of population based health and economic impacts. Models can account for and potentially predict the effect of marketing an MRTP on initiation, cessation, or relapse. Simulation models that use mathematical formulas need to account for population dynamics, because initiation and cessation rates can depend on demographic differences and social behaviors.

The synthesis of all of this information will be challenging because it involves explicit or implicit weightings of the various possible outcomes. No well-defined cut scores are available for gauging benefit, and interrelations of variables may be complex. For instance, an MRTP should be compared with one or more NRTs in RCTs (Kotlyar et al., 2007); however, note that the MRTP need not necessarily be "better" or even equivalent to the NRT in order to exert a public health benefit. An MRTP that is inferior to NRTs (more toxicants, less effective at boosting cessation of smoking conventional cigarettes) could still exert a net public health benefit if its modest effects were additive, meaning they occurred on top of those of NRTs. For example, while not being very effective at helping smokers quit when used as a sole product, it is possible that the combination of NRT plus the MRTP yields additive (or even positive synergistic) effects on smoking cessation when in combination. This is entirely possible because combinations of NRT medications are more effective than single medications (Fiore et al., 2008; Piper et al., 2009; Smith et al., 2009). Another possibility is that dual use reduces the rate of cigarette use and exposure to toxicants and therefore results in a net benefit to both individual and public health. Conversely, the net public health impact of the MRTP may be compromised to the extent that it reduced use of NRTs that ultimately led to smoking cessation. Or, the MRTP might benefit a different population of smokers than do NRTs. Ideally, an experimental design should permit the testing of a broad range of MRTPs and MRTP effects.

REFERENCES

al'Absi, M., D. Hatsukami, G. L. Davis, and L. E. Wittmers. 2004. Prospective examination of effects of smoking abstinence on cortisol and withdrawal symptoms as predictors of early smoking relapse. *Drug and Alcohol Dependence* 73(3):267-278.

Baker, T. B., T. H. Brandon, and L. Chassin. 2004a. Motivational influences on cigarette smoking. *Annual Review of Psychology* 55:463-491.

Baker, T. B., M. E. Piper, D. E. McCarthy, M. R. Majeskie, and M. C. Fiore. 2004b. Addiction motivation reformulated: An affective processing model of negative reinforcement. *Psychological Review* 111(1):33-51.

Benowitz, N. L., P. Jacob, and B. Herrera. 2006. Nicotine intake and dose response when smoking reduced-nicotine content cigarettes. *Clinical Pharmacology and Therapeutics* 80(6):703-714.

Benowitz, N. L., K. M. Dains, D. Dempsey, B. Herrera, L. Yu, and P. Jacob. 2009. Urine nicotine metabolite concentrations in relation to plasma cotinine during low-level nicotine exposure. *Nicotine & Tobacco Research* 11(8):954-960.

Bickel, W. K., and G. J. Madden. 1999a. *The behavioral economics of smoking*. Chicago: University of Chicago Press.

Bickel, W. K., and G. J. Madden. 1999b. A comparison of measures of relative reinforcing efficacy and behavioral economics: Cigarettes and money in smokers. *Behavioural Pharmacology* 10(6-7):627-637.

Bickel, W. K., R. Yi, E. T. Mueller, B. A. Jones, and D. R. Christensen. 2010. The behavioral economics of drug dependence: Towards the consilience of economics and behavioral neuroscience. In *Behavioral neuroscience of drug addiction*. Vol. 3, *Current topics in behavioral neurosciences*, edited by D. W. Self, and J. K. Staley Gottschalk. Berlin, Germany: Springer Berlin Heidelberg. Pp. 319-341.

Bigelow, G. E. 1991. Human drug abuse liability assessment: Opioids and analgesics. *British Journal of Addiction* 86(12):1615-1628.

Blondal, T., M. Franzon, and A. Westin. 1997. A double-blind randomized trial of nicotine nasal spray as an aid in smoking cessation. *European Respiratory Journal* 10(7):1585-1590.

Bolt, D. M., M. E. Piper, D. E. McCarthy, S. J. Japuntich, M. C. Fiore, S. S. Smith, and T. B. Baker. 2009. The Wisconsin Predicting Patients' Relapse questionnaire. *Nicotine & Tobacco Research* 11(5):481-492.

Brendryen, H., and P. Kraft. 2008. Happy ending: A randomized controlled trial of a digital multi-media smoking cessation intervention. *Addiction* 103(3):478-484.

Carpenter, M. J., J. R. Hughes, L. J. Solomon, and P. W. Callas. 2004. Both smoking reduction with nicotine replacement therapy and motivational advice increase future cessation among smokers unmotivated to quit. *Journal of Consulting and Clinical Psychology* 72(3):371-381.

Carter, B. L., and S. T. Tiffany. 1999. Cue-reactivity and the future of addiction research. *Addiction* 94(3):349-351.

Carter, L. P., and R. R. Griffiths. 2009. Principles of laboratory assessment of drug abuse liability and implications for clinical development. *Drug and Alcohol Dependence* 105(Suppl. 1):S14-S25.

CDC (Centers for Disease Control and Prevention). 2005. Annual smoking-attributable mortality, years of potential life lost, and productivity losses—United States, 1997-2001. *Morbidity and Mortality Weekly Report* 54(25):625-628.

CDC. 2009. Cigarette smoking among adults and trends in smoking cessation—United States, 2008. *Morbidity and Mortality Weekly Report* 58(44):1227-1232.

CDC and National Center for Health Statistics. 2008. *National Health and Nutrition Examination Survey 2007-2008*. http://www.cdc.gov/nchs/nhanes.htm (accessed October 5, 2010).

Chan, S. S., D. Y. Leung, A. S. Abdullah, V. T. Wong, A. J. Hedley, and T. H. Lam. 2011. A randomized controlled trial of a smoking reduction plus nicotine replacement therapy intervention for smokers not willing to quit smoking. *Addiction* 106(6):1155-1163.

Colby, S. M., S. T. Tiffany, S. Shiffman, and R. S. Niaura. 2000. Measuring nicotine dependence among youth: A review of available approaches and instruments. *Drug and Alcohol Dependence* 59:23-39.

Collins, L. M., T. B. Baker, R. J. Mermelstein, M. E. Piper, D. E. Jorenby, S. S. Smith, B. A. Christiansen, T. R. Schlam, J. W. Cook, and M. C. Fiore. 2011. The multiphase optimization strategy for engineering effective tobacco use interventions. *Annals of Behavioral Medicine* 41(2):208-226.

Comer, S. D., J. B. Ashworth, R. W. Foltin, C. E. Johanson, J. P. Zacny, and S. L. Walsh. 2008. The role of human drug self-administration procedures in the development of medications. *Drug and Alcohol Dependence* 96(1-2):1-15.

Cramer, J. A., V. L. Ouellette, and R. H. Mattson. 1990. Effect of microelectronic observation of compliance. *Epilepsia* 31:617-618.

Dewit, H., S. Dudish, and J. Ambre. 1993. Subjective and behavioral effects of diazepam depend on its rate of onset. *Psychopharmacology* 112(2-3):324-330.

Fant, R. V., K. J. Schuh, and M. L. Stitzer. 1995. Response to smoking as a function of prior smoking amounts. *Psychopharmacology* 119(4):385-390.

Fiore, M. C., C. R. Jaen, T. B. Baker, W. C. Bailey, N. Benowitz, S. J. Curry, S. F. Dorfman, E. S. Froelicher, M. G. Goldstein, C. G. Healton, P. N. Henderson, R. B. Heyman, H. K. Koh, T. E. Kottke, H. A. Lando, R. E. Mecklenburg, R. J. Mermelstein, P. D. Mullen, C. T. Orleans, L. Robinson, M. Stitzer, A. C. Tommasello, L. Villejo, and M. E. Wewers. 2008. *Treating tobacco use and dependence: 2008 update*. Rockville, MD: U.S. Department of Health and Human Services, U.S. Public Health Service.

Foltin, R. W., and M. W. Fischman. 1991. Assessment of abuse liability of stimulant drugs in humans: A methodological survey. *Drug and Alcohol Dependence* 28(1):3-48.

Gloria, R., L. Angelos, H. S. Schaefer, J. M. Davis, M. Majeskie, B. S. Richmond, J. J. Curtin, R. J. Davidson, and T. B. Baker. 2009. An fMRI investigation of the impact of withdrawal on regional brain activity during nicotine anticipation. *Psychophysiology* 46(4):681-693.

Griffiths, R. R., J. R. Troisi, II, K. Silvermian, and G. K. Miumford. 1993. Multiple-choice procedure: An efficient approach for investigating drug reinforcement in humans. *Behavioural Pharmacology* 4(1):3-14.

Griffiths, R. R., G. E. Bigelow, and N. A. Ator. 2003. Principles of initial experimental drug abuse liability assessment in humans. *Drug and Alcohol Dependence* 70(3):S41-S54.

Haney, M. 2009. Self-administration of cocaine, cannabis and heroin in the human laboratory: Benefits and pitfalls. *Addiction Biology* 14(1):9-21.

Haney, M., and R. Spealman. 2008. Controversies in translational research: Drug self-administration. *Psychopharmacology* 199(3):403-419.

Hatsukami, D. K., N. L. Benowitz, S. I. Rennard, C. Oncken, and S. S. Hecht. 2006. Biomarkers to assess the utility of potential reduced exposure tobacco products. *Nicotine & Tobacco Research* 8(2):169-191.

Hatsukami, D. K., K. Hanson, A. Briggs, M. Parascandola, J. M. Genkinger, R. O'Connor, and P. G. Shields. 2009. Clinical trials methods for evaluation of potential reduced exposure products. *Cancer Epidemiology, Biomarkers & Prevention* 18(12):3143-3195.

Hatsukami, D. K., J. Jensen, A. Anderson, B. Broadbent, S. Allen, Y. Zhang, and H. Severson. 2011. Oral tobacco products: Preference and effects among smokers. *Drug and Alcohol Dependence* 118(2-3):230-236.

Hawkins, J., W. Hollingworth, and R. Campbell. 2010. Long-term smoking relapse: A study using the British household panel survey. *Nicotine & Tobacco Research* 12(12):1228-1235.

Heffner, J. L., T. C. Lee, C. Arteaga, and R. M. Anthenelli. 2010. Predictors of post-treatment relapse to smoking in successful quitters: Pooled data from two phase III varenicline trials. *Drug and Alcohol Dependence* 109(1-3):120-125.

Heishman, S. J., B. A. Kleykamp, and E. G. Singleton. 2010. Meta-analysis of the acute effects of nicotine and smoking on human performance. *Psychopharmacology* 210(4):453-469.

Hughes, J. R., and D. Hatsukami. 1986. Signs and symptoms of tobacco withdrawal. *Archives of General Psychiatry* 43(3):289-294.

Hughes, J. R., S. W. Gust, K. Skoog, R. M. Keenan, and J. W. Fenwick. 1991. Symptoms of tobacco withdrawal—a replication and extension. *Archives of General Psychiatry* 48(1):52-59.

Hughes, J. R., A. H. Oliveto, and S. Y. Terry. 1996. Saying versus doing and other methodological issues in the study of human drug self-administration. *Experimental and Clinical Psychopharmacology* 4(2):234-238.

Hughes, J. R., J. P. Keely, R. S. Niaura, D. J. Ossip-Klein, R. L. Richmond, and G. E. Swan. 2003. Measures of abstinence in clinical trials: Issues and recommendations. *Nicotine & Tobacco Research* 5(1):13-25.

Hughes, J. R., E. N. Peters, and S. Naud. 2008. Relapse to smoking after 1 year of abstinence: A meta-analysis. *Addictive Behaviors* 33(12):1516-1520.

Hughes, J. R., T. Baker, N. Breslau, L. Covey, and S. Shiffman. 2011. Applicability of DSM criteria to nicotine dependence. *Addiction* 106(5):894-895.

Hursh, S. R., and A. Silberberg. 2008. Economic demand and essential value. *Psychological Review* 115(1):186-198.

Jasinski, D. 1977. Assessment of the abuse potentiality of morphine-like drugs (methods used in man). In *Handbook of experimental pharmacology*. Vol. 45, edited by W. Martin. Heidelberg: Springer. Pp. 197-258.

Johnson, M. W., and W. K. Bickel. 2006. Replacing relative reinforcing efficacy with behavioral economic demand curves. *Journal of the Experimental Analysis of Behavior* 85(1):73-93.

Johnson, M. W., W. K. Bickel, and A. P. Kirshenbaum. 2004. Substitutes for tobacco smoking: A behavioral economic analysis of nicotine gum, denicotinized cigarettes, and nicotine-containing cigarettes. *Drug and Alcohol Dependence* 74(3):253-264.

Kleykamp, B. A., J. M. Jennings, and T. Eissenberg. 2011. Effects of transdermal nicotine and concurrent smoking on cognitive performance in tobacco-abstinent smokers. *Experimental and Clinical Psychopharmacology* 19(1):75-84.

Kotlyar, M., M. I. Mendoza-Baumgart, Z. Li, P. R. Pentel, B. C. Barnett, R. M. Feuer, E. A. Smith, and D. K. Hatsukami. 2007. Nicotine pharmacokinetics and subjective effects of three potential reduced exposure products, moist snuff and nicotine lozenge. *Tobacco Control* 16(2):138.

Krall, E. A., A. J. Garvey, and R. I. Garcia. 2002. Smoking relapse after 2 years of abstinence: Findings from the VA normative aging study. *Nicotine & Tobacco Research* 4:95-100.

Lam, T. H., A. S. M. Abdullah, S. S. C. Chan, A. J. Hedley, and Hong Kong Council on Smoking and Health Smoking Cessation Health Centre Steering Group. 2005. Adherence to nicotine replacement therapy versus quitting smoking among Chinese smokers: A preliminary investigation. *Psychopharmacology* 177(4):400-408.

Lancaster, T., and L. F. Stead. 2005. Individual behavioural counselling for smoking cessation. *Cochrane Database of Systematic Reviews* (2):CD001292.

Liu, H. H., C. E. Golin, L. G. Miller, R. D. Hays, K. Beck, S. Sanandaji, J. Christian, T. Maldonado, D. Duran, A. H. Kaplan, and N. S. Wenger. 2001. A comparison study of multiple measures of adherence to HIV protease inhibitors. *Annals of Internal Medicine* 134(10):968-977.

MacKinnon, D. P., C. M. Lockwood, J. M. Hoffman, S. G. West, and V. Sheets. 2002. A comparison of methods to test mediation and other intervening variable effects. *Psychological Methods* 7(1):83-104.

Madden, G. J., and W. K. Bickel. 1999. Abstinence and price effects on demand for cigarettes: A behavioral-economic analysis. *Addiction* 94(4):577-588.

Matsui, D., C. Hermann, M. Braudo, S. Ito, N. Olivieri, and G. Koren. 1992. Clinical use of the medication event monitoring system: A new window into pediatric compliance. *Clinical Pharacology and Therapeutics* 52:102-103.

McClernon, F. J., R. V. Kozink, A. M. Lutz, and J. E. Rose. 2009. 24-h smoking abstinence potentiates fMRI-bold activation to smoking cues in cerebral cortex and dorsal striatum. *Psychopharmacology* 204(1):25-35.

Metry, J. M. 1999. Measuring compliance in clinical trials and ambulatory care. In *Drug regimen compliance: Issues in clinical trials and patient management*. 1st ed, edited by J.-M. Metry, and U. A. Meyer. Chichester: John Wiley & Sons. Pp. 1-21.

Miller, C. L., and V. Sedivy. 2009. Using a quitline plus low-cost NRT to help disadvantaged smokers quit. *Tobacco Control* 18(2):144-149.

Mumford, G. K., S. M. Evans, J. C. Fleishaker, and R. R. Griffiths. 1995. Alprazolam absorption kinetics affects abuse liability. *Clinical Pharmacology and Therapeutics* 57(3):356-365.

Murphy, J. G., J. MacKillop, J. R. Skidmore, and A. A. Pederson. 2009. Reliability and validity of a demand curve measure of alcohol reinforcement. *Experimental and Clinical Psychopharmacology* 17(6):396-404.

Murphy, J. G., J. MacKillop, J. W. Tidey, L. A. Brazil, and S. M. Colby. 2011. Validity of a demand curve measure of nicotine reinforcement with adolescent smokers. *Drug and Alcohol Dependence* 113(2-3):207-214.

Perkins, K. A. 2009. Acute responses to nicotine and smoking: Implications for prevention and treatment of smoking in lower SES women. *Drug and Alcohol Dependence* 104(Suppl. 1):S79-S86.

Perkins, K. A., L. H. Epstein, J. Grobe, and C. Fonte. 1994a. Tobacco abstinence, smoking cues, and the reinforcing value of smoking. *Pharmacology Biochemistry and Behavior* 47(1):107-112.

Perkins, K. A., J. E. Grobe, C. Fonte, J. Goettler, A. R. Caggiula, W. A. Reynolds, R. L. Stiller, A. Scierka, and R. G. Jacob. 1994b. Chronic and acute tolerance to subjective, behavioral and cardiovascular effects of nicotine in humans. *Journal of Pharmacology and Experimental Therapeutics* 270(2):628-638.

Perkins, K. A., J. Grobe, and C. Fonte. 1997. Influence of acute smoking exposure on the subsequent reinforcing value of smoking. *Experimental and Clinical Psychopharmacology* 5(3):277-285.

Perkins, K. A., E. Donny, and A. R. Caggiula. 1999. Sex differences in nicotine effects and self-administration: Review of human and animal evidence. *Nicotine & Tobacco Research* 1(4):301-315.

Perkins, K. A., L. Jacobs, L. Clark, C. A. Conklin, M. Sayette, and A. Wilson. 2004a. Instructions about nicotine dose influence acute responses to nasal spray. *Nicotine & Tobacco Research* 6(6):1051-1060.

Perkins, K. A., L. Jacobs, M. Sayette, and A. Caggiula. 2004b. The influence of instructions and nicotine dose on the subjective and reinforcing effects of smoking. *Experimental and Clinical Psychopharmacology* 12(2):91-101.

Perkins, K. A., M. Ciccocioppo, C. A. Conklin, M. E. Milanak, A. Grottenthaler, and M. A. Sayette. 2008. Mood influences on acute smoking responses are independent of nicotine intake and dose expectancy. *Journal of Abnormal Psychology* 117(1):79-93.

Perkins, K. A., S. B. Coddington, J. L. Karelitz, C. Jetton, J. A. Scott, A. S. Wilson, and C. Lerman. 2009. Variability in initial nicotine sensitivity due to sex, history of other drug use, and parental smoking. *Drug and Alcohol Dependence* 99(1-3):47-57.

Perkins, K. A., J. L. Karelitz, C. A. Conklin, M. A. Sayette, and G. E. Giedgowd. 2010. Acute negative affect relief from smoking depends on the affect situation and measure but not on nicotine. *Biological Psychiatry* 67(8):707-714.

Piasecki, T. M., D. E. Jorenby, S. S. Smith, M. C. Fiore, and T. B. Baker. 2003. Smoking withdrawal dynamics: II. Improved tests of withdrawal-relapse relations. *Journal of Abnormal Psychology* 112(1):14-27.

Piasecki, T. M., M. E. Piper, and T. B. Baker. 2010. Refining the tobacco dependence phenotype using the Wisconsin Inventory of Smoking Dependence Motives: II. Evidence from a laboratory self-administration assay. *Journal of Abnormal Psychology* 119(3):513-523.

Pierce, J. P., K. Messer, M. M. White, D. W. Cowling, and D. P. Thomas. 2011. Prevalence of heavy smoking in California and the United States, 1965-2007. *JAMA* 305(11):1106-1112.

Piper, M. E., T. M. Piasecki, E. B. Federman, D. M. Bolt, S. S. Smith, M. C. Fiore, and T. B. Baker. 2004. A multiple motives approach to tobacco dependence: The Wisconsin Inventory of Smoking Dependence Motives (WISDM-68). *Journal of Consulting and Clinical Psychology* 72(2):139.

Piper, M. E., E. B. Federman, D. E. McCarthy, D. M. Bolt, S. S. Smith, M. C. Fiore, and T. B. Baker. 2008. Using mediational models to explore the nature of tobacco motivation and tobacco treatment effects. *Journal of Abnormal Psychology* 117(1):94-105.

Piper, M. E., S. S. Smith, T. R. Schlam, M. C. Fiore, D. E. Jorenby, D. Fraser, and T. B. Baker. 2009. A randomized placebo-controlled clinical trial of 5 smoking cessation pharmacotherapies. *Archives of General Psychiatry* 66(11):1253-1262.

Powell, J., L. Dawkins, R. West, and A. Pickering. 2010. Relapse to smoking during unaided cessation: Clinical, cognitive and motivational predictors. *Psychopharmacology* 212(4):537-549.

Reid, R. D., A. L. Pipe, B. Quinlan, and J. Oda. 2007. Interactive voice response telephony to promote smoking cessation in patients with heart disease: A pilot study. *Patient Education and Counseling* 66(3):319-326.

Robinson, M. L., E. J. Houtsmuller, E. T. Moolchan, and W. B. Pickworth. 2000. Placebo cigarettes in smoking research. *Experimental and Clinical Psychopharmacology* 8(3):326-332.

Rogers, S., T. Gould, and T. Baker. 2009. Mouse models and the genetics of nicotine dependence. In *Phenotypes and endophenotypes: Foundations for genetic studies of nicotine use and dependence. Tobacco Control Monograph No. 20.* Bethesda, MD: U.S. Department of Health and Human Services, National Institutes of Health, National Cancer Institute.

Rose, J. E., E. C. Westman, F. M. Behm, M. P. Johnson, and J. S. Goldberg. 1999. Blockade of smoking satisfaction using the peripheral nicotinic antagonist trimethaphan. *Pharmacology Biochemistry and Behavior* 62(1):165-172.

Rose, J. E., F. M. Behm, E. C. Westman, J. E. Bates, and A. Salley. 2003. Pharmacologic and sensorimotor components of satiation in cigarette smoking. *Pharmacology, Biochemistry and Behavior* 76(2):243-250.

Sayette, M. A., and M. R. Hufford. 1994. Effects of cue exposure and deprivation on cognitive resources in smokers. *Journal of Abnormal Psychology* 103(4):812-818.

Sayette, M. A., C. S. Martin, J. G. Hull, J. M. Wertz, and M. A. Perrott. 2003. Effects of nicotine deprivation on craving response covariation in smokers. *Journal of Abnormal Psychology* 112(1):110-118.

Schnoll, R. A., F. Patterson, E. P. Wileyto, R. F. Tyndale, N. Benowitz, and C. Lerman. 2009. Nicotine metabolic rate predicts successful smoking cessation with transdermal nicotine: A validation study. *Pharmacology Biochemistry and Behavior* 92(1):6-11.

Schulz, K. F., D. G. Altman, and D. Moher. 2010. CONSORT 2010 statement: Updated guidelines for reporting parallel group randomized trials. *Annals of Internal Medicine* 152(11):726-732.

Schuurmans, M. M., A. H. Diacon, X. van Biljon, and C. T. Bolliger. 2004. Effect of pretreatment with nicotine patch on withdrawal symptoms and abstinence rates in smokers subsequently quitting with the nicotine patch: A randomized controlled trial. *Addiction* 99(5):634-640.

Shaham, Y., H. Rajabi, and J. Stewart. 1996. Relapse to heroin-seeking in rats under opioid maintenance: The effects of stress, heroin priming, and withdrawal. *The Journal of Neuroscience* 16(5):1957-1963.

Shaham, Y., U. Shalev, L. Lu, H. De Wit, and J. Stewart. 2003. The reinstatement model of drug relapse: History, methodology and major findings. *Psychopharmacology* 168(1-2):3-20.

Shiffman, S., and M. A. Sayette. 2005. Validation of the nicotine dependence syndrome scale (NDSS): A criterion-group design contrasting chippers and regular smokers. *Drug and Alcohol Dependence* 79(1):45-52.

Shiffman, S., W. G. Shadel, R. Niaura, M. A. Khayrallah, D. E. Jorenby, C. F. Ryan, and C. L. Ferguson. 2003. Efficacy of acute administration of nicotine gum in relief of cue-provoked cigarette craving. *Psychopharmacology* 166(4):343-350.

Shiffman, S., A. Waters, and M. Hickcox. 2004. The nicotine dependence syndrome scale: A multidimensional measure of nicotine dependence. *Nicotine & Tobacco Research* 6(2):327-348.

Shiffman, S., S. G. Ferguson, J. Rohay, and J. G. Gitchell. 2008a. Perceived safety and efficacy of nicotine replacement therapies among US smokers and ex-smokers: Relationship with use and compliance. *Addiction* 103(8):1371-1378.

Shiffman, S., C. T. Sweeney, S. G. Ferguson, M. A. Sembower, and J. G. Gitchell. 2008b. Relationship between adherence to daily nicotine patch use and treatment efficacy: Secondary analysis of a 10-week randomized, double-blind, placebo-controlled clinical trial simulating over-the-counter use in adult smokers. *Clinical Therapeutics* 30(10):1852-1858.

Smith, S. S., D. E. McCarthy, S. J. Japuntich, B. Christiansen, M. E. Piper, D. E. Jorenby, D. L. Fraser, M. C. Fiore, T. B. Baker, and T. C. Jackson. 2009. Comparative effectiveness of 5 smoking cessation pharmacotherapies in primary care clinics. *Archives of Internal Medicine* 169(22):2148-2155.

Smith, S. S., M. E. Piper, D. M. Bolt, M. C. Fiore, D. W. Wetter, P. M. Cinciripini, and T. B. Baker. 2010. Development of the brief Wisconsin Inventory of Smoking Dependence Motives. *Nicotine & Tobacco Research* 12(5):489-499.

Stead, L. F., and T. Lancaster. 2007. Interventions to reduce harm from continued tobacco use. *Cochrane Database of Systematic Reviews* (3):CD005231.

Strasser, A. A., C. Lerman, P. M. Sanborn, W. B. Pickworth, and E. A. Feldman. 2007. New lower nicotine cigarettes can produce compensatory smoking and increased carbon monoxide exposure. *Drug and Alcohol Dependence* 86(2-3):294-300.

Strasser, A. A., K. Z. Tang, P. M. Sanborn, J. Y. Zhou, and L. T. Kozlowski. 2009. Behavioral filter vent blocking on the first cigarette of the day predicts which smokers of light cigarettes will increase smoke exposure from blocked vents. *Experimental and Clinical Psychopharmacology* 17(6):405-412.

Sutherland, G., J. A. Stapleton, M. A. Russell, M. J. Jarvis, P. Hajek, M. Belcher, and C. Feyerabend. 1992. Randomised controlled trial of nasal nicotine spray in smoking cessation. *Lancet* 340(8815):324-329.

Tinkelman, D., S. M. Wilson, J. Willett, and C. T. Sweeney. 2007. Offering free NRT through a tobacco quitline: Impact on utilisation and quit rates. *Tobacco Control* 16(Suppl. 1):i42-i46.

Vogt, F., S. Hall, and T. M. Marteau. 2008. Understanding why smokers do not want to use nicotine dependence medications to stop smoking: Qualitative and quantitative studies. *Nicotine & Tobacco Research* 10(8):1405.

Weiss, R. B., T. B. Baker, D. S. Cannon, A. von Niederhausern, D. M. Dunn, N. Matsunami, N. A. Singh, L. Baird, H. Coon, W. M. McMahon, M. E. Piper, M. C. Fiore, M. B. Scholand, J. E. Connett, R. E. Kanner, L. C. Gahring, S. W. Rogers, J. R. Hoidal, and M. F. Leppert. 2008. A candidate gene approach identifies the CHRNA5-A3-B4 region as a risk factor for age-dependent nicotine addiction. *PLoS Genetics* 4(7):e1000125.

Welsch, S. K., S. S. Smith, D. W. Wetter, D. E. Jorenby, M. C. Fiore, and T. B. Baker. 1999. Development and validation of the Wisconsin Smoking Withdrawal Scale. *Experimental and Clinical Psychopharmacology* 7(4):354-361.

Westman, E. C., F. M. Behm, and J. E. Rose. 1996. Dissociating the nicotine and airway sensory effects of smoking. *Pharmacology, Biochemistry and Behavior* 53(2):309-315.

Zinser, M. C., M. C. Fiore, R. J. Davidson, and T. B. Baker. 1999. Manipulating smoking motivation: Impact on an electrophysiological index of approach motivation. *Journal of Abnormal Psychology* 108(2):240-254.

5

Methods for Studying Risk Perception and Risk Communication

According to the Family Smoking Prevention and Tobacco Control Act of 2009 (FSPTCA),[1] consumer perceptions of labels or marketing statements for modified risk tobacco products (MRTPs) should be tested to show that they will not mislead the consumer to believe that the product is less harmful or demonstrates less risk than is actually true. As such, on an annual basis, pre- and postmarket studies should be conducted to demonstrate that current and potential consumers of each MRTP understand the actual and relative risks of the product. As discussed in Chapter 1, the FSPTCA articulates a public health standard whereby product sponsors must conduct studies on the effect of the product on the population as a whole. As outlined in the law, this evaluation of the health of the population must include studies demonstrating that (1) perceptions of less risk from the MRTP do not result in nontobacco users initiating tobacco use, (2) existing tobacco users who would otherwise consider quitting all tobacco products do not switch to this new MRTP, and (3) usage of tobacco products does not increase as a result of this new product.

This chapter begins with a brief review of how users and nonusers perceive tobacco-related outcomes, including perceptions of epidemiologic data, short- and long-term risks to the individual, addiction, and potential benefits. Careful attention is given regarding perceptions of different types of tobacco products, as well as how perceptions of tobacco

[1] *Family Smoking Prevention and Tobacco Control Act of 2009*, Public Law 111-31, 123 Stat. 1776 (June 22, 2009).

use outcomes vary by age and demographics. Next, the chapter outlines the standards for studies on risk perceptions, including the questions that should be addressed through the studies, standards for the research designs, participant recruitment, measurement, and analysis.

BACKGROUND AND RATIONALE: IMPORTANCE OF RISK PERCEPTIONS

Judgments about risk, otherwise known as risk perceptions, are viewed as a fundamental element of most theoretical models of health behavior and behavioral decision making, including social cognitive theory (Bandura, 2001), the health belief model (Rosenstock, 1974), the theory of reasoned action (Fishbein and Ajzen, 1975), the theory of planned behavior (Ajzen, 1985), self-regulation theory (Kanfer, 1970), and subjective culture and interpersonal relations theory (Triandis, 1977). In general, these models argue that individuals' perceptions about the value and likelihood of behavior-related positive and negative consequences and their vulnerability to those consequences play a key role in behavioral choices. As such, understanding individuals' perceptions of tobacco-related products, including MRTPs, whether such perceptions change over time with the introduction of MRTPs, and whether such perceptions play a role in tobacco use behavior, is critical. The committee also acknowledges, as the 2007 Institute of Medicine (IOM) report articulated, that perceptions of risk (and benefit) may have differing implications for product use among different consumers. It is important to understand both the risk (and benefit) perceptions of the consumer and the value that is placed upon these perceptions.

In the next few sections, the committee provides an overview of the literature on tobacco-related perceptions, followed by methodological considerations to design studies to determine perceptions and behavioral implications of MRTPs.

PERCEPTIONS OF EPIDEMIOLOGIC DATA FOR TOBACCO USE

It is critical to first understand the extent to which both tobacco users and nonusers understand the actual risks of tobacco use, compared to epidemiologic data. Much of the literature comparing perceptions to actual data suggests that, on average, smokers overestimate the risks of smoking (Borland, 1997; Johnson et al., 2002; Kristiansen et al., 1983; Viscusi, 1990, 1991, 1992), while other studies show that smokers underestimate them (Arnett, 2000; Hansen and Malotte, 1986; Leventhal et al., 1987; Schoenbaum, 1997; Sutton, 1998; Virgili et al., 1991). Among adolescents and young adults (ages 18–22), Jamieson and Romer (2001) found that

70 percent of smokers and 79 percent of nonsmokers overestimated the risk of contracting lung cancer from smoking. Just over a third of the smokers and more than 40 percent of nonsmokers overestimated the risk of death from smoking, and 41 percent of smokers and 27 percent of nonsmokers either underestimated or did not know this rate (Jamieson and Romer, 2001). About a quarter of the nonsmoking participants and 21 percent of the smokers also underestimated the number of years of life that would be lost due to smoking, and they inaccurately perceived more deaths caused by gunshots, car accidents, alcohol, and other drug use than by smoking cigarettes (Jamieson and Romer, 2001). Given people's limited understanding of tobacco-related risk, MRTP labels and advertisements should be careful to convey information on tobacco-related risks accurately and in a manner that can be fully comprehended by the general population.

PERCEPTIONS OF TOBACCO-RELATED RISKS AND BENEFITS TO THE INDIVIDUAL

A great number of studies have examined both smokers' and nonsmokers' perceptions of tobacco-related outcomes, including perceived short- and long-term health risks, social risks, risks of becoming addicted, risks from secondhand smoke, and cumulative risks. Findings on these tobacco-related perceptions as well as the important relationship between perceptions and tobacco use are reviewed next.

Historically, studies of tobacco-related perceptions were largely focused on perceptions of long-term health risks associated with smoking, such as heart attack and lung cancer. More recently, there has been an emphasis on short-term health and social risks that are more pertinent to adolescents and even adults, such as the smell of cigarettes, the yellowing of teeth, and the possibility of getting into trouble (Gritz et al., 2003; Halpern-Felsher et al., 2004; IOM, 2007; Prokhorov et al., 2002).

Studies have also examined whether such tobacco-related perceptions are related to actual tobacco use. There have been a number of studies that have relied on cross-sectional data to test the relationship between adolescents' perceived tobacco risk and actual tobacco use. The bulk of these findings indicate that adolescents who have smoked hold lower perceptions of risk than adolescents who have not smoked (Jamieson and Romer, 2001; Romer and Jamieson, 2001).

Using prospective, longitudinal data to examine whether perceptions actually predict the initiation of tobacco use, Song and colleagues (2009b) showed that, compared to adolescents with the highest perceptions of tobacco-related risks, adolescents with the lowest perceptions of tobacco-related long-term risks were 3.64 times more likely to initiate tobacco use. The same relationship was observed with perceptions of short-term risks, whereby the adolescent participants who believed that tobacco-related short-term risks were unlikely were 2.68 times more likely to initiate smoking compared to adolescents with higher perceptions of short-term risks (Song et al., 2009b).

In addition to understanding the extent to which adolescent and adult smokers and nonsmokers perceive tobacco-related risks and whether these risk perceptions deter tobacco use, it is critical to learn the extent to which perceived tobacco-related benefits motivate individuals to use a tobacco product. Indeed, studies provide support that perceived benefits are an equally, if not more important, component of the decision equation. For example, Prokhorov and colleagues (2002) found that scores on a smoking-related pros or benefits scale increased and con scores decreased as adolescents became more susceptible to smoking. Pallonen et al. (1998) showed that nonsmokers were more likely to initiate tobacco use if they perceived more smoking benefits, whereas perceived smoking risks were less related to smoking onset.

Halpern-Felsher et al. (2004), as well as Goldberg et al. (2002), found that participants who have smoked perceive benefits more likely to occur, and risks less likely to occur, compared to adolescents who have not smoked. Results from more recent longitudinal studies have demonstrated that adolescents who report the highest perceptions of smoking-related benefits are as much as 3.3 times more likely to initiate smoking (Song et al., 2009b), and that adolescents who have experimented with as little as one puff of cigarette have greater perceptions of benefits compared to those who have never smoked (Morrell et al., 2010).

In summary, adolescents' perceptions of the risks and benefits of cigarette smoking play an important role in adolescents' decisions to smoke, and adolescents with lower perceptions of tobacco risks are more likely to initiate tobacco use. It is therefore essential that studies of consumer perceptions examine whether the information about MRTPs that is provided to consumers affects the perceived risks and benefits of the products, and what implications these perceptions have for subsequent use of the MRTP in relation to pre-existing tobacco products. Given that adolescence is a period of heightened vulnerability for the initiation of tobacco use, it is particularly important to evaluate whether adolescents accurately understand the purported benefits of an MRTP. The ethical considerations for studies involving populations at high risk for tobacco initiation, such as adolescents, are discussed in Chapter 2 and Chapter 6.

Other aspects of tobacco-associated risks that are not fully understood by many adolescents and young adults include misunderstandings about nicotine addiction and the ability to quit using tobacco products. Studies suggest that smokers and nonsmokers are not fully aware of the addictive nature of smoking (Arnett, 2000; DiFranza et al., 2011; Halpern-Felsher et al., 2004; Leventhal et al., 1987; Slovic, 1998, 2001). It is argued that adolescent smokers may be less concerned about the long-term risks of smoking partly because they believe that they can stop smoking easily and at any time (Arnett, 2000; Halpern-Felsher et al., 2004; IOM, 2007; Slovic, 1998).

Perceptions of addiction go beyond the physical need to smoke, and include fulfilling an emotional or social need, such as avoiding unpleasant mood states or wanting to socially relate to others (Johnson et al., 2003). Rugaska et al. (2001) concluded that youth perceive dependence risks to be associated solely with adult smoking; the authors found that adolescents believe their underage smoking for social settings was safe, in contrast to adults who smoke to cope with everyday life stress.

Weinstein et al. (2004) examined smokers' beliefs concerning the ease of quitting and the nature of addiction. They found that more than 96 percent of the adolescents and adults in their study agreed with the statement, "the longer you smoke, the harder it is to quit," and most believed that addiction develops quickly. Other analyses have found that smokers are relatively optimistic about the idea of addiction, believing that smoking cessation is not that difficult (Jamieson and Romer, 2001) and overestimating the ease with which a smoker can quit (Weinstein et al., 2004).

When inquired about the ease of quitting smoking, adolescents with smoking experience believed they will find it easier to quit and will be more likely to quit smoking compared to adolescents without smoking experience (Halpern-Felsher et al., 2004). Arnett (2000) found that 60 percent of the adolescents and almost half of the adults in their study endorsed the idea that they could smoke for a few years and then quit if and when they wanted. Weinstein et al. (2005) found differences in perceptions of risks between smokers who did and did not plan to quit smoking, with those planning to quit recognizing higher risks of lung cancer.

In addition to examining perceptions of personal risk from smoking, a few studies have examined perceptions of risk from secondhand smoke, including risk to others if you smoke, and personal risk from others' smoke. Glantz and Jamieson (2000) found that youth who smoked were less likely than nonsmoking youth to believe that secondhand smoke leads to thousands of deaths each year. They also showed that adolescents who planned to quit smoking were more aware of the effects of secondhand smoke than were smokers without quit intentions. Romer and Jamieson (2001) found that knowledge of secondhand smoke harm was indirectly related to intentions to quit because of its relationship with perceived risk of smoking overall. Kurtz and colleagues (2001) showed that elementary, middle, and high school students with smoking experience were less knowledgeable about and had less negative views of secondhand smoke compared to students without smoking experience. Similarly, Halpern-Felsher and Rubinstein (2005) found that adolescents with smoking experience perceived less risk from secondhand smoke than did adolescents without smoking experience. In a follow-up study, Song et al. (2009a) showed that perceptions of risk from secondhand smoke predicted smoking initiation, with adolescents with the lowest perceived risk of secondhand smoke being the most likely to subsequently try smoking.

Taken together, this set of literature demonstrates the need to understand and describe perceptions of tobacco-related outcomes, including perceptions of short- and long-term risks, addiction, and potential benefits. It is also important to understand perceptions concerning secondhand smoke as well as other tobacco products. These studies aid in identifying critical perceptions held by smokers and nonsmokers; perceptions are also instrumental in predicting subsequent tobacco use and changes in patterns of use that are important to capture. Data from these studies should be included in the portfolio of evidence submitted to the Food and Drug Administration (FDA) when applying for a modified risk claim on a tobacco product.

Differences in Perceptions of Risks and Benefits by Type of Tobacco Product

A small set of literature has examined whether perceptions of risks and benefits vary by the type of tobacco product. Most of this research has examined perceptions of so called "light," "ultra light," and "low-tar" cigarettes. The studies show that adults have misperceptions about the health risks associated with smoking light and ultra light cigarettes; most adult smokers believe these cigarettes deliver less tar and nicotine, produce milder sensations, and result in less health consequences (Etter et al., 2003; Shiffman et al., 2001; Slovic, 2001). Studies have also shown that smokers have switched to these so-called lighter cigarettes to reduce the health risks of smoking (Slovic, 2001). Shiffman et al. (2001) examined the perceptions of light, ultra light, and regular cigarettes among adult daily smokers; participants believed that lights and ultra lights were less risky compared to regular cigarettes and that the ultra light cigarettes were the least harmful. Similarly, Etter et al. (2003) quantified the perceptions of smoking different cigarettes, showing that participants believed they needed to smoke two light cigarettes or four ultra light cigarettes to inhale the same amount of nicotine as one would inhale from a single regular cigarette. Etter and colleagues (2003) also found that current adult light cigarette smokers believed they were at less risk of developing lung cancer than did smokers of regular cigarettes.

Kropp and Halpern-Felsher (2004) extended these studies to examine adolescents' perceptions of light cigarettes. In their study, adolescents believed they were significantly less likely to have a heart attack, get lung cancer, have trouble breathing, get a bad cough, and die from a smoking-related disease if smoking light cigarettes compared with smoking regular cigarettes. The participants also believed that light cigarettes have less tar and nicotine than regular cigarettes and that it would be easier to quit smoking light compared to regular cigarettes.

A study of Norwegian older adolescents and young adults (aged 16–20 years) examined perceptions of different tobacco products, including roll-your-own tobacco, factory-made cigarettes, low-tar factory-made cigarettes, pipe tobacco, cigars or cigarillos, loose snus, prepackaged snus, and nicotine replacement therapies (NRTs). Participants rated roll-your-own tobacco as most harmful and NRTs less harmful (Øverland et al., 2008). In a direct comparison, snus was considered less harmful than cigarettes on average, and participants who used snus rated it less harmful than did nonusers of snus (Øverland et al., 2008). Callery and colleagues (2011) examined the relative health risk beliefs among a group of adult Canadian smokers (aged 18–30 years). They found that between 30 percent and 47 percent of the participants wrongly believed that smokeless tobacco and cigarettes are equally harmful, and some wrongly noted that smokeless tobacco is more harmful than cigarettes (Callery et al., 2011).

Other studies have examined whether smokers believe there are differences in harm based on type, brand, or color packaging of tobacco products. Mutti and colleagues (2011) showed that adult smokers attributed differential risks based on cigarette brands and packaging color (e.g., gold or silver compared to red or black). Smokers of light and mild cigarettes perceived their cigarette brand to be less harmful compared to others, as did smokers of cigarettes found in gold, silver, purple, or blue packages. Similarly, Bansal-Travers et al. (2011) perceived differences in risk based on color of the cigarette package, with white coloring denoting less risk.

These studies confirm that adults and adolescents, as well as smokers and nonsmokers, harbor misconceptions about tobacco products based on the packaging coloring or descriptors. As noted by a previous IOM committee (2007), "such perceptions are likely the result, in part, of the tobacco industry's marketing of light cigarettes as the healthier smoking choice, a safer alternative to cessation, and a first step toward quitting smoking altogether." More favorable perceptions of light, ultra light, and low-tar cigarettes are important to note, since many smokers have made the choice to smoke light cigarettes because they believe such cigarettes are less addictive or safer than regular cigarettes (Etter et al., 2003). Further, adults who smoke light or ultra light cigarettes might be less likely to attempt to quit smoking, believing that their cigarette choices provide a safer alternative to regular cigarette smoking (Etter et al., 2003; Shiffman et al., 2001).

Demographic Differences in Tobacco-Related Perceptions

With the exception of identifying age differences, there are surprisingly few studies that have examined differences in tobacco-related perceptions by other demographic variables, such as gender, race/ethnicity, or socioeconomic status. The small literature on these topics is reviewed next.

Previous studies have found limited gender-specific differences among smokers with regards to benefit perceptions of smoking. Among adults, women are more likely than men to be concerned about post-cessation weight gain, women are more likely to identify weight gain as the cause for relapse to smoking, and women are less likely to be motivated to quit smoking if they fear subsequent weight gain (Swan et al., 1993; Weekley, 1992). McKee et al. (2005) showed that adult females perceived both greater risk and greater benefits from smoking than did adult males. Others have found that women are less likely to acknowledge the health benefits of smoking cessation (Sorensen and Pechacek, 1987) and that men are more likely to quit smoking in order to have better health (Curry et al., 1997). Adolescent males report fewer health concerns than females, and they perceive fewer risks and greater benefits associated with a variety of health-related risky behaviors (Millstein and Halpern-Felsher, 2002). Taken together, these studies provide evidence to support the existence of gender-based differences in perceptions of the risks and benefits of smoking. These differences may also relate to why females have poorer smoking cessation outcomes as compared to males (Perkins, 2001). Thus, consumer perceptions of tobacco products applying for the modified risk claim should be explored separately for males and females in adolescent and adult samples.

Surprisingly few studies have examined cultural variation (including race, ethnicity, country of origin, acculturation, language usage, and social class) in perceptions, especially related to tobacco use. As described in a previous IOM report (2007),

> it is possible that the level of perceived risk (and benefit) may differ across groups of individuals, possibly as a factor of culture, socioeconomic status, or differences in exposure to behavior-related outcomes, for example. Alternatively, groups of adolescents or young adults might perceive the same level of risk, but these perceptions might have different implications for their smoking, in part due to differences in perceived control, risk-reducing strategies used, or value placed on the negative outcome (e.g., bad breath or trouble breathing) compared to the value placed on the benefit (e.g., looking cool) of smoking.

Future studies are needed.

Adolescents' Reasons for Smoking

Qualitative studies have used methods such as one-on-one interviewing or focus groups to understand the motivations for smoking (IOM, 2007). Based on these studies, the most commonly identified reasons for smoking include: to satisfy curiosity, to fit in with peers, to relieve stress and boredom, to decrease appetite, to increase the high from alcohol and drugs, and because parents smoke (Clark et al., 2002; Dunn and Johnson, 2001; Gittelsohn et al., 2001; Kegler and Cleaver, 2000; Nichter et al., 1997; Vuckovic et al., 2003). A previous IOM committee (2007) noted that "adolescents form perceptions of smoking images, such as nonsmokers being more mature (Lloyd et al., 1997), and adolescents recognize that different types of smoking identities (beyond the usual categories of nonsmokers, experimenters, and smokers) exist for adolescents (Johnson et al., 2003)." A number of studies indicate that such images have an impact on adolescents' smoking. Gerrard and colleagues' (2008) Prototype Willingness Model of adolescent risk behavior postulates that an adolescent's image of a typical smoker or nonsmoker will influence his or her willingness to smoke, and ultimately his or her actual smoking behavior. Research confirms that adolescents who hold more favorable images of a typical smoker are more willing to smoke and accept the consequences of smoking (Gerrard et al., 2008).

Advertisements for tobacco products have targeted reasons for smoking across a variety of groups defined by demographic characteristics such as age (adolescents, young adults, and adults), gender, race, socioeconomic status, and psychosocial needs; they have also been directed at creating favorable images of smokers in order to increase sales (Anderson et al., 2005; Balbach et al., 2003; Carpenter et al., 2005; Cook et al., 2003; Cummings et al., 2002; Landrine et al., 2005; Ling and Glantz, 2004; Wakefield et al., 2002; Wayne and Connolly, 2002). Pre- and postmarket studies should show that perceptions of MRTPs do not cause consumers to increase use of harmful tobacco products or lead to dual use of MRTPs and conventional tobacco products.

SCIENTIFIC STANDARDS FOR STUDIES ON RISK PERCEPTION AND RISK COMMUNICATION

Study Questions to Address the Risk Perceptions of Modified Risk Tobacco Products

With reference to each MRTP, it will be important to identify consumers' perceptions of disease risk, likelihood of addiction, likelihood of reducing or increasing others' exposure to potentially hazardous compounds (e.g., secondhand smoke), and perceptions of risk compared to other products that are already on the market. Perceptions of general harm, such as overall risk of harm or addiction, as well as perceptions of specific harm, such as risk of lung cancer or heart disease, should be studied. It is also important to establish consumers' intentions of using the product, both for consumers who do and do not currently use any other tobacco product. Of particular importance are adolescents' perceptions of the risks and benefits of using the product, and whether they intend to initiate tobacco use with the MRTP rather than a traditional tobacco product because they believe the latter is a "safe" alternative. These issues should be addressed in both pre- and postmarket studies.

Studies of risk perception should also include comprehensive questions that address the many aspects of risk perceptions, including areas which researchers may ordinarily regard as self-evident. For example, it is important to include questions about perceived risks of secondhand smoke to nonusers for all MRTPs, regardless if the product is inhaled or non-inhaled. Such a comprehensive approach will allow researchers and regulators to better understand all components of perceived risk reduction. In addition, longitudinal postmarket studies should address whether differences in perceptions and/or intentions among different age, racial, socioeconomic status, or education groups predict later product use, change in product use, or progression to dual use of MRTPs and traditional tobacco products.

Research Designs

This section outlines the committee's review of research designs for use in pre- and postmarket studies of consumer perceptions of MRTPs. The focus of the discussion is on specific issues related to ethical procedures, target population selection and recruitment, construct measurement, and analysis.

To determine perceptions of MRTPs, as well as whether such perceptions influence tobacco use behavior, studies will need to occur both pre- and postmarket for each MRTP. Premarket research will play an essential role in developing the messages that the tobacco industry can use to communicate information about the MRTPs to consumers. This research will determine consumers' ability to accurately understand messages that communicate information about the risks, benefits, and conditions of use pertaining to the MRTP itself and compared to existing tobacco products. Studies should also test how these messages influence consumers' perceptions of the risks, benefits, and likelihood of addiction related to the MRTP. Clearly, no message developed can result in any significant misunderstanding, misinterpretation, or generalization of what exactly the MRTP is supposedly modifying. For example, if the tobacco company claims that the product contains less nicotine, then the consumer or potential consumer cannot believe that the product also reduces the risk of lung cancer. Thus, the perceived influence of the new product on health and other outcomes should match the actual difference in health effects.

The first stage of premarket research will involve formative work using focus groups. Focus groups are useful for offering depth and insight from similar groups of people, especially when the intent is to gather general themes and ideas on topics not yet well studied. Focus groups are particularly useful when no existing research can provide the information, and they are an ideal way to generate new ideas that will be relevant for subsequent larger-scale studies, surveys, and future research (Krueger, 2000). These focus groups should consist of the target populations described below. The first phase of focus group research should include discussions with various groups of individuals regarding the best, most effective, and most comprehensible messaging that should be used to market and to label the product if the product is later approved as an MRTP. That is, what is the most accurate and easily comprehended message? The second phase should include discussions with groups of similar individuals to assess how the messages that were developed in phase 1 are received by consumers. Specifically, do potential consumers understand the messages correctly? Do the messages change intentions to use this MRTP or any other tobacco product?

Once messages that communicate potential risks and benefits of use are developed using the focus groups, the effects of these messages on consumer perceptions should be tested. Statements to be tested should include not only product labels or inserts intended to convey health information about the product, but also marketing statements that will appear on any form of advertisement of the MRTP. Nonverbal messages should be tested as well. For example, when banned from using labels such as "light" or "mild" on cigarette packages in countries other than the United States, the industry switched to using lighter colors to indicate "lighter" cigarettes. As a result, smokers perceived cigarettes in the lighter colored packs to be less harmful and easier to quit (Hammond and Parkinson, 2009; Hammond et al., 2009); this phenomenon has been replicated in a recent U.S. study as well (Bansal-Travers et al., 2011). Therefore, if the industry decides to use imagery, color-coding, or any other visual (but nonverbal) means of conveying information about the MRTP, then they should also test the influence of this type of messaging on consumer perceptions in pre- and postmarket studies, as well as its influence on use of the MRTP in postmarket studies.

The minimum standards to test consumer perceptions and understanding of messages about MRTPs include showing these messages to participants in randomized order and then evaluating participants' understanding of the messages and health outcomes affected by the message (see later section on measuring risk communication) and their subsequent perceptions of the product in terms of its potential risks, benefits, and likelihood of being addictive (see later section on measuring perceptions). Techniques such as eye-tracking could also be employed by researchers to study how research participants react to and understand warning labels, texts, or advertisements. It will be important to compare consumers' perceptions of the MRTP to selected comparison products that are currently on the market, using experimental designs. Additionally, it will be informative to investigate how perceptions are linked to product use by the consumer. The relevance of behavioral economic self-administration studies in evaluating the reinforcement potency of a product is discussed in Chapter 4.

It will also be important to test consumers' intentions to use the MRTP in general and compared to current products on the market (see later section on measuring intentions). That is, given information about a specific MRTP, questions to be investigated include (1) do participants plan to start using tobacco for the first time by using the MRTP, (2) do they intend to use it to help them quit smoking regular cigarettes or other traditionally available tobacco products, (3) do they intend to use both products concurrently, or (4) do they not intend to use the MRTP at all?

The studies required by FDA for products applying to switch from a prescription to over-the-counter (OTC) product may be useful in setting standards for studies on risk perceptions and risk communication. Under this requirement, prescription drug sponsors must conduct labeling comprehension studies to provide data on how the candidate OTC product label can inform the consumer about the product, including how the consumer can understand and apply the information presented on the drug label. The product itself does not need to be administered to the research participants. Although label comprehension studies may not fully predict consumer behavior once a prescription drug reaches the market as an OTC product, they can assist in creating a label that communicates effectively. The committee believes that the standards for the label comprehension studies required for a prescription-to-OTC switch can be useful in the regulation of MRTPs.

After the product has been approved by the FDA as an MRTP and released for general sale, it is vital to continue monitoring consumer perceptions and behavior related to that product via ongoing postmarket research. Conducting nationally representative cohort-sequential longitudinal surveys a minimum of three times per year (every 4 months) will be useful, with longitudinal studies ongoing until sufficient time has passed to be able to observe changes in tobacco use patterns. The longitudinal aspect of the design will allow researchers to track changes in consumer perceptions and intentions over time; it will also determine how these perceptions influence subsequent usage of the new MRTP, initiation of other tobacco use, and changes in overall patterns of tobacco usage. The cohort-sequential aspect of the design will allow researchers to control for historical or age effects that may affect real and perceived outcomes (e.g., effects on perceived health risks, addiction risks, and actual usage). How long a particular cohort should be followed depends on the age group of the cohort. Ideally, children and adolescents should be followed at least through young adulthood (e.g., age 25) because this is the period in which most people begin to use tobacco products. Adults who begin the survey after age 25 may be followed for a shorter period of time, perhaps 3 to 5 years. The next section will provide more information on participants and sampling.

Populations to Be Studied

In a preceding section, a number of questions about consumer perceptions that should be addressed were outlined. Each of these questions should be asked and answered across a variety of important study populations.

Based on the scientific literature discussed earlier in this chapter, perceptions of MRTPs, including interpretation of marketing and health messages regarding particular MRTPs, and whether such perceptions influence changes in tobacco use, are likely to differ depending on whether or not consumers are current tobacco users, and whether or not current users desire to quit. Therefore, perceptions should be studied among people who have never used a tobacco product; people who have used any tobacco product in the past, but not currently; people who currently use a tobacco product and do not intend to quit; and people who currently use a tobacco product and do intend to quit, either with or without the use of NRT or other approved smoking cessation aids. Assessment of tobacco use with items from previously validated measures and surveys is standard. A list of sample items and their sources are listed in Box 5-1. Tobacco use should be assessed for each category of tobacco product separately: cigarettes, cigars, chewing tobacco, snuff, and pipe tobacco.

Among tobacco users, level of nicotine dependence should also be assessed and included as a potential predictor of differential perceptions toward the MRTP. Levels of nicotine dependence can be investigated by employing widely used and well-validated measures of nicotine dependence, such as the Fagerstrom Test for Nicotine Dependence and the Hooked on Nicotine Checklist (HONC) (DiFranza et al., 2002; Heatherton

BOX 5-1
Sample Items to Assess Tobacco Use

- Have you smoked at least 100 cigarettes in your entire life? (CDC, 2010)
- Do you now smoke cigarettes every day, some days, or not at all? (CDC, 2010)
- How long has it been since you last smoked part or all of a cigarette? (during the past 30 days, more than 30 days ago but within the past 12 months, more than 12 months ago but within the past 3 years, more than 3 years ago) (SAMHSA, 2009)
- What is your best estimate of the number of days you smoked part or all of a cigarette during the past 30 days? (1 or 2 days, 3 to 5 days, 6 to 9 days, 10 to 19 days, 20 to 29 days, all 30 days) (SAMHSA, 2009)
- How many cigarettes per day do you smoke? (Heatherton et al., 1991)
- During your entire life, about how many times have you smoked a few puffs of a cigarette? (Lee and Halpern-Felsher, 2011; Song et al., 2009a)
- During your entire life, about how many times have you smoked a whole cigarette? (Lee and Halpern-Felsher, 2011; Song et al., 2009a)
- Have you ever smoked part or all of any type of cigar (including big cigars, cigarillos, or even little cigars that look like cigarettes)? (SAMHSA, 2009)
- Have you ever smoked tobacco in a pipe, even once? (SAMHSA, 2009)
- Have you ever used chewing tobacco, even once? (SAMHSA, 2009)
- Have you ever used snuff, even once? (SAMHSA, 2009)

et al., 1991). Additional measures of nicotine dependence include the nicotine dependence criteria of the *Diagnostic and Statistical Manual of Mental Disorders*, 4th edition (DSM-IV), as well as the Nicotine Dependence Syndrome Scale, Minnesota Withdrawal Scale, and Shiffman-Jarvik Withdrawal Scale (American Psychiatric Association, 2000; Hughes and Hatsukami, 1986; Shiffman and Jarvik, 1976; Shiffman et al., 2004). Many of these measures, such as the HONC, DSM-IV measures, and the modified Fagerstrom Tolerance Questionnaire, were developed for or have been adapted for use among adolescents (Prokhorov et al., 1996). At present, the only known reliable measures of nicotine dependence for smokeless tobacco use are the HONC and the Autonomy Over Smoking Scale, and they have only been tested for reliability in adolescents (DiFranza et al., 2011).

Smoking behavior can be characterized through an assessment of the frequency, timing, and duration of prior quit attempts; this should be incorporated into the minimum standards. Having experienced an unsuccessful quit attempt versus never having tried to stop smoking may differentially influence smokers' perceptions of an MRTP, and thus have an impact on their responses to marketing messages and subsequent product use. For example, tobacco users who were unsuccessful in their quit attempts may perceive an MRTP as a potential cessation aid, even if the product is not marketed as such, which would have important implications for use. Alternately, this type of smoker may believe he or she will never be able to quit using tobacco, and therefore view the MRTP as an option to continue using tobacco with less risk. Having experienced more than one failed attempt at smoking cessation may serve to solidify any beliefs smokers may have about their likelihood of success in the future, and affect their perceptions of the MRTP and their behavior accordingly. Finally, quit attempts made more recently may have a stronger effect on perceptions and behavioral outcomes than those made in the more distant past because of the salience of the event (see Box 5-2 for sample questions to assess prior cessation attempts).

Perceptions of and intentions to use a given MRTP are also likely to differ by age group. Thus, it is critical that studies include participants in the following age groups: children (≤ 12 years old), adolescents (13–17 years old), young or emerging adults (18–25 years old), and adults (≥ 25 years old). Studies should compare perceptions, intentions, and actual tobacco use patterns within and across the age groups.

Research has shown that tobacco use and perceptions of tobacco-related risks/benefits may also differ by race and ethnicity, thus placing certain ethnic groups at increased risk for tobacco use and subsequent disease. Evaluation of differences in perceptions by racial or ethnic categories is standard in all studies of consumer perception. The basic racial or ethnic categories recommended by the IOM Subcommittee on Standardized Collection of Race/Ethnicity Data for Healthcare Quality Improvement is appropriate for these studies: Asian, Black or African American, Native Hawaiian or Other Pacific Islander, White, American Indian or Alaska Native, and Hispanic or Latino, plus additional categories for Other and Two or More Races (IOM, 2009).

BOX 5-2
Sequence of Questions to Assess Prior Smoking Cessation Attempts

- How many times in the past have you made a serious attempt to quit smoking?
- What was the longest period of time that you were able to quit smoking?
- When was your most recent serious attempt to quit smoking?
- How long were you able to quit smoking during your most recent quit attempt?

SOURCE: Abrams et al. (2003).

Studies also show that individuals with low socioeconomic status are more likely to use tobacco and carry a disproportionate amount of the health burden associated with tobacco use. As a result, it is imperative that the potential influence of socioeconomic status on consumer perceptions and use of MRTPs is understood. The most recent reported estimates of family income and poverty thresholds published by the U.S. Census Bureau can help researchers to understand the influence of socioeconomic status.[2]

Finally, numerous studies show that individuals with less education are more likely to use tobacco; thus, they are more likely to suffer the health consequences of tobacco use. Researchers investigating these tobacco products should evaluate potential differences in consumer perceptions of MRTPs by level of education. For studies on consumer perceptions of MRTPs, it is standard to include the use of the Substance Abuse and Mental Health Services Administration education categories: less than high school, high school graduate, some college, and college graduate. Adding a category for individuals who have completed graduate school will strengthen these studies.

Participant Recruitment

Study participants should be recruited such that there are satisfactory numbers of participants falling into each tobacco use, age, racial/ethnic, socioeconomic status, or educational category described above. A sample size will be considered satisfactory based on a priori statistical power analyses to ensure that the sample adequately reflects the demographic characteristics of the population of interest. Study participants should not have any affiliation with the tobacco industry, the FDA, or any tobacco control agency.

For focus groups and experiments, the samples should be drawn from multiple sites across the United States because of the regional differences in tobacco use and exposure to pro- and antitobacco marketing and campaigns. Each focus group should contain 8–12 participants, with participants within a given focus group having similar age, racial/ethnic, socioeconomic status, or educational category described above. Multiple focus groups should be conducted, each one representing different demographic characteristics, to ensure the results are generalizable across each group. For experimental designs, participants should be randomly assigned to each group, with numbers of each demographic group equally represented. For surveys, the samples should be nationally representative; however, certain groups may be oversampled because of low prevalence

[2] These estimates can be found on the U.S. Census Bureau website: http://www.census.gov/hhes/www/poverty/methods/definitions.html.

rates in the general population, such as minority racial or ethnic groups. Participants should be recruited for surveys using the random digit dialing method.

Measurement

Specific information on measuring tobacco use and sample demographic characteristics was discussed above. Here details on measuring perceptions, risk communication, and tobacco use intentions are provided.

Perceptions

Inclusion of conditional risk assessments is standard for evaluations of participants' perceptions of risks, benefits, and likelihood of addiction associated with a given MRTP. This type of risk assessment uses scenarios to explicitly place the outcomes under investigation in the context of a specific behavior. Previous research shows that conditional risk assessments more closely reflect health risk behavior models and are stronger predictors of behavior than unconditional risk assessments, which do not place outcomes in a precise behavioral context (Halpern-Felsher et al., 2001; Ronis, 1992; Van Der Velde et al., 1996). As an example, for evaluating short- and long-term risks and benefits, the committee suggests using a conditional risk scenario such as the following: "Imagine that you just began smoking. You smoke about 2 or 3 cigarettes each day. Sometimes you smoke alone, and sometimes you smoke with friends. If you smoke about 2 or 3 cigarettes each day, what is the chance that . . . ?" (Halpern-Felsher et al., 2004). The second scenario for evaluating long-term risks can include "Imagine that you continue to smoke about 2 or 3 cigarettes each day for the rest of your life. What is the chance that . . . ?" (Halpern-Felsher et al., 2004).

After reading the scenarios, research participants should then be evaluated on their perceptions of relevant outcomes occurring to them, others, and to smokers and nonsmokers. Research participants can be asked about perceptions of general harm or specific harm (such as lung cancer, heart attack, etc.). The committee believes that researchers should ask about specific tobacco-related outcomes (rather than more general perceptions of harm), given that generalized outcomes can be vague, can lead to misperceptions, or can produce results that are difficult to interpret. Inquiring about specific outcomes reduces misinterpretation of the questions, and allows the investigator to determine the domains in which the misperceptions of the MRTP are most likely to occur.

In addition to considering the types of risks to assess, it is important to utilize the best response set for each question. Perceptions can be assessed using a number of scales, including likelihood scales (such as: "what is the likelihood or chance that [a specific outcome] will occur?"), log linear scales, lexical scales (such as: "very likely to not very likely" or "small chance to large chance"), comparative scales (such as: "compared to [another product], is this MRTP more or less likely to cause [a specific outcome]"). The committee supports the use of likelihood estimates assessed through numerical scales (such as: "please estimate the chance that [a specific outcome] will occur using any number between 0 percent and 100 percent") or comparative risk assessments (Biehl and Halpern-Felsher, 2001; Halpern-Felsher et al., 2004).

Risk Communication

Numeracy, or the ability to understand and use numbers, is very low in the general population (Gigerenzer et al., 2007; Reyna et al., 2009). As a result, a large proportion of the population, including health professionals and educated laypeople, has difficulty comprehending numerical information about risks and benefits to health (Gigerenzer et al., 2007; Reyna et al., 2009). It is important that the tobacco product sponsor communicates the risks and benefits associated with a given MRTP accurately and in a way that the general population can clearly understand. Thus, it is essential that the product sponsor carefully crafts messages about the risks and benefits of any MRTP and then demonstrates through rigorous testing that people correctly understand and interpret such messages.

There is a significant public health interest in ensuring that consumers accurately understand the risks if they use the MRTP. This includes understanding their increased level of risk if they are not current tobacco users, or their presumed decreased level of risk if they are already tobacco users (this requires comparisons of risk between the MRTP and specific tobacco products that are currently on the market). It also includes understanding changes in specific types of risk, such as risk of carbon monoxide exposure, risk of heart attack, or risk of specific types of cancer. In addition, consumers should comprehend what conditions of use are associated with the stated risks and benefits of the product (e.g., daily use, hourly use, or as indicated on package). Finally, consumers should be able to understand how these risks and benefits relate to groups of people similar to themselves. For example, what are the risks for female, African-American adolescents? Conveying such information of course assumes the tobacco product sponsor has already completed the appropriate and scientifically sound research that will allow it to make claims about the risks and benefits associated with using the MRTP under specific conditions of use and across a variety of relevant populations.

Research indicates that the best way to promote an accurate understanding of risk is to report absolute rather than relative risks (Gigerenzer et al., 2007). For instance, it is better to state that 5 in 100 people will develop shortness of breath when using the MRTP as compared to 10 in 100 people who smoke a traditional cigarette, than to state that there is a 50 percent reduction in risk for shortness of breath when using the MRTP compared to smoking a traditional cigarette.

Based on studies of exposure and risk, the industry should first generate statements that communicate the risks and/or benefits of using the MRTP and include the following elements:

- Use statements of absolute rather than relative risk.
- Clearly state what type of risk and outcome is being addressed.
- Clearly state under what conditions of use the risks/benefits are incurred.
- Clearly state what comparison is being made (i.e., among which alternative products).
- Clearly state what populations incur the risks/benefits (e.g., people who do versus do not use tobacco, males versus females, certain age or racial/ethnic groups).

Once these statements of risk are generated, they should be presented to research participants, and the participants' understanding of the statements should be assessed using the research designs discussed earlier.

Intentions

Several types of intentions to engage in MRTP use should be assessed, including intent to try the new product, intent to use the MRTP to aid in tobacco use cessation (related to intent to quit tobacco use), and intent to use the MRTP while continuing to use current tobacco product(s). The general format of the questions may include, but is not limited to the sample questions in Box 5-3.

Outcome Expectancies

In addition to the methods for assessing risk perceptions outlined above, other factors that may relate to the likelihood of trying or adopting use of an MRTP should be considered. A number of research groups have examined outcome expectancies as predictors of both smoking uptake and relapse after cessation. Expectancies are a class of attitude, formed from previous knowledge, beliefs, and experiences, that serve to guide behavior (Del Boca et al., 2002). The most widely used measure for ciga-

BOX 5-3
Sample Questions for Measuring Intentions to Use an MRTP

- What is the chance that you will try [the MRTP] sometime in the next 6 months?
- What is the chance that you will try [the MRTP] in your life?
- What is the chance that you will ever use [the MRTP]?
- What is the chance that you will use [the MRTP] to help you quit smoking cigarettes/chewing tobacco/etc.?
- What is the chance that you will use [the MRTP] in addition to other tobacco products that you already use?
- If one of your best friends were to offer you [the MRTP] in the next 6 months, would you use it? (for adolescents)

rette expectancies has been the Smoking Consequences Questionnaire (Brandon and Baker, 1991) and its various derivatives for adults, adolescents, and children (Copeland et al., 1995; Lewis-Esquerre et al., 2005; Rash and Copeland, 2008). Broadly, expectancies can be divided into positive outcomes (i.e., anticipated benefits) and negative outcomes (i.e., anticipated harms). Wetter and colleagues (1994) established that positive expectancies (positive reinforcement, negative reinforcement, and appetite-weight control) predicted withdrawal severity while negative expectancies predicted cessation success. Studies in adolescent populations have shown outcome expectancies (those relating to negative affect relief in particular) are related to smoking uptake, behavior, and nicotine dependence (Heinz et al., 2010; Wahl et al., 2005). One study that examined expectancies in relation to modified tobacco products showed that positive expectancies predicted interest in trying both Quest and Eclipse, regardless of level of smoking experience (O'Connor et al., 2007). The committee suggests that studies of MRTP perceptions include a measure of outcome expectancies.

Affective Responses

Evidence that has emerged over the past decade points to the importance of affect in shaping decisions about a wide array of health behaviors, including tobacco use (Keer et al., 2010; Kiviniemi et al., 2007; Lawton et al., 2009). While judgment and decision making have most widely been regarded as rational processes, accumulating evidence suggests an important role for affective processes and emotions in guiding decisions, primarily through heuristics (Greifeneder et al., 2011; Slovic et al., 2005). According to this model, affective reactions to stimuli, which are often automatic, can become salient in guiding decisions based on individual and situational conditions, particularly those requiring complex analysis or under time pressure. Broadly speaking, activities viewed positively are seen as low risk/high benefit, whereas those viewed negatively are seen as high risk/low benefit. Other evidence suggests that messages that evoke emotional responses are better remembered (Lang and Dhillon, 1995) and promote higher order cognitive processing (Donohew et al., 1998; Keller and Block, 1996). Thus, affective heuristics, and emotional factors more broadly, can be important to consider in user and nonuser reactions to MRTPs. A number of measures have been developed to assess affective reactions. Some measures are scales that ask participants to rate their feelings by responding to descriptive statements or words along unipolar or bipolar (semantic differential) axes, either numeric or visual-analog. Validated clinical measures such as the Profile of Mood States or Positive and Negative Affect Schedule can also be employed to measure current feelings among participants (McNair et al., 1971; Watson et al., 1988). Affect can also be measured using pictograms to assess affect valence, arousal, and dominance brought about by a particular stimulus (Bradley and Lang, 1994). This measure has been validated against the International Affective Picture System (IAPS) (Lang et al., 1997). The IAPS images cover five domains: pleasant-aroused, pleasant-calm, neutral, unpleasant-calm, and unpleasant-aroused. The committee suggests that studies examining affective reactions to MRTP-related stimuli (e.g., advertising, packaging, marketing) include a set of IAPS images from each domain for comparative purposes.

Consistent with the importance of affect and outcome expectancies, Wakefield and colleagues have been working to evaluate youth reactions to smoking messaging (Wakefield et al., 2003, 2005). They have noted five key considerations to understand ad impact and facilitate comparison of different ads: (1) previous exposure and reactions to general antismoking information and to test ads, (2) comprehension, (3) specific ad appraisals, (4) relative utility of target ad compared to generic antismoking information, and (5) recall of the test ad within 1 week. The proposed metrics are broadly applicable across media types (e.g., video, print, Internet) and include both cognitive and emotional responses. A sample questionnaire used by the Wakefield et al. research team to assess youth responses to anti-smoking ads is provided in Figure 5-1.

Analyses

The industry should hire independent, professional biostatisticians to aid in initial measurement design and all analyses following completion of data collection. The biostatisticians should conduct a priori power

What is the MAIN point that this ad is trying to make?

What ELSE is it trying to say?

How well do the following phrases describe this ad? **(Circle one number for each phrase)**

This ad....	**Strongly Disagree**	**Disagree**	**Neither Disagree nor Agree**	**Agree**	**Strongly Agree**
...was clear	1	2	3	4	5
...had a message that is important to me	1	2	3	4	5
...said things that were hard to believe	1	2	3	4	5
...made me stop and think	1	2	3	4	5
...made me curious to know if what the ad says is true	1	2	3	4	5
...is one that I would talk to other people about	1	2	3	4	5
...told me something new	1	2	3	4	5
...talked down to me	1	2	3	4	5
This ad made me feel...					
...sad	1	2	3	4	5
...angry	1	2	3	4	5
...happy	1	2	3	4	5
...scared	1	2	3	4	5
This ad was...					
...funny	1	2	3	4	5
...powerful	1	2	3	4	5
...boring	1	2	3	4	5
...emotional	1	2	3	4	5
Overall, I thought this ad was a very good anti-smoking advertisement...					
	1	2	3	4	5

What makes it that way?

Have you seen this ad on TV before today?

Yes No Not Sure

FIGURE 5-1 Sample advertisement rating questions.
SOURCE: Wakefield et al., "Assessment of youth responses to anti-smoking ads: Description of a research protocol," Practice Ad Coding Scheme. Chicago: ImpacTeen, 2002. Reprinted with permission.

analyses for all studies in order to determine appropriate sample size and level of acceptable power.

In general, the analysis of the focus group data should involve a continuum from the raw data to descriptive statements to interpretation. Analysis of the data should occur in three steps: (1) identification of participants' concepts, (2) organization of participants' concepts into a hierarchy (a model), and (3) quantitative analysis of frequency of participants' concepts. Analyses should first identify concepts (e.g., health) used by participants in the course of the interview using a variety of techniques drawn from grounded theory (Strauss and Corbin, 1997). Next, the identified concepts should be organized into a hierarchy, making use of diagrams and other comparative analytic techniques. Once a hierarchy of participants' concepts is completed, the entire dataset should be coded for participants' concepts using an appropriate software package, such as NVivo (NUD*IST). Participant concept data should then be exported into a database, allowing for analyses of whether frequency of participant concepts varies by individual-level characteristics, and so on.

The issues discussed in this chapter are relevant for the interpretation of data generated from scientific studies and for the evaluation of product applications at the FDA. The next chapter discusses the cross-cutting issues presented in this chapter as well as earlier chapters. The next chapter will also focus on methods to integrate information, and present the committee's findings and recommendations.

REFERENCES

Abrams, D. B., R. Niaura, R. A. Brown, K. M. Emmons, M. G. Goldstein, P. M. Monti, and L. A. Linnan. 2003. *The tobacco dependence treatment handbook: A guide to best practices*. New York, NY: The Guilford Press.

Ajzen, I. 1985. From intentions to actions: A theory of planned behavior. In *Action control, from cognition to behavior*, edited by J. Kuhl, and J. Beckmann. Berlin: Springer-Verlag.

American Psychiatric Association. 2000. *Diagnostic and statistical manual of mental disorders: DSM-IV-TR*. 4th ed. Washington, DC: American Psychiatric Publishing, Inc.

Anderson, S. J., S. A. Glantz, and P. Ling. 2005. Emotions for sale: Cigarette advertising and women's psychosocial needs. *Tobacco Control* 14(2):127.

Arnett, J. J. 2000. Optimistic bias in adolescent and adult smokers and nonsmokers. *Addictive Behaviors* 25(4):625-632.

Balbach, E. D., R. J. Gasior, and E. M. Barbeau. 2003. R. J. Reynolds' targeting of African Americans: 1988-2000. *American Journal of Public Health* 93(5):822-827.

Bandura, A. 2001. Social cognitive theory of mass communication. *Media Psychology* 3(3):265-299.

Bansal-Travers, M., R. O'Connor, B. V. Fix, and K. M. Cummings. 2011. What do cigarette pack colors communicate to smokers in the U.S.? *American Journal of Preventive Medicine* 40(6):683-689.

Biehl, M., and B. L. Halpern-Felsher. 2001. Adolescents' and adults' understanding of probability expressions. *Journal of Adolescent Health* 28(1):30-35.

Borland, R. O. N. 1997. Tobacco health warnings and smoking-related cognitions and behaviours. *Addiction* 92(11):1427-1435.

Bradley, M. M., and P. J. Lang. 1994. Measuring emotion: The self-assessment manikin and the semantic differential. *Journal of Behavior Therapy and Experimental Psychiatry* 25(1):49-59.

Brandon, T. H., and T. B. Baker. 1991. The smoking consequences questionnaire: The subjective expected utility of smoking in college students. *Psychological Assessment: A Journal of Consulting and Clinical Psychology* 3(3):484.

Callery, W. E., D. Hammond, R. J. O'Connor, and G. T. Fong. 2011. The appeal of smokeless tobacco products among young Canadian smokers: The impact of pictorial health warnings and relative risk messages. *Nicotine & Tobacco Research* 13(5):373-383.

Carpenter, C. M., G. F. Wayne, and G. N. Connolly. 2005. Designing cigarettes for women: New findings from the tobacco industry documents. *Addiction* 100(6):837-851.

CDC (Centers for Disease Control and Prevention). 2010. *Healthy people 2010 operational definition.* ftp://ftp.cdc.gov/pub/health_statistics/nchs/datasets/data2010/focusarea27/o2701a.pdf (accessed October 5, 2011).

Clark, V. L. P., D. L. Miller, J. W. Creswell, K. McVea, R. McEntarffer, L. M. Harter, and W. T. Mickelson. 2002. In conversation: High school students talk to students about tobacco use and prevention strategies. *Qualitative Health Research* 12(9):1264.

Cook, B. L., G. F. Wayne, L. Keithly, and G. Connolly. 2003. One size does not fit all: How the tobacco industry has altered cigarette design to target consumer groups with specific psychological and psychosocial needs. *Addiction* 98(11):1547-1561.

Copeland, A. L., T. H. Brandon, and E. P. Quinn. 1995. The smoking consequences questionnaire—adult: Measurement of smoking outcome expectancies of experienced smokers. *Psychological Assessment* 7(4):484.

Cummings, K. M., C. Morley, J. Horan, C. Steger, and N. R. Leavell. 2002. Marketing to America's youth: Evidence from corporate documents. *Tobacco Control* 11(Suppl. 1):i5.

Curry, S. J., L. Grothaus, and C. McBride. 1997. Reasons for quitting: Intrinsic and extrinsic motivation for smoking cessation in a population-based sample of smokers. *Addictive Behaviors* 22(6):727-739.

Del Boca, F. K., J. Darkes, M. S. Goldman, and G. T. Smith. 2002. Advancing the expectancy concept via the interplay between theory and research. *Alcoholism: Clinical and Experimental Research* 26(6):926-935.

DiFranza, J. R., J. A. Savageau, K. Fletcher, J. K. Ockene, N. A. Rigotti, A. D. McNeill, M. Coleman, and C. Wood. 2002. Measuring the loss of autonomy over nicotine use in adolescents: The DANDY (Development and Assessment of Nicotine Dependence in Youths) study. *Archives of Pediatrics and Adolescent Medicine* 156(4):397.

DiFranza, J. R., M. Sweet, J. A. Savageau, and W. Ursprung. 2011. The assessment of tobacco dependence in young users of smokeless tobacco. *Tobacco Control.* Published Online June 28, 2011.

Donohew, L., E. P. Lorch, and P. Palmgreen. 1998. Applications of a theoretic model of information exposure to health interventions. *Human Communication Research* 24(3):454-468.

Dunn, D. A., and J. L. Johnson. 2001. Choosing to remain smoke-free: The experiences of adolescent girls. *Journal of Adolescent Health* 29(4):289-297.

Etter, J.-F., L. T. Kozlowski, and T. V. Perneger. 2003. What smokers believe about light and ultralight cigarettes. *Preventive Medicine* 36(1):92-98.

Fishbein, M., and I. Ajzen. 1975. *Belief, attitude, intention, and behavior: An introduction to theory and research.* Reading, MA: Addision-Wesley.

Gerrard, M., F. X. Gibbons, A. E. Houlihan, M. L. Stock, and E. A. Pomery. 2008. A dual-process approach to health risk decision making: The prototype willingness model. *Developmental Review* 28(1):29-61.

Gigerenzer, G., W. Gaissmaier, E. Kurz-Milcke, L. M. Schwartz, and S. Woloshin. 2007. Helping doctors and patients make sense of health statistics. *Psychological Science in the Public Interest* 8(2):53-96.

Gittelsohn, J., K. M. Roche, C. S. Alexander, and P. Tassler. 2001. The social context of smoking among African-American and white adolescents in Baltimore City. *Ethnicity and Health* 6(3-4):211-225.

Glantz, S. A., and P. Jamieson. 2000. Attitudes toward secondhand smoke, smoking, and quitting among young people. *Pediatrics* 106(6):e82.

Goldberg, J. H., B. L. Halpern-Felsher, and S. G. Millstein. 2002. Beyond invulnerability: The importance of benefits in adolescents' decision to drink alcohol. *Health Psychology* 21(5):477.

Greifeneder, R., H. Bless, and M. T. Pham. 2011. When do people rely on affective and cognitive feelings in judgment? A review. *Personality and Social Psychology Review* 15(2):107.

Gritz, E. R., A. V. Prokhorov, K. S. Hudmon, M. M. Jones, C. Rosenblum, C.-C. Chang, R. M. Chamberlain, W. C. Taylor, D. Johnston, and C. de Moor. 2003. Predictors of susceptibility to smoking and ever smoking: A longitudinal study in a triethnic sample of adolescents. *Nicotine & Tobacco Research* 5(4):493-506.

Halpern-Felsher, B. L., and M. L. Rubinstein. 2005. Clear the air: Adolescents' perceptions of the risks associated with secondhand smoke. *Preventive Medicine* 41(1):16-22.

Halpern-Felsher, B. L., S. G. Millstein, J. M. Ellen, N. E. Adler, J. M. Tschann, and M. Biehl. 2001. The role of behavioral experience in judging risks. *Health Psychology* 20(2):120-126.

Halpern-Felsher, B. L., M. Biehl, R. Y. Kropp, and M. L. Rubinstein. 2004. Perceived risks and benefits of smoking: Differences among adolescents with different smoking experiences and intentions. *Preventive Medicine* 39(3):559-567.

Hammond, D., and C. Parkinson. 2009. The impact of cigarette package design on perceptions of risk. *Journal of Public Health* 31(3):345-353.

Hammond, D., M. Dockrell, D. Arnott, A. Lee, and A. McNeill. 2009. Cigarette pack design and perceptions of risk among UK adults and youth. *European Journal of Public Health* 19(6):631-637.

Hansen, W. B., and C. K. Malotte. 1986. Perceived personal immunity: The development of beliefs about susceptibility to the consequences of smoking. *Preventive Medicine* 15(4):363-372.

Heatherton, T. F., L. T. Kozlowski, and R. C. Frecker. 1991. The Fagerstrom test for nicotine dependence: A revision of the Fagerstrom tolerance questionnaire. *British Journal of Addiction* 86(9):1119-1127.

Heinz, A. J., J. D. Kassel, M. Berbaum, and R. Mermelstein. 2010. Adolescents' expectancies for smoking to regulate affect predict smoking behavior and nicotine dependence over time. *Drug and Alcohol Dependence* 111(1-2):128-135.

Hughes, J. R., and D. Hatsukami. 1986. Signs and symptoms of tobacco withdrawal. *Archives of General Psychiatry* 43(3):289-294.

IOM (Institute of Medicine). 2007. *Ending the tobacco problem: A blueprint for the nation*. Washington, DC: The National Academies Press.

IOM. 2009. *Race, ethnicity, and language data: Standardization for health care quality improvement.* Washington, DC: The National Academies Press.

Jamieson, P., and D. Romer. 2001. What do young people think they know about the risks of smoking. In *Smoking: Risk, perception & policy*, edited by P. Slovic. Thousand Oaks, CA: Sage Publications.

Johnson, J. L., J. L. Bottorff, B. Moffat, P. A. Ratner, J. A. Shoveller, and C. Y. Lovato. 2003. Tobacco dependence: Adolescents' perspectives on the need to smoke. *Social Science and Medicine* 56(7):1481-1492.

Johnson, R. J., K. D. McCaul, and W. M. Klein. 2002. Risk involvement and risk perception among adolescents and young adults. *Journal of Behavioral Medicine* 25(1):67-82.

Kanfer, F. H. 1970. *Self-regulation: Research, issues, and speculations*. Edited by Institute for Research in Clinical Psychology, C. Neuringer, and J. L. Michael, *Behavior modification in clinical psychology*. New York: Appleton-Century-Crofts.

Keer, M., B. Putte, and P. Neijens. 2010. The role of affect and cognition in health decision making. *British Journal of Social Psychology* 49(1):143-153.

Kegler, M., and V. Cleaver. 2000. The social context of experimenting with cigarettes: American Indian "start stories." *American Journal of Health Promotion* 15(2):89-92.

Keller, P. A., and L. G. Block. 1996. Increasing the persuasiveness of fear appeals: The effect of arousal and elaboration. *Journal of Consumer Research* 22(4):448-459.

Kiviniemi, M. T., A. M. Voss-Humke, and A. L. Seifert. 2007. How do I feel about the behavior? The interplay of affective associations with behaviors and cognitive beliefs as influences on physical activity behavior. *Health Psychology* 26(2):152.

Kristiansen, C. M., C. M. Harding, and J. R. Eiser. 1983. Beliefs about the relationship between smoking and causes of death. *Basic and Applied Social Psychology* 4(3):253-261.

Krueger, R. A. 2000. *Focus groups: A practical guide for applied research*. Thousand Oaks, CA: Sage Publications.

Kurtz, M. E., J. Kurtz, S. M. Johnson, and W. Cooper. 2001. Sources of information on the health effects of environmental tobacco smoke among African-American children and adolescents. *Journal of Adolescent Health* 28(6):458-464.

Landrine, H., E. A. Klonoff, S. Fernandez, N. Hickman, K. Kashima, B. Parekh, K. Thomas, C. R. Brouillard, M. Zolezzi, and J. A. Jensen. 2005. Cigarette advertising in black, latino, and white magazines, 1998-2002: An exploratory investigation. *Ethnicity and Disease* 15(1):63-67.

Lang, A., and K. Dhillon. 1995. The effects of emotional arousal and valence on television viewers' cognitive capacity and memory. *Journal of Broadcasting & Electronic Media* 39(3):313.

Lang, P., M. Bradley, and B. Cuthbert. 1997. *International affective picture system (IAPS): Technical manual and affective ratings*. Gainesville, FL: National Institute of Mental Health Center for the Study of Emotion and Attention.

Lawton, R., M. Conner, and R. McEachan. 2009. Desire or reason: Predicting health behaviors from affective and cognitive attitudes. *Health Psychology* 28(1):56.

Lee, J., and B. L. Halpern-Felsher. 2011. What does it take to be a smoker? Adolescents' characterization of different smoker types. *Nicotine & Tobacco Research* 13(11):1106-1113.

Leventhal, H., K. Glynn, and R. Fleming. 1987. Is the smoking decision an "informed choice"? *JAMA* 257(24):3373.

Lewis-Esquerre, J. M., J. R. Rodrigue, and C. W. Kahler. 2005. Development and validation of an adolescent smoking consequences questionnaire. *Nicotine & Tobacco Research* 7(1):81.

Ling, P. M., and S. A. Glantz. 2004. Tobacco industry research on smoking cessation. *Journal of General Internal Medicine* 19(5p1):419-426.

Lloyd, B., K. Lucas, and M. Fernbach. 1997. Adolescent girls' constructions of smoking identities: Implications for health promotion. *Journal of Adolescence* 20(1):43-56.

McKee, S. A., S. S. O'Malley, P. Salovey, S. Krishnan-Sarin, and C. M. Mazure. 2005. Perceived risks and benefits of smoking cessation: Gender-specific predictors of motivation and treatment outcome. *Addictive Behaviors* 30(3):423-435.

McNair, D. M., M. Lorr, and L. F. Droppleman. 1971. *Manual for the profile of mood states*. San Diego, CA: Educational and Industrial Testing Service.

Millstein, S. G., and B. L. Halpern-Felsher. 2002. Judgments about risk and perceived invulnerability in adolescents and young adults. *Journal of Research on Adolescence* 12(4):399-422.

Morrell, H. E. R., A. V. Song, and B. L. Halpern-Felsher. 2010. Predicting adolescent perceptions of the risks and benefits of cigarette smoking: A longitudinal investigation. *Health Psychology* 29(6):610-617.

Mutti, S., D. Hammond, R. Borland, M. K. Cummings, R. J. O'Connor, and G. T. Fong. 2011. Beyond light and mild: Cigarette brand descriptors and perceptions of risk in the international tobacco control (ITC) four country survey. *Addiction* 106:1166-1175.

Nichter, M., N. Vuckovic, G. Quintero, and C. Ritenbaugh. 1997. Smoking experimentation and initiation among adolescent girls: Qualitative and quantitative findings. *Tobacco Control* 6(4):285.

O'Connor, R. J., R. L. Ashare, B. V. Fix, L. W. Hawk, K. M. Cummings, and W. C. Schmidt. 2007. College students' expectancies for light cigarettes and potential reduced exposure products. *American Journal of Health Behavior* 31(4):402-410.

Øverland, S., J. Hetland, and L. E. Aarø. 2008. Relative harm of snus and cigarettes: What do Norwegian adolescents say? *Tobacco Control* 17(6):422-425.

Pallonen, U. E., J. O. Prochaska, W. F. Velicer, A. V. Prokhorov, and N. F. Smith. 1998. Stages of acquisition and cessation for adolescent smoking: An empirical integration. *Addictive Behaviors* 23(3):303-324

Perkins, K. A. 2001. Smoking cessation in women: Special considerations. *CNS Drugs* 15(5):391-411.

Prokhorov, A. V., U. E. Pallonen, J. L. Fava, L. Ding, and R. Niaura. 1996. Measuring nicotine dependence among high-risk adolescent smokers. *Addictive Behaviors* 21(1):117-127.

Prokhorov, A. V., C. A. de Moor, K. S. Hudmon, S. Hu, S. H. Kelder, and E. R. Gritz. 2002. Predicting initiation of smoking in adolescents: Evidence for integrating the stages of change and susceptibility to smoking constructs. *Addictive Behaviors* 27(5):697-712.

Rash, C. J., and A. L. Copeland. 2008. The Brief Smoking Consequences Questionnaire–Adult (BSCQ-A): Development of a short form of the SCQ-A. *Nicotine & Tobacco Research* 10(11):1633.

Reyna, V. F., W. L. Nelson, P. K. Han, and N. F. Dieckmann. 2009. How numeracy influences risk comprehension and medical decision making. *Psychological Bulletin* 135(6):943-973.

Romer, D., and P. Jamieson. 2001. Do adolescents appreciate the risks of smoking? Evidence from a national survey. *Journal of Adolescent Health* 29(1):12-21.

Ronis, D. L. 1992. Conditional health threats—health beliefs, decisions, and behaviors among adults. *Health Psychology* 11(2):127-134.

Rosenstock, I. M. 1974. The health belief model and preventive health behavior. In *The health belief model and personal health behavior*, edited by M. H. Becker. Thorofare, NJ: Slack.

Rugkåsa, J., B. Knox, J. Sittlington, O. Kennedy, M. P. Treacy, and P. S. Abaunza. 2001. Anxious adults vs. cool children: Children's views on smoking and addiction. *Social Science and Medicine* 53(5):593-602.

SAMHSA (Substance Abuse and Mental Health Services Administration). 2009. *2010 national survey on drug use and health: CAI specifications for programming*. Rockville, MD: Substance Abuse and Mental Health Services Administration.

Schoenbaum, M. 1997. Do smokers understand the mortality effects of smoking? Evidence from the Health and Retirement Survey. *American Journal of Public Health* 87(5):755-759.

Shiffman, S. M., and M. E. Jarvik. 1976. Smoking withdrawal symptoms in two weeks of abstinence. *Psychopharmacology* 50(1):35-39.

Shiffman, S., J. L. Pillitteri, S. L. Burton, J. M. Rohay, and J. G. Gitchell. 2001. Smokers' beliefs about "light" and "ultra light" cigarettes. *Tobacco Control* 10:I17-I23.

Shiffman, S., A. Waters, and M. Hickcox. 2004. The nicotine dependence syndrome scale: A multidimensional measure of nicotine dependence. *Nicotine & Tobacco Research* 6(2):327-348.

Slovic, P. 1998. Do adolescent smokers know the risks? *Duke Law Journal* 47:1133-1141.

Slovic, P. 2001. *Smoking: Risk, perception, & policy*. Thousand Oaks, CA: Sage Publications.

Slovic, P., E. Peters, M. L. Finucane, and D. G. MacGregor. 2005. Affect, risk, and decision making. *Health Psychology* 24(4, Suppl. 1):S35-S40.

Song, A. V., S. A. Glantz, and B. L. Halpern-Felsher. 2009a. Perceptions of second-hand smoke risks predict future adolescent smoking initiation. *Journal of Adolescent Health* 45(6):618-625.

Song, A. V., H. E. R. Morrell, J. L. Cornell, M. E. Ramos, M. Biehl, R. Y. Kropp, and B. L. Halpern-Felsher. 2009b. Perceptions of smoking-related risks and benefits as predictors of adolescent smoking initiation. *American Journal of Public Health* 99(3):487-492.

Sorensen, G., and T. F. Pechacek. 1987. Attitudes toward smoking cessation among men and women. *Journal of Behavioral Medicine* 10(2):129-137.

Strauss, A. L., and J. M. Corbin. 1997. *Grounded theory in practice*. Thousand Oaks, CA: Sage Publications, Inc.

Sutton, S. 1998. How ordinary people in Great Britain perceive the health risks of smoking *Journal of Epidemiology and Community Health* 52(5):338-339.

Swan, G. E., M. M. Ward, D. Carm Elli, and L. M. Jack. 1993. Differential rates of relapse in subgroups of male and female smokers. *Journal of Clinical Epidemiology* 46(9):1041-1053.

Triandis, H. C. 1977. Subjective culture and interpersonal relations across cultures. *Annals of the New York Academy of Sciences* 285(1):418-434.

Van Der Velde, F. W., C. Hooykaas, and J. Van Der Pligt. 1996. Conditional versus unconditional risk estimates in models of AIDS-related risk behaviour. *Psychology & Health* 12(1):87-100.

Virgili, M., N. Owen, and H. H. Sverson. 1991. Adolescents' smoking behavior and risk perceptions. *Journal of Substance Abuse* 3(3):315-324.

Viscusi, W. K. 1990. Do smokers underestimate risks? *Journal of Political Economy* 98(6):1253-1269.

Viscusi, W. K. 1991. Age variations in risk perceptions and smoking decisions. *The Review of Economics and Statistics* 73(4):577-588.

Viscusi, W. K. 1992. *Smoking: Making the risky decision*. New York: Oxford University Press.

Vuckovic, N., M. R. Polen, and J. F. Hollis. 2003. The problem is getting us to stop: What teens say about smoking cessation. *Preventive Medicine* 37(3):209-218.

Wahl, S. K., L. R. Turner, R. J. Mermelstein, and B. R. Flay. 2005. Adolescents' smoking expectancies: Psychometric properties and prediction of behavior change. *Nicotine & Tobacco Research* 7(4):613.

Wakefield, M., C. Morley, J. K. Horan, and K. M. Cummings. 2002. The cigarette pack as image: New evidence from tobacco industry documents. *Tobacco Control* 11(Suppl. 1):i73-i80.

Wakefield, M., R. Durrant, Y. Terry-McElrath, E. Ruel, G. Balch, S. Anderson, G. Szczypka, S. Emery, and B. Flay. 2003. Appraisal of anti-smoking advertising by youth at risk for regular smoking: A comparative study in the United States, Australia, and Britain. *Tobacco Control* 12(Suppl. 2):ii82-ii86.

Wakefield, M., G. I. Balch, E. Ruel, Y. Terry McElrath, G. Szczypka, B. Flay, S. Emery, and K. Clegg Smith. 2005. Youth responses to anti smoking advertisements from tobacco control agencies, tobacco companies, and pharmaceutical companies. *Journal of Applied Social Psychology* 35(9):1894-1910.

Watson, D., L. A. Clark, and A. Tellegen. 1988. Development and validation of brief measures of positive and negative affect: The PANAS scales. *Journal of Personality and Social Psychology* 54(6):1063-1070.

Wayne, G. F., and G. N. Connolly. 2002. How cigarette design can affect youth initiation into smoking: Camel cigarettes 1983-93. *Tobacco Control* 11(Suppl. 1):i32-i39.

Weekley, C. K. 1992. Smoking as a weight-control strategy and its relationship to smoking status. *Addictive Behaviors* 17(3):259-271.

Weinstein, N. D., P. Slovic, and G. Gibson. 2004. Accuracy and optimism in smokers' beliefs about quitting. *Nicotine & Tobacco Research* 6(Suppl. 3):S375-S380.

Weinstein, N. D., S. E. Marcus, and R. P. Moser. 2005. Smokers' unrealistic optimism about their risk. *Tobacco Control* 14(1):55-59.

Wetter, D. W., S. S. Smith, S. L. Kenford, D. E. Jorenby, M. C. Fiore, R. D. Hurt, K. P. Offord, and T. B. Baker. 1994. Smoking outcome expectancies: Factor structure, predictive validity, and discriminant validity. *Journal of Abnormal Psychology* 103(4):801.

6

Decision Making and Oversight of MRTP Studies: Findings and Recommendations

In the previous three chapters, the committee describes the evidence domains for potential modified risk tobacco products (MRTPs), including studies of health effects, addictive potential, and risk perception and communication. The committee discussed the governance of those studies in Chapter 2. The Food and Drug Administration (FDA) will have to integrate the evidence from those diverse domains when making regulatory decisions about MRTPs. Many of the same issues and concerns related to the governance and design of studies arise regardless of the evidence domain. The committee's findings and recommendations, therefore, cut across the evidence domains and focus on the types of evidence and studies that the FDA should use in making regulatory decisions about MRTPs, the design of studies of MRTPs for the FDA's decision making, and the governance of those studies. This chapter focuses on those cross-cutting issues, ways the FDA can integrate information from those studies as part of its regulatory decisions, and the committee's overarching findings and recommendations.

This chapter begins with a brief summary of the evidence domains discussed in more detail in the previous chapters, a discussion of the scientific and ethical issues in studying MRTPs, and the governance issues that accompany such studies. The committee then discusses (1) the integration of all the evidence for the FDA's regulatory decisions about these products, including mathematical modeling and simulation techniques, and (2) the comparative nature of the harm reduction claims that the FDA will be evaluating. The committee then presents its findings and recommendations.

ISSUES WITH EVIDENCE

Evidence Domains

The evidence to support the marketing authorization of an MRTP will come from studies of health effects, addictive potential, and risk perception. In this section the committee provides a brief summary of the studies within those evidence domains.

Health Effects

Laboratory analysis of the performance and of the constituents of tobacco products will be the first step in the evaluation of any new product. These analyses involve standard methods of extraction, sample cleanup, analytic identification, and quantitation. There are important limitations to laboratory analysis of product performance and composition. Laboratory analysis of constituents may not reflect constituent uptake under conditions of use. Smoking machines do not replicate human smoking conditions, and there is no proven way to replicate the many ways humans use tobacco. It is crucial, therefore, to describe the smoking regimen or other extraction methods employed.

The second step in the evaluation of MRTPs will be preclinical studies of toxicity. These assays are essential in identifying particularly risky or toxic products that should not be tested in humans, as well as products that have reasonable potential to reduce risk and, therefore, should proceed to clinical evaluation. Evaluation of products *in vitro* should precede *in vivo* assays. Although it is not possible to make laboratory animals use tobacco products the way humans do, and there are inherent interspecies differences that prevents meaningful extrapolation of human effects, it is still informative to observe the effect of tobacco products in live animal models. The number of animal studies required to characterize an MRTP preclinically could potentially be reduced by setting composition standards or limits or establishing standards for certain categories of MRTPs. Assays of toxicity in humans will also be essential, in particular assays of urinary mutagenicity and sister chromatid exchange in peripheral lymphocytes.

Biomarkers of exposure measure human exposure to constituents of tobacco and could include the constituents themselves, their metabolites, or protein (or DNA) binding products of the constituents or their metabolites. These biomarkers have the potential to bypass many of the uncertainties in product composition analysis and provide a realistic and direct assessment of carcinogen and toxicant dose in an individual. Biomarkers of exposure should be validated before their use.

When the FDA evaluates studies, it is important that it ensures the constituents of a product are accurately and precisely measured, that exposure methods are appropriate, and that any biomarkers of exposure have been validated.

Experimental designs, in particular randomized controlled trials (RCTs), provide data that can support the strong inferences about the effect of an MRTP on human health relative to conventional tobacco products. The use of appropriately designed clinical trials will be important to establish whether the use of the MRTP reduces exposure to toxicants or induces positive changes in surrogate markers. An RCT is an effective means of examining acceptability and use of the MRTP, the ability of the MRTP to increase cessation in users of conventional tobacco products, and the likelihood that MRTP availability will lead to dual use. RCT methods can also produce evidence on whether and how much individuals use an MRTP after they have used it to help them quit conventional products, changes in perception of the MRTP with its continued use, and the MRTP's ability to suppress tobacco withdrawal symptoms. It is important to recognize that no single RCT can address all of these areas, and each study should have a focused objective with a primary endpoint.

Postmarket studies of marketed MRTPs will be critical to evaluating the effect of the MRTP on both individuals and on the public's health. In particular, the prospective cohort design will be an essential tool to validating anticipated or claimed effects of marketed MRTPs. These studies can assess baseline tobacco and MRTP exposures; summarize product use as the study is ongoing, including any changes in product use habits; document and verify outcomes as they occur; and evaluate a wide variety of outcomes, including both intermediate and clinical outcomes.

In addition, other study designs will be necessary to provide evidence on the public health effects of MRTPs, including retrospective cohort studies, case-control studies, crossover or case-crossover designs, and comparative effectiveness research methods. Different study designs will be necessary depending on the circumstances and the research question.

Addictive Potential

Evaluation of the likelihood of initiation, maintenance, and persistence of use in both conventional tobacco users and nonusers is critical to estimating the public health effect of marketing MRTPs. Specifically, evaluation of the MRTP's ability to promote initiation and continuation of its regular use, switching to its use and cessation of the consumption of more harmful products, dual use, and to promote relapse back to more harmful tobacco use are all essential. All of these outcomes are related to the reinforcing value of the MRTP (that is, how rewarding it is).

There is a continuum of reinforcement value. In theory, the MRTP should be somewhat more reinforcing than nicotine replacement therapies but perhaps less reinforcing than conventional cigarettes. Ideally, an MRTP would be sufficiently reinforcing so as to attract smokers away from conventional cigarettes but not encourage the widespread dependent use of the product by individuals who were previously nonusers or who would have quit smoking. Data from all relevant measures should be integrated, taking into account not only the effects of the MRTP on important health outcomes, but also the prevalence of use and projected public health impact.

Evaluation of the abuse and addiction potential of a product can be accomplished with a variety of experimental designs and in a variety of contexts. Those include subjective evaluations in laboratory contexts, acute self-administration studies in laboratory contexts, use in extended residence facilities, and natural environment contexts where long-term use can be studied in real-world circumstances via RCTs, cross-sectional survey studies, and longitudinal cohort studies.

Evaluation of reinforcement value in a laboratory setting is particularly important because the results of these studies reliably correspond to an agent's addictive potential in real-world use. One standard with regard to human abuse liability drug testing are acute dose-effect comparison studies, because of the correspondence between subjective ratings of drug effects and real-world abuse potential. Behavioral economic self-administration studies will also be important to evaluating the reinforcement potency of a product. The usefulness of all studies in forecasting the risk for initiation and abuse of a product depends on study design factors. Important design considerations include the size of the sample, the nature of the sample (whether the sample includes heavy smokers or light smokers, smokers who want to quit, and nonsmokers), the characterization of the sample (age, sex, gender, ethnicity, educational attainment, socioeconomic status, etc.), and the nature of the comparison product.

Risk Perception and Communication

Judgments about risk, otherwise known as risk perceptions, are a fundamental element to most theoretical models of health behavior and behavioral decision making. In general, those models argue that individuals' perceptions about the value and likelihood of behavior-related positive and negative consequences and their vulnerability to those consequences play a key role in behavioral choices. As such, understanding individuals' perceptions of tobacco-related products, including MRTPs, whether such perceptions change over time, and whether such perceptions play a role in tobacco use behavior, is critical. It will be important to identify consumers' perceptions of disease risk, likelihood of addiction, likelihood of reducing or increasing others' exposure to potentially hazardous compounds, and perceptions of risk compared to other products already on the market. It is also important to assess intentions of using the product. It is essential that industry carefully crafts messages about risks and benefits of any MRTP and demonstrates through rigorous testing that people correctly understand and interpret the risks.

Studies evaluating risk perceptions and risk communication must be performed both before the marketing of an MRTP and after the MRTP has been marketed. Premarket research will play an essential role in developing the messages that the tobacco industry can use to communicate information about MRTPs to consumers. This research will determine consumers' ability to accurately understand messages that communicate information about the risks, benefits, and conditions of use pertaining to the MRTP itself and compared to existing tobacco products. Studies should also test how these messages influence consumers' perceptions of the risks, benefits, and likelihood of addiction related to the MRTP. The first stage of premarket research will involve formative work using focus groups. The second stage should include discussions with groups of similar individuals to assess how the messages that were developed in the first stage are received by consumers. Finally, the effects of these messages on consumer perceptions should be tested. It will be important to evaluate consumer understanding and to compare consumer perceptions of the MRTP to conventional products. After the product is released on the market, it is vital to continue monitoring consumer perceptions and behavior related to that product. Conducting nationally representative cohort-sequential longitudinal surveys will be essential.

Cross-Cutting Issues with Studies of MRTPs

When evaluating the studies of MRTPs, there are a number of considerations that are relevant regardless of whether a study is looking at health effects, addictive potential, or risk perception. Those cross-cutting issues, which include issues related to study design and governance, are discussed in this section.

Study Design

Issues that could affect the FDA's evaluation of a study on MRTPs include the generalizability of the study and how well the study is conducted.

Studies should be designed appropriately to create an evidence base that can support a finding of public health benefit. The ultimate goal of studying the effect of MRTPs on human health and behavior is to be able to accurately predict the public health effects of allowing an MRTP to be marketed. In other words, the ultimate goal of scientific studies it to produce generalizable data. The "generalizability" of data, or the reliability of predictions that can be made about the real world based on scientific observations, will depend on the design of the studies.

The FDA should carefully evaluate the size and nature of the sample to assess the generalizability of the study data. Sample sizes should be carefully determined and tailored to the study design and the effects studied. Statistically underpowered studies cannot support inferences or projections about the effects of a product. The nature of the study sample is critical to the usefulness of study results. Results from studies conducted in one population may not be applicable to other populations, because the characteristics that define the study population either are related to or cause the responses to the product. As such, it is important to study a wide range of populations. It is particularly important to include populations that have a high risk of using tobacco and populations that will be affected by the marketing of the product.

Study designs must also carefully consider the degree of control imposed on experimental designs. Internal and external validity should be balanced not only within studies but also across studies of the same product. Highly controlled experimental designs can eliminate many variables and confounders and support strong inferences, but simultaneously lose relevance to the FDA's decisions because the conditions of product use do not reflect real-world circumstances and behaviors. Experimental designs that are less controlled can emulate circumstances that reflect real-world conditions and behaviors, and therefore may be more relevant in predicting real-world effects, but uncontrolled variables may confound meaningful associations or inferences.

Regardless of the type of study design, the planning and conduct of the study should meet good research practice standards. As discussed earlier in this report, there are minimum standards for studies in other settings, such as studies of pharmaceuticals. Consensus statements such as Consolidated Standards of Reporting Trials (CONSORT) for clinical trials, Strengthening the Reporting of Observational Studies in Epidemiology (STROBE) for observational studies, publication criteria from the International Council of Medical Journal Editors, and reporting criteria of the International Conference on Harmonization have been implemented to ensure that the design, conduct, and reporting of studies are consistent with the state-of-the-art scientific standards. Studies of MRTPs should meet those or similar criteria to help ensure the overall quality and integrity of the studies. Certain types of studies will have specific criteria to ensure quality research. Some of those specific criteria were discussed in the preceding chapters.

Governance

Research on MRTPs will require some oversight or governance to ensure the research is free from bias and conflicts of interest and that appropriate controls are in place for human subject research. Such governance will also ensure the disclosure of data to ensure transparency and instill confidence in the research findings.

The tobacco industry has a history of hiding and misrepresenting information about the risks of tobacco products (Cummings, 2003; Cummings et al., 2002, 2007).[1] This history has lead to profound public distrust in both the tobacco industry and in the research it sponsors, and the absence of governance in the tobacco industry has created an isolated industry that lacks not only the expertise to produce the necessary range of credible and reliable data, but also the trustworthiness to acquire external expertise and avenues to disseminate acquired data (Ashley and Cohen, 2003; Harris Interactive, 2003, 2010; NCI, 2008). The production of reliable and credible data depends upon building rigor, oversight, and independence into the entire research process. Data problems often cannot be detected after study completion, and therefore integrity and accountability need to be built into the research throughout the study's execution. There is neither an established set of regulatory practices for the review of MRTPs nor an established set of federal research standards for the design, conduct, analysis, monitoring, and completion of studies in support of MRTPs.

There are also a number of ethical issues associated with conducting human subjects research involving MRTPs. The first issue is the risk of conducting clinical trials of MRTPs or other tobacco products in populations with a high risk for tobacco initiation and addiction, including but not limited to adolescents, certain ethnic minorities, and individuals with mental health disorders. Randomization of participants to a product known to be potentially addictive and hazardous is ethically problematic. The second issue is the risk of improperly disclosing the substance abuse of a minor to the minor's parents or guardians in the process of obtaining parental consent for research. While the assent of minors is always necessary, investigators should also be aware of circumstances where waiver of parental consent is warranted because obtaining parental consent will violate the confidence of the study participant. The third issue is the inclusion of individuals from high-risk groups with reduced decision-making capacity. Some populations at a high risk for tobacco use, such as

[1] The history of research conducted, funded, or supported by the tobacco industry is not raised to be retributive or punitive, but simply to acknowledge that past actions reflect on the credibility of the industry's current research, which may pose a problem for regulators, particularly in the contentious area of MRTPs.

adolescents and populations with mental health issues, may have a higher prevalence of individuals with reduced decision-making capacity. When investigators are conducting research involving these high-risk groups, they should be particularly cautious about the inclusion of individuals who lack the capacity to provide meaningful consent.

The tobacco industry is currently limited in its ability to produce credible and comprehensive data. The challenges created by that lack of credibility are augmented by the fact that it is inevitable that product sponsors will need to collect extensive data on the effect of products in populations vulnerable to high use rates. Because of those challenges, at least part of the research base in support of an MRTP may need to be generated by researchers and organizations independent of the sponsors of the MRTP in question (Rees et al., 2009a, 2009b). The creation of a third party or third parties for the conduct and oversight of studies of MRTPs could help overcome some of the issues discussed above. The Health Effects Institute (HEI), a nonprofit corporation with approximately one-half of its funding coming from the automobile industry and the other half coming from the federal government and other government sources, has successfully managed the boundary between industry and government, and between the research community in health effects and the research community in air quality (Keating, 2001). HEI, however, does not fund projects in support of specific marketing applications, but rather it funds projects that contribute to general knowledge. The Reagan-Udall Foundation (RUF) advises the FDA on modernizing regulatory science, and it conducts and oversees studies on regulatory science, particularly in the emerging fields of pharmacogenomics and genomic-based prediction of drug response and adverse event risk. The RUF has in place controls for bias and conflict of interest that are noteworthy.

Public access to the totality of the data on MRTPs is critical to the credibility of MRTP research. Registering studies funded by the National Institutes of Health or used in drug approvals on the National Library of Medicine website Clinicaltrials.gov has greatly improved the transparency of those studies. The FDA may have to balance the need for public access to key information with the need to protect proprietary information and promote innovation. Public availability of data provided in an MRTP application is discussed in the Family Smoking Prevention and Tobacco Control Act of 2009 (FSPTCA).[2]

---

[2] *Family Smoking Prevention and Tobacco Control Act of 2009*, Public Law 111-31, 123 Stat. 1776 (June 22, 2009).

INTEGRATION OF EVIDENCE AND DECISION MAKING

Table 6-1 presents the evidence domains and example considerations for using evidence from the different domains. A key challenge facing the FDA will be integrating the various domains and levels of evidence provided by sponsors in support of an MRTP application. A systematic, explicit approach that weights outcomes in terms of their public health importance, identifies the measures and data most relevant to those outcomes, and combines the available evidence in a manner that is psychometrically sound, objective, and reproducible would be helpful. This effort could be informed by decision theory concepts and techniques such as expected utility, Bayesian, and improper linear model approaches. Mathematical modeling and simulation analysis, such as discussed earlier in Chapter 3, can also be used to predict population-level effects.

It is clear that no single class of evidence (e.g., preclinical, RCTs, consumer perception, epidemiologic) in itself will be sufficient to support an MRTP application. Studies offered in support of an MRTP application should address all key research domains needed to prognosticate the product's likely public health impact, including the following:

- Product content (including but not limited to harmful and potentially harmful constituents), performance, and quality assurance regarding product consistency
- Self-administration and subjective evaluation
- Exposure assessment by state-of-the-art methods and measurements that focus on human exposure to harmful and potentially harmful constituents
- Consumer and nonconsumer perceptions; and assessment of biomarkers of exposure, biomarkers of risk, and where feasible, disease outcomes

The research should use designs that are properly powered, balance internal and external validity, and comprise multiple populations appropriate to the experimental questions being addressed. Target populations of special relevance include (but are not limited to) users of tobacco products, both those interested and uninterested in quitting; nonsmokers; former smokers; beginning smokers; and adolescents. In addition, study samples must permit inferences to populations with significant smoking prevalence such as those low in socioeconomic status and educational attainment, and certain ethnic minorities.

Beyond merely amassing evidence in support of the modified risk claims, higher-level processing of the evidence is needed to inform decision making. It is probable that, depending on their construction, intended uses, and desired claims, different MRTP applications could require dif-

TABLE 6-1 Evidence Domains Relevant to an MRTP Application and Examples of Types of Findings

Class of Evidence	Examples of Types of Finding That May Be Required
Preclinical	• Assurance of manufacturing quality control • Significant and substantial reduction in toxicant and carcinogen content in product • Significant reduction in exposure to toxicants and carcinogens in limited human study • No significant evidence for offsetting increases in content of or exposure to other toxicants
Clinical trial	• Significant reduction in exposure to toxicants and carcinogens in relation to continued use of traditional product, preferably approaching nonsmoker levels • Significant rates of cessation of conventional tobacco product use, or significant decrease in the rates of conventional tobacco product use • Significant reduction in biomarkers or surrogates of disease
Abuse potential	• No more liable for abuse than currently marketed products • No significant evidence of attractiveness to nonusers of tobacco
Epidemiology	• Clear and consistent evidence of reduction in disease risk (e.g., cancer, cardiovascular disease, chronic obstructive pulmonary disease) or intermediate endpoint thereof • No significant evidence of offsetting increased risk for other diseases • No significant evidence of uptake among nonusers or relapse among former users (postmarketing)
Consumer and nonconsumer perceptions	• Evidence for accurate understanding of product claim • No significant evidence that consumers equate reduced exposure with reduced risk • No significant evidence of intention to use product among nonusers (especially adolescents) • No significant evidence of switching from MRTP to other tobacco product usage
Populations at high risk for tobacco use	• No significant evidence of risk of initiation among nonusers (especially adolescents) • Consistency of findings across relevant subpopulations of interest (e.g., low socioeconomic status, racial/ethnic minorities)
Modeling and synthesis	• Population predictions show reduction in smoking-related morbidity and mortality following the introduction of an MRTP with no significant evidence of uptake by nonusers (especially adolescents)

NOTE: This table is not comprehensive and is not intended to be a guideline or framework for the evaluation of MRTP applications.

ferent balances of the various classes of evidence. As such, prescribing a universally applicable portfolio of evidence is difficult to envision. However, it remains possible to frame four broad scenarios for an MRTP application, as shown in the Table 6-2.

A product sponsor may file an MRTP application for a product that is already on the market or for a new product (meaning one not available on the U.S. market). For each of these, a sponsor could request a Modified Risk claim or a modified exposure claim under the Special Rule for Certain Products.

Existing products for which sponsors wish to make a risk/exposure claim raise particular concerns because of conscious and unconscious associations built over time with branding and other marketing messages. This is particularly true for MRTPs that are cobranded with other tobacco products. In the case of an existing product wishing to make a claim of modified risk, it would not be unreasonable of the FDA to expect epidemiologic evidence (e.g., case-control or cohort studies of users and nonusers of the MRTP about which the claim is made, with clinical or validated surrogate endpoints) and to have this evidence weigh heavily in its decision making. Supporting evidence for such an application could include switching studies and RCTs that conclusively demonstrate substantially reduced biomarkers of toxicant exposure or biomarkers of risk. It would be anticipated that preclinical evidence from human, animal, *in vitro*, or *in vivo* studies would play relatively minor roles (e.g., providing mechanistic context) in justifying a claim for a product that is already on the market. However, all MRTP products would benefit from having supportive preclinical studies completed prior to human studies, even if they are already on the market. Lastly, significant emphasis would need to be placed on extensive consumer and nonconsumer testing of the proposed advertising and marketing materials, product packaging, and design to determine (1) whether the typical consumer understands the message and is unlikely to be misled and (2) if the product is minimally attractive to nonusers of tobacco.

In the case of a product not currently available to consumers, claims of reduced risk (which require epidemiologic evidence) would be difficult to support. However, if the identical product were available in other

TABLE 6-2 Four Broad Scenarios for an MRTP Application

	Modified Risk	Modified Exposure (Special Rule)
Product already on market	A	B
Product not on market	C	D

markets, then epidemiologic evidence from users in those markets could be informative as to demonstrating disease risk reduction. In such a case, cautions must be applied, because emerging evidence suggests that clinical trial results from non-U.S. populations can differ substantially from those obtained in U.S. populations, even when identical drugs are given and identical study designs are employed (Glickman et al., 2009). The FDA should articulate guidance for the acceptance of data produced in foreign studies.

In the case of a truly new product (where nothing similar is sold elsewhere) an Investigational New Drug application model could be the most appropriate approach. In such cases, then, the FDA could require that preclinical laboratory testing be completed before moving to animal or human studies (e.g., Phase I), which would have to be justified by significant findings in the laboratory work. If preclinical findings pointed to potential reduced exposure, then the FDA could authorize Phase II or III trials to explore the experience of reduced exposure in larger human populations under controlled conditions (e.g., RCTs). Concurrent or subsequent to these trials, consumer perception studies would be required to understand how potential users and nonusers would respond to product claims. To gather important population-level data on truly novel products, the FDA should consider authorizing a limited test market (perhaps requiring sponsors to track a cohort of MRTP purchasers for later follow-up) to gather data on real-world use.

Mathematical Modeling and Simulation Analysis

Mathematical modeling and simulation analyses can be useful to estimate exposures, health effects, and broad public health effects, providing information to the FDA for its decisions. The role of modeling in evaluating the effect of MRTPs and considerations for the conduct and reporting of model-based analyses are discussed below.

The Role of Modeling

Empirical studies employing a variety of designs, such as prospective, retrospective, randomized, or cohort designs, are expected to provide fundamental information about the societal impact of MRTPs. Considerations for the design and conduct of such studies are discussed in other sections of this report. However, the evidence empirical studies can provide has several limitations, especially in the context of assessing societal impact. First, empirical studies typically address streamlined questions. For example, they may compare only two interventions, have limited follow-up, or involve narrow subsets of potential users and nonusers of tobacco products.

As a result, generalizable evidence from empirical studies may be slow to accumulate and may also need updating with future studies. Second, empirical studies often require significant resources and time to launch and carry out. The magnitude of the required resources (sample size, length of follow-up) may be very large for studies intended to derive generalizable assessments of the short- and long-term societal effects of MRTPs.

Mathematical modeling and simulation analysis provide a complementary approach to the conduct of empirical studies that can be useful at each stage of the regulatory process for MRTPs. Modeling has already been used in the evaluation of smoking behaviors and related interventions (Mendez et al., 2008; TPSAC, 2011). Modeling is also widely used in several areas of health care research and public policy, such as the analysis of the economic impact of therapeutic and diagnostic interventions, the health implications of the introduction of disease screening programs, and the health impact of environmental conditions and exposures (Rutter et al., 2011; Weinstein et al., 2003).

According to the National Research Council (NRC) report *Improving Information for Social Policy Decisions: The Uses of Microsimulation Modeling*, a simulation model is "a replicable, objective sequence of computations used for generating estimates of quantities of concern" (NRC, 1991). In the context of this report the committee adopts the NRC definition with the following clarifications that adapt well-established criteria for models used in health care evaluation (Weinstein et al., 2003) to the regulation of MRTPs:

a. Models account for smoking behaviors and outcomes (health, behavioral, economic) over time and across various cohorts of individuals.
b. Models are based on information from empirical studies.
c. Models are used to estimate the effects of the introduction and use of MRTPs on outcomes of interest to the regulatory process.

The literature on the methodology and uses of modeling is by now extensive and addresses modeling both in general and for particular domains. Decision trees are some of the simpler and most widely used models in health care evaluation (Sonnenberg and Beck, 1993). Cohort Markov models describe the trajectory of groups of individuals through time and conditions of the process being modeled (Sonnenberg and Beck, 1993). Micro-simulation models (discrete or continuous time, agent based, etc.) describe the trajectories of individuals in a population, incorporating information on exposures and other events and outcomes (NRC, 1991; Rutter et al., 2011). A survey of recently used methods in tobacco modeling is provided in the proceedings of the 2008 Tobacco Modelers Conference (2008).

Modeling analyses have multiple potential uses in the assessment of the societal impact of MRTPs, as required by the regulatory process. Those aspects can help the FDA in its decision-making process. First, model-based analyses can synthesize the available information from empirical studies of MRTPs. In doing so, models can help clarify the logical and scientific relations between the structural assumptions and information sources used in the model (inputs) and the final conclusions and recommendations from the analysis. In addition to combining available information, models can help clarify and possibly reconcile seemingly contradictory evidence from previous studies. Second, models can enable researchers and decision makers to explore complex interactions and systems that may be impractical to evaluate in empirical studies. In particular, models can include components relating to quantities or processes that are not directly observable in empirical studies. Thus, model-based analyses can inform and augment current knowledge from clinical practice and empirical observation for the underlying mechanisms of MRTP utilization, smoking behavior, and outcomes. Third, model-based analyses allow researchers and decisions makers to explore "what if" questions relevant to decision making, which would not be practical to assess in empirical studies. For example, models can be used to assess the potential impact of modifications in MRTP structure, delivery, and scheduling of intervention. The vast multitude of such potential modifications makes it impractical to expect, for example, that comparative studies will be conducted for each possible combination of changes. Fourth, and of major importance to the regulatory process, models can be used to make projections about the short- and long-term effects of the introduction of MRTPs. These projections can be used not only for current decision making but also for planning future studies designed to fill particular gaps in scientific knowledge or to reduce uncertainty in the evidence about key determinants of the outcome.

Considerations for the Conduct and Reporting of Model-Based Analyses

The adoption and use of modeling in the regulation of MRTPs will represent a relatively new dimension in the regulatory process. The decades-long experience of modeling in health care evaluation and other areas of policy and decision making can provide a solid foundation for the introduction of model-based analyses in the regulatory process. Of particular value in the eventual formulation of a regulatory framework could be the results of completed (Weinstein et al., 2003) and ongoing work on the development of good research practices for modeling.[3]

---

[3] ISPOR-SMDM Modeling Good Research Practices Task Force Working Group, Report Parts 1-7, unpublished data, 2011.

As with models in general, transparency in all aspects of the model, validation of the model, and proper assessment of the uncertainty regarding model parameters and results are major aspects of modeling for the societal impact of MRTPs.

Transparency Transparency refers to the availability of detailed information on all aspects of model structure, sources of evidence used, computational approach, and the construction of summaries and reporting of the results. This information is essential for a proper scientific understanding of the modeling and for enabling model critique and validation by researchers and other participants in the regulatory process. Transparency is also essential for the proper interpretation of the results of modeling analyses and their usefulness to decision making at the individual and policy levels. A practical goal of transparency in modeling used for MRTP regulation is to enable others to reproduce the results of the model-based analysis.

Validation Validation of the model examines its structure and performance from several perspectives, including internal, face, cross, and predictive validity.[4] The internal validity of a model refers to whether the components of the model perform their intended tasks, the programming is correct, and the summaries of the results are accurate. Thus internal validation requires verification of the correct functioning of each component of the model.

The external validity of a model refers to the model's ability to simulate events and outcomes that have actually occurred in settings that are close to those captured in the assumptions of the model. Thus external validation requires its developers to select appropriate databases, simulate the outcomes for individuals in these databases, and compare the results of the simulation to the actual outcomes. When models are to be used in settings that are not very close to those used for their development, careful calibration is needed to ensure applicability of models to these new settings (Vanni et al., 2011).

The face validity of a model is a more subjective attribute than the previous two, because it refers to the degree to which the structure, inputs, and methods for summarizing the results correspond to currently accepted scientific knowledge and practice, as judged by experts in tobacco studies. Standard components of the regulatory process, such as advisory panels, can take on a significant role in the assessment of face validity for models used on MRTP regulation.

---

[4] David M. Eddy et al., ISPOR-SMDM Modeling Good Research Practices Task Force Working Group Part 4, unpublished data, 2011.

The cross-validity of a model refers to whether the model's results agree with those of other models developed for the same purpose. For example, the Cancer Intervention and Surveillance Modeling Network (CISNET) initiative funded by the National Cancer Institute involves the development and comparison of the results of alternative models for assessing the impact of technologies for the early detection of cancer (CISNET Breast Cancer Collaborators, 2006). Cross-validation can be time and resource intensive, because it requires the development of several alternative models and the comparison of their results.

The predictive validity of a model refers to the model's ability to accurately predict future outcomes. This aspect of model validity is particularly important for the intended use of modeling in regulating MRTPs. For example, a model developed on the basis of currently available information can be linked to future empirical studies of the impact of an MRTP and validated partly or fully using data from those studies.

Uncertainty Uncertainty accounting refers to the systematic examination, assessment, and reporting of the uncertainty in model inputs and assumptions, estimates of model parameters, and summary measures of the results (Bilcke et al., 2011). There are three important sources of uncertainty in model-based analyses. First, there can be uncertainty about the structural assumptions of the model, such as assumptions about variables to include and causation and prediction pathways. The uncertainty about such assumptions is often difficult to quantify, but their impact on the conclusions of the analysis may be substantial (Bojke et al., 2009). Second, there can be uncertainty about model parameter estimates derived from available studies. Formal statistical measures for this uncertainty can be derived using meta-analysis methods. However, the assessment of how this uncertainty propagates through the modeling computations to the final results continues to be a challenge. Third, there can be variability among individuals in the population, which can be explained on the basis of characteristics of these individuals (systematic or explained variation) or is considered to be random.

The impact of the various forms of uncertainty on the final results of the model is typically assessed by deterministic or probabilistic sensitivity analysis. However, because of the extent and variety of uncertainty in a modeling analysis, a realistic account of its impact on the results of the analysis continues to be challenging. For example, most sensitivity analyses in practice do not account for the multivariate structure of inputs and the correlations in this structure. Guidelines on how the results of sensitivity analyses should be reported are currently in development[5] (Bilcke et al., 2011; Stout et al., 2009).

Selection of Comparison Products

The amount of harm reduction claimed by an MRTP sponsor in an application is a critical issue in deciding whether to issue an order for the marketing of the MRTP. Harm reduction is inherently relative; a reduction claim is by definition relative to a comparison product. Selection of an appropriate comparison product is essential for informed and accurate decision making. The FSPTCA recognizes this, giving the Secretary of the Department of Health and Human Services authority to require product sponsors to compare their product to a commercially marketed representative product. The choice of appropriate comparison products will be driven by the type of MRTP being tested, the anticipated claim, and the study design. And, indeed, the comparison products may differ between different classes of evidence. However, two reference products come to the forefront in terms of integration and synthesis of evidence: leading brands and smoking cessation products.

Leading Brands

Those products that are most commonly used by consumers are likely to provide a good comparison for products that claim to demonstrate reduced health risk. "Leading brands" represent a set of products that accounts for a significant portion of the market and could capture subgroups of interest (e.g., low socioeconomic status, who tend to use discount brands, and racial/ethnic minorities, who tend toward menthols). Using leading brands increases the likelihood that the findings will have broader applicability to the population, which is crucial given the public health standard against which MRTPs are evaluated. Using leading brands as a comparator also avoids potential mischief in comparing an MRTP to a product that is little used but may inflate the apparent risk reduction of the MRTP. In some cases, when desiring a reduced exposure or risk claim, the comparison product will be a product within the same product class (e.g., cigarette-like MRTP versus leading brand cigarette), and thus the comparison is relatively straightforward. In other cases, an MRTP may make a reduced risk/exposure claim across product classes (e.g., smokeless MRTP versus leading brand cigarettes). In this case, the product should also be compared to leading products in its own class

[5] ISPOR-SMDM Modeling Good Research Practices Task Force Working Group Part 3, unpublished data, 2011.

(e.g., smokeless MRTP versus moist snuff). Following this example, a smokeless MRTP that succeeds in both within- and cross-class comparisons against leading brands could reduce risk/exposure for both smokers and traditional smokeless users. However, it is also possible that a smokeless MRTP does pose less risk or exposure than cigarettes but is no different than other smokeless products not seeking a claim.

Smoking Cessation

On the opposite end of the spectrum of exposure and risk reduction is the "gold standard" of smoking cessation (or tobacco cessation in the case of smokeless tobacco users). This provides an aspirational goal for risk and exposure for MRTPs—in principle, the closer risks and exposures from the MRTP are to cessation products, the more confident a regulator can be in the chances for net public health benefit. Note that the use of this comparison product is not the same as studying whether the MRTP acts as an aid to smoking cessation. Rather, the goal is to compare how the risk or exposure reduction attained with use of the MRTP compares to smoking cessation of similar duration. It is also important to consider that for some health conditions, such as acute cardiovascular outcomes and lung function decline, the benefits of cessation accrue more quickly than for cancer.

FINDINGS AND RECOMMENDATIONS

In the committee's view, the fundamental problem that confronts the FDA is a shortage of credible and reliable evidence about the effects of MRTPs on both individual and public health. The history of deceptive behavior by the tobacco industry undermined the trust of the public as well as the public's confidence in the industry's ability to rigorously conduct studies that will generate the data needed to evaluate these products. Therefore, the committee's recommendations are designed to articulate the minimum standards for producing credible and reliable evidence to demonstrate that the marketing of an MRTP is consistent with the protection of public health. The committee articulates a strategy for the production of scientific evidence by making recommendations in three areas:

1. Types of evidence and studies
2. Design and integration of studies
3. Governance of studies

Types of Evidence and Studies

Finding 1: Types of Evidence. The public health standard articulated by the FSPTCA requires collection of scientific evidence from a wide range of disciplines and research domains. Although the committee respects the FDA's independence and discretion in regulating MRTPs, the committee maintains there is a minimum range of research domains required to evaluate the effect of MRTPs on individuals and public health. Individual methods may change as the technology or state of the science may evolve, but the minimum standards for the domains of evidence will be relevant regardless of the state of the science in the future.

Recommendation 1: The FDA should require that studies submitted in support of an MRTP application address all key research domains needed to forecast and monitor the product's public health impact, including:

- **product composition and performance;**
- **addiction potential and likelihood for initiation or persistence of use;**
- **human exposure to harmful and potentially harmful constituents;**
- **perceptions about the product's effects and likelihood of addiction; and**
- **effects of the product on human health and surrogates of human health.**

Finding 2: Phased Approach to New MRTPs. Many novel MRTPs are likely to be developed for marketing in the near future. There are inherent uncertainties and risks with new products that should be addressed. Risks should be minimized before new products are tested in humans. To address the risk of new products, a phased approach, similar to the New Drug Application framework for the regulation and control of new drugs, is appropriate for the evaluation of new MRTPs. A phased approach will help the FDA ensure that only products that are unlikely to be unsafe and have a reasonable expectation of reducing harm relative to conventional tobacco products will be used in human studies.

Recommendation 2: The FDA should establish guidance that conveys an expected sequencing of studies, such that preclinical work is completed and submitted to the FDA before clinical (human subjects) work commences, and that there is a reasonable expectation based on preclinical work that a reduction or lack of harm will be seen in humans.

Finding 3: Clinical Trial Studies. Although the use of randomized controlled trial methods will be constrained for a number of reasons (including the practical limitations of study cost, size, and follow-up, and ethical constraints on randomizing study participants to harmful exposures), they will continue to play an essential role in creating an evidence base on the public health effects of MRTPs. Randomized controlled trial methods can provide highly reliable data on the likelihood of addiction and initiation or cessation of product use. Also, these methods can provide reliable evidence on human exposure.

Recommendation 3: The FDA should require randomized controlled trials in the following domains:

- **Exposure reduction**
- **Self-administration of the MRTP**
- **Effects on use of conventional tobacco products**

These randomized controlled trials should include multiple comparison products (such as nicotine replacement products, conventional cigarettes or smokeless tobacco, placebo preparations, and alternative nicotine delivery systems). These trials should also assess the effect of the MRTP on human exposure and on human health and surrogates of human health.

Finding 4: Requirement for Postmarket, Prospective Epidemiologic Studies. Postmarket studies of MRTPs will be critical to evaluating the effect of MRTPs on both individuals and the public's health. In particular, the prospective cohort design will be an essential tool to validating anticipated or claimed effects of marketed MRTPs. These studies have several important strengths: (1) biochemical tobacco and MRTP exposure can be assessed at baseline, offering "unbiased" exposure assessment before health outcomes occur; (2) there is less of a problem with retrospective recall of product use, beause this information can be summarized at the start of the study and followed prospectively; (3) changing product use habits can be monitored as the study progresses; (4) outcomes can be documented as they occur, and verification is more efficient; and (5) a wide variety of outcomes can be evaluated in the same study, particularly outcomes that are more common. Furthermore, cohort studies allow assessment of overall health status and outcomes.

Recommendation 4: The FDA should require prospective epidemiologic studies to commence upon issuance of a marketing order to confirm reduced exposure and reduced risk claims, and to examine

effects of MRTP availability on the population as a whole, including the likelihood of initiation and cessation. The FDA should issue guidance on the design, conduct, and analysis of such studies.

Finding 5: Modeling of Public Health Outcomes. Mathematical modeling and simulation analysis provides a complementary approach to the conduct of empirical studies that can be useful at each stage of the regulatory process for MRTPs. Model-based analyses can (1) synthesize the available information from empirical studies of MRTPs; (2) enable researchers and decision makers to explore complex interactions and systems that may be impractical to evaluate in empirical studies; (3) allow researchers and decisions makers to explore "what if" questions relevant to decision making, which would not be practical to assess in empirical studies; and (4) be used to make projections about the short- and long-term effects of the introduction of MRTPs.

Recommendation 5: The FDA should issue guidance on the development and use of simulation and modeling approaches to predict public health impact through the systematic integration of information about relevant assumptions and influences. Such approaches should be tested for robustness with regard to results and assumptions, they should be public and transparent, and they should be validated against postmarketing epidemiologic research.

Design and Integration of Studies

Finding 6: Standards for Sampling in MRTP Studies. To have regulatory usefulness, studies of MRTPs must be generalizable to the overall population of interest and to specific populations, including populations at high risk for tobacco use. Failure to include relevant populations in studies will result in incomplete evidence on the effect of an MRTP on the public's health and, therefore, will be inadequate to support regulatory decisions about the marketing of MRTPs.

Recommendation 6: The FDA should require studies to include populations of special relevance, including (but not limited to)

- **users of tobacco products, including users who are and are not interested in quitting;**
- **in certain circumstances, nonusers of tobacco products;**
- **former smokers;**
- **beginning smokers;**
- **adolescents; and**

- **populations at a high risk for tobacco use, including, but not limited to, those low in socioeconomic status and educational attainment, and certain ethnic minorities.**

Finding 7: Quality of Studies. The usefulness of a study to inform a regulatory decision hinges on the quality and appropriateness of the design. In many cases, complementary studies might be needed to provide a breadth of evidence for an informed regulatory decision with appropriate control of confounders and internal and external validity.

> **Recommendation 7: For all studies of the effects of MRTPs on human health and behavior, the FDA should require a range of designs that are properly powered, balance internal and external validity, and comprise multiple populations appropriate to the experimental questions being addressed.**

Finding 8: Standards for Good Research Practice. A significant amount of guidance on minimum standards for scientific studies directly relevant to the evaluation of MRTPs has already been developed. Guidelines for formatting, design, conduct, and reporting of science are articulated in consensus statements, such as the Consolidated Standards of Reporting Trials (CONSORT) reporting criteria for clinical trials, the Strengthening the Reporting of Observational Studies in Epidemiology (STROBE) statement for observational studies, the publication criteria of the International Council of Medical Journal Editors, and the reporting criteria of the International Conference on Harmonization. These existing guidelines represent robust standards for the conduct of science across many of the research domains relevant to the evaluation of MRTPs.

> **Recommendation 8: The FDA should issue guidance to the industry regarding the format, design, conduct, and reporting of studies in support of MRTP applications that is based upon current generally accepted principles for scientific investigation.**

Finding 9: Standards for Integration of Evidence. Regulatory decisions regarding MRTPs will be based on a wide range and variety of scientific evidence, and the integration of scientific evidence will play a pivotal role in that decision making. The assessment of MRTPs will typically require the evaluation and integration of evidence on risks and benefits across multiple diverse outcomes, such as measures of toxicity, biomarkers, addictiveness, and disease endpoints. Modeling and simulation approaches are relevant to estimating public health effects of tobacco and, therefore, the FDA will likely engage in various methods of data integration, synthesis, and analysis, including, but not limited to, simulation and modeling. It is critical that these approaches are transparent and reproducible.

Recommendation 9: The FDA should develop and use an approach to data integration that is explicit and transparent with regard to the importance of the different outcomes, that uses optimal available evidence, and that employs objective and reproducible methods for data integration.

Governance of Studies

Finding 10: Independent Oversight and Conduct of Studies. It has been established in public records and as a matter of law that the tobacco industry has engaged in illegal and improper practices, including the destruction and manipulation of scientific data. As a result, the tobacco industry is profoundly isolated from the mainstream scientific community. Many major universities have policies against acceptance of tobacco funding, and many high-impact scientific and medical journals will not accept tobacco industry-supported manuscripts. The consequence of this isolation is a lack of the expertise and the resources necessary to produce high-quality science across the range of disciplines to support an application to market an MRTP. Use of a trusted third party, particularly for products developed by the tobacco industry, could provide an avenue for the production of credible evidence needed by the FDA to evaluate tobacco products. Ultimately, such a research structure could encourage and support the production and dissemination of credible and reliable evidence about the effects of tobacco products on the public's health.

Recommendation 10: MRTP sponsors should consider use of independent third parties to undertake one or more key functions, including the design and conduct of research, the oversight of specific studies, and the distribution of sponsor funds for research. Such independent third parties should be approved by the FDA in advance of the research.

Finding 11: Public Disclosure of Research. Public availability of data not only builds credibility and public trust, but also benefits the public because it allows for independent analysis of study methods and data. The model of Clinicaltrials.gov is particularly compelling and relevant, and a similar model of public accounting and open disclosure should be expected of the tobacco industry.

Recommendation 11: The FDA should require all MRTP sponsors to place all data generated in the development and marketing of the MRTP in a public repository selected by the FDA.

Finding 12: Proper Conduct of Research. Standards for the conduct of science and the protection of human research participants have been established for biomedical research enterprises not only in academics but also in commercial research. The FDA has the tools to ensure studies adhere to established standards in the drug development framework, which can be applied to the development of MRTPs. Those standards not only protect human participants, but also build credibility into any data that are provided to the FDA, particularly by the tobacco industry. Institutional credibility and trustworthiness is particularly relevant in this context, given the history of unethical and illegal practices of the tobacco industry.

Recommendation 12: The FDA should require studies offered in support of an MRTP application to adhere to established standards and principles of good research governance, including appropriately qualified investigators, transparency, independent institutional review board or ethical review, and adherence to the Common Rule (21 CFR parts 50 and 56).

REFERENCES

Ashley, M. J., and J. E. Cohen. 2003. What the public thinks about the tobacco industry and its products. *Tobacco Control* 12(4):396-400.

Bilcke, J., P. Beutels, M. Brisson, and M. Jit. 2011. Accounting for methodological, structural, and parameter uncertainty in decision-analytic models: A practical guide. *Medical Decision Making* 31(4):675-692.

Bojke, L., K. Claxton, M. Sculpher, and S. Palmer. 2009. Characterizing structural uncertainty in decision analytic models: A review and application of methods. *Value Health* 12(5):739-749.

CISNET Breast Cancer Collaborators. 2006. The impact of mammography and adjuvant therapy on U.S. breast cancer mortality (1975-2000): Collective results from the cancer intervention and surveillance modeling network. *JNCI Monographs* 2006(36):1-126.

Cummings, K. M. 2003. A promise is a promise. *Tobacco Control* 12(2):117-118.

Cummings, K. M., C. Morley, and A. Hyland. 2002. Failed promises of the cigarette industry and its effect on consumer misperceptions about the health risks of smoking. *Tobacco Control* 11(Suppl. 1):i110-i117.

Cummings, K. M., A. Brown, and R. O'Connor. 2007. The cigarette controversy. *Cancer Epidemiology, Biomarkers & Prevention* 16(6):1070-1076.

Glickman, S. W., J. G. McHutchison, E. D. Peterson, C. B. Cairns, R. A. Harrington, R. M. Califf, and K. A. Schulman. 2009. Ethical and scientific implications of the globalization of clinical research. *NEJM* 360(8):816-823.

Harris Interactive. 2003. *Attitudes to government regulation vary greatly for different industries; The Harris Poll #19*. http://www.harrisinteractive.com/vault/Harris-Interactive-Poll-Research-Attitudes-to-Government-Regulation-Vary-Greatly-for-Different-Industries-2003-04.pdf (accessed December 5, 2011).

Harris Interactive. 2010. *Oil, pharmaceutical, health insurance, and tobacco top the list of industries that people think should be more regulated; The Harris Poll #149*. http://www.harrisinteractive.com/vault/HI-Harris-Poll-Industry-Regulation-2010-12-02.pdf (accessed December 5, 2010).

Keating, T. J. 2001. Lessons from the recent history of the Health Effects Institute. *Science, Technology, & Human Values* 26(4):409-430.

Mendez, D., S. P. Marcus, P. I. Clark, S. J. Leischow, and T. F. Pechacek. 2008. Tobacco modelers conference proceedings. In *Tobacco modelers conference*. Ann Arbor: University of Michigan.

NCI (National Cancer Institute). 2008. *The role of the media in promoting and reducing tobacco use*. Edited by R. M. Davis, *NCI tobacco control monograph series*. Bethesda, MD: U.S. Department of Health and Human Services, National Institutes of Health, National Cancer Institute.

NRC (National Research Council). 1991. *Improving information for social policy decisions: The uses of microsimulation modeling*. Vol. 1. Washington, DC: National Academy Press.

Rees, V. W., J. M. Kreslake, K. M. Cummings, R. J. O'Connor, D. K. Hatsukami, M. Parascandola, P. G. Shields, and G. N. Connolly. 2009a. Assessing consumer responses to potential reduced-exposure tobacco products: A review of tobacco industry and independent research methods. *Cancer Epidemiology, Biomarkers & Prevention* 18(12):3225-3240.

Rees, V. W., J. M. Kreslake, R. J. O'Connor, K. M. Cummings, M. Parascandola, D. Hatsukami, P. G. Shields, and G. N. Connolly. 2009b. Methods used in internal industry clinical trials to assess tobacco risk reduction. *Cancer Epidemiology, Biomarkers & Prevention* 18(12):3196-3208.

Rutter, C. M., A. M. Zaslavsky, and E. J. Feuer. 2011. Dynamic microsimulation models for health outcomes. *Medical Decision Making* 31(1):10-18.

Sonnenberg, F. A., and J. R. Beck. 1993. Markov models in medical decision making: A practical guide. *Medical Decision Making* 13(4):322-338.

Stout, N. K., A. B. Knudsen, C. Y. J. Kong, P. M. McMahon, and G. S. Gazelle. 2009. Calibration methods used in cancer simulation models and suggested reporting guidelines. *Pharmacoeconomics* 27(7):533-545.

TPSAC (Tobacco Products Scientific Advisory Committee). 2011. *Menthol cigarettes and public health: Review of the scientific evidence and recommendations.* Tobacco Products Scientific Advisory Committee.

Vanni, T., J. Karnon, J. Madan, R. G. White, W. J. Edmunds, A. M. Foss, and R. Legood. 2011. Calibrating models in economic evaluation: A seven-step approach. *Pharmacoeconomics* 29(1):35-49.

Weinstein, M. C., B. O'Brien, J. Hornberger, J. Jackson, M. Johannesson, C. McCabe, and B. R. Luce. 2003. Principles of good practice for decision analytic modeling in health-care evaluation: Report of the ispor task force on good research practices—modeling studies. *Value Health* 6(1):9-17.

Appendix A

Section 911 of the Family Smoking Prevention and Tobacco Control Act of 2009

21 USC 387k. **'SEC. 911. MODIFIED RISK TOBACCO PRODUCTS.**

"(a) In General- No person may introduce or deliver for introduction into interstate commerce any modified risk tobacco product unless an order issued pursuant to subsection (g) is effective with respect to such product.

"(b) Definitions- In this section:

"(1) MODIFIED RISK TOBACCO PRODUCT- The term 'modified risk tobacco product' means any tobacco product that is sold or distributed for use to reduce harm or the risk of tobacco-related disease associated with commercially marketed tobacco products.

"(2) SOLD OR DISTRIBUTED-

"(A) IN GENERAL- With respect to a tobacco product, the term 'sold or distributed for use to reduce harm or the risk of tobacco-related disease associated with commercially marketed tobacco products' means a tobacco product--

"(i) the label, labeling, or advertising of which represents explicitly or implicitly that--

"(I) the tobacco product presents a lower risk of tobacco-related disease or is less harmful than one or more other commercially marketed tobacco products;

"(II) the tobacco product or its smoke contains a reduced level of a substance or presents a reduced exposure to a substance; or

"(III) the tobacco product or its smoke does not contain or is free of a substance;

"(ii) the label, labeling, or advertising of which uses the descriptors 'light', 'mild', or 'low' or similar descriptors; or

"(iii) the tobacco product manufacturer of which has taken any action directed to consumers through the media or otherwise, other than by means of the tobacco product's label, labeling, or advertising, after the date of enactment of the Family Smoking Prevention and Tobacco Control Act, respecting the product that would be reasonably expected to result in consumers believing that the tobacco product or its smoke may present a lower risk of disease or is less harmful than one or more commercially marketed tobacco products, or presents a reduced exposure to, or does not contain or is free of, a substance or substances.

"(B) LIMITATION- No tobacco product shall be considered to be 'sold or distributed for use to reduce harm or the risk of tobacco-related disease associated with commercially marketed tobacco products', except as described in subparagraph (A).

"(C) SMOKELESS TOBACCO PRODUCT- No smokeless tobacco product shall be considered to be 'sold or distributed for use to reduce harm or the risk of tobacco-related disease associated with commercially marketed tobacco products' solely because its label, labeling, or advertising uses the following phrases to describe such product and its use: 'smokeless tobacco', 'smokeless tobacco product', 'not consumed by smoking', 'does not produce smoke', 'smokefree', 'smoke-free', 'without smoke', 'no smoke', or 'not smoke'.

"(3) EFFECTIVE DATE- The provisions of paragraph (2)(A)(ii) shall take effect 12 months after the date of enactment of the Family Smoking Prevention and Tobacco Control Act for those products whose label, labeling, or advertising contains the terms described in such paragraph on such date of enactment. The effective date shall be with respect to the date of manufacture, provided that, in any case, beginning 30 days after such effective date, a manufacturer shall not introduce into the domestic commerce of the United States any product, irrespective of the date of manufacture, that is not in conformance with paragraph (2)(A)(ii).

"(c) Tobacco Dependence Products- A product that is intended to be used for the treatment of tobacco dependence, including smoking cessation, is not a modified risk tobacco product under this section if it has been approved as a drug or device by the Food and Drug Administration and is subject to the requirements of chapter V.

"(d) Filing- Any person may file with the Secretary an application for a modified risk tobacco product. Such application shall include--

"(1) a description of the proposed product and any proposed advertising and labeling;

"(2) the conditions for using the product;

"(3) the formulation of the product;

"(4) sample product labels and labeling;

"(5) all documents (including underlying scientific information) relating to research findings conducted, supported, or possessed by the tobacco product manufacturer relating to the effect of the product on tobacco-related diseases and health-related conditions, including information both favorable and unfavorable to the ability of the product to reduce risk or exposure and relating to human health;

"(6) data and information on how consumers actually use the tobacco product; and

"(7) such other information as the Secretary may require.

"(e) Public Availability- The Secretary shall make the application described in subsection (d) publicly available (except matters in the application which are trade secrets or otherwise confidential, commercial information) and shall request comments by interested persons on the information contained in the application and on the label, labeling, and advertising accompanying such application.

"(f) Advisory Committee-

"(1) IN GENERAL- The Secretary shall refer to the Tobacco Products Scientific Advisory Committee any application submitted under this section.

"(2) RECOMMENDATIONS- Not later than 60 days after the date an application is referred to the Tobacco Products Scientific Advisory Committee under paragraph (1), the Advisory Committee shall report its recommendations on the application to the Secretary.

"(g) Marketing-

"(1) MODIFIED RISK PRODUCTS- Except as provided in paragraph (2), the Secretary shall, with respect to an application submitted under this section, issue an order that a modified risk product may be commercially marketed only if the Secretary determines that the applicant has demonstrated that such product, as it is actually used by consumers, will--

"(A) significantly reduce harm and the risk of tobacco-related disease to individual tobacco users; and

"(B) benefit the health of the population as a whole taking into account both users of tobacco products and persons who do not currently use tobacco products.

"(2) SPECIAL RULE FOR CERTAIN PRODUCTS-

"(A) IN GENERAL- The Secretary may issue an order that a tobacco product may be introduced or delivered for introduction into interstate commerce, pursuant to an application under this section, with respect to a tobacco product that may not be commercially marketed under paragraph (1) if the Secretary makes the findings required under this paragraph and determines that the applicant has demonstrated that--

"(i) such order would be appropriate to promote the public health;

"(ii) any aspect of the label, labeling, and advertising for such product that would cause the tobacco product to be a modified risk tobacco product under subsection (b) is limited to an explicit or implicit representation that such tobacco product or its smoke does not contain or is free of a substance or contains a reduced level of a substance, or presents a reduced exposure to a substance in tobacco smoke;

"(iii) scientific evidence is not available and, using the best available scientific methods, cannot be made available without conducting long-term epidemiological studies for an application to meet the standards set forth in paragraph (1); and

"(iv) the scientific evidence that is available without conducting long-term epidemiological studies demonstrates that a measurable and substantial reduction in morbidity or mortality among individual tobacco users is reasonably likely in subsequent studies.

"(B) ADDITIONAL FINDINGS REQUIRED- To issue an order under subparagraph (A) the Secretary must also find that the applicant has demonstrated that--

"(i) the magnitude of the overall reductions in exposure to the substance or substances which are the subject of the application is substantial, such substance or substances are harmful, and the product as actually used exposes consumers to the specified reduced level of the substance or substances;

"(ii) the product as actually used by consumers will not expose them to higher levels of other harmful substances compared to the similar types of tobacco products then on the market unless such increases are minimal and the reasonably likely overall impact of use of the product remains a substantial and measurable reduction in overall morbidity and mortality among individual tobacco users;

"(iii) testing of actual consumer perception shows that, as the applicant proposes to label and market the product, consumers will not be misled into believing that the product--

"(I) is or has been demonstrated to be less harmful; or

"(II) presents or has been demonstrated to present less of a risk of disease than 1 or more other commercially marketed tobacco products; and

"(iv) issuance of an order with respect to the application is expected to benefit the health of the population as a whole taking into account both users of tobacco products and persons who do not currently use tobacco products.

"(C) CONDITIONS OF MARKETING-

"(i) IN GENERAL- Applications subject to an order under this paragraph shall be limited to a term of not more than 5 years, but may be renewed upon a finding by the Secretary that the requirements of this paragraph continue to be satisfied based on the filing of a new application.

"(ii) AGREEMENTS BY APPLICANT- An order under this paragraph shall be conditioned on the applicant's agreement to conduct postmarket surveillance and studies and to submit to the Secretary the results of such surveillance and studies to determine the impact of the order on consumer perception, behavior, and health and to enable the Secretary to review the accuracy of the determinations upon which the order was based in accordance with a protocol approved by the Secretary.

"(iii) ANNUAL SUBMISSION- The results of such postmarket surveillance and studies described in clause (ii) shall be submitted annually.

"(3) BASIS- The determinations under paragraphs (1) and (2) shall be based on--

"(A) the scientific evidence submitted by the applicant; and

"(B) scientific evidence and other information that is made available to the Secretary.

"(4) BENEFIT TO HEALTH OF INDIVIDUALS AND OF POPULATION AS A WHOLE- In making the determinations under paragraphs (1) and (2), the Secretary shall take into account--

"(A) the relative health risks to individuals of the tobacco product that is the subject of the application;

"(B) the increased or decreased likelihood that existing users of tobacco products who would otherwise stop using such products will switch to the tobacco product that is the subject of the application;

"(C) the increased or decreased likelihood that persons who do not use tobacco products will start using the tobacco product that is the subject of the application;

"(D) the risks and benefits to persons from the use of the tobacco product that is the subject of the application as compared to the use of products for smoking cessation approved under chapter V to treat nicotine dependence; and

"(E) comments, data, and information submitted by interested persons.

"(h) Additional Conditions for Marketing-

"(1) MODIFIED RISK PRODUCTS- The Secretary shall require for the marketing of a product under this section that any advertising or labeling concerning modified risk products enable the public to comprehend the information concerning modified risk and to understand the relative significance of such information in the context of total health and in relation to all of the diseases and health-related conditions associated with the use of tobacco products.

"(2) COMPARATIVE CLAIMS-

"(A) IN GENERAL- The Secretary may require for the marketing of a product under this subsection that a claim comparing a tobacco product to 1 or more other commercially marketed tobacco products shall compare the tobacco product to a commercially marketed tobacco product that is representative of that type of tobacco product on the market (for example the average value of the top 3 brands of an established regular tobacco product).

"(B) QUANTITATIVE COMPARISONS- The Secretary may also require, for purposes of subparagraph (A), that the percent (or fraction) of change and identity of the reference tobacco product and a quantitative comparison of the amount of the substance claimed to be reduced shall be stated in immediate proximity to the most prominent claim.

"(3) LABEL DISCLOSURE-

"(A) IN GENERAL- The Secretary may require the disclosure on the label of other substances in the tobacco product, or substances that may be produced by the consumption of that tobacco product, that may affect a disease or health-related condition or may increase the risk of other diseases or health-related conditions associated with the use of tobacco products.

"(B) CONDITIONS OF USE- If the conditions of use of the tobacco product may affect the risk of the product to human health, the Secretary may require the labeling of conditions of use.

"(4) TIME- An order issued under subsection (g)(1) shall be effective for a specified period of time.

"(5) ADVERTISING- The Secretary may require, with respect to a product for which an applicant obtained an order under subsection (g)(1), that the product comply with requirements relating to advertising and promotion of the tobacco product.

"(i) Postmarket Surveillance and Studies-

"(1) IN GENERAL- The Secretary shall require, with respect to a product for which an applicant obtained an order under subsection (g)(1), that the applicant conduct postmarket surveillance and studies for such a tobacco product to determine the impact of the order issuance on consumer perception, behavior, and health, to enable the Secretary to review the accuracy of the determinations upon which the order was based, and to provide information that the Secretary determines is otherwise necessary regarding the use or health risks involving the tobacco product. The results of postmarket surveillance and studies shall be submitted to the Secretary on an annual basis.

"(2) SURVEILLANCE PROTOCOL- Each applicant required to conduct a surveillance of a tobacco product under paragraph (1) shall, within 30 days after receiving notice that the applicant is required to conduct such surveillance, submit, for the approval of the Secretary, a protocol for the required surveillance. The Secretary, within 60 days of the receipt of such protocol, shall determine if the principal investigator proposed to be used in the surveillance has sufficient qualifications and experience to conduct such surveillance and if such protocol will result in collection of the data or other information designated by the Secretary as necessary to protect the public health.

"(j) Withdrawal of Authorization- The Secretary, after an opportunity for an informal hearing, shall withdraw an order under subsection (g) if the Secretary determines that--

"(1) the applicant, based on new information, can no longer make the demonstrations required under subsection (g), or the Secretary can no longer make the determinations required under subsection (g);

"(2) the application failed to include material information or included any untrue statement of material fact;

"(3) any explicit or implicit representation that the product reduces risk or exposure is no longer valid, including if--

"(A) a tobacco product standard is established pursuant to section 907;

"(B) an action is taken that affects the risks presented by other commercially marketed tobacco products that were compared to the product that is the subject of the application; or

"(C) any postmarket surveillance or studies reveal that the order is no longer consistent with the protection of the public health;

"(4) the applicant failed to conduct or submit the postmarket surveillance and studies required under subsection (g)(2)(C)(ii) or subsection (i); or

"(5) the applicant failed to meet a condition imposed under subsection (h).

"(k) Chapter IV or V- A product for which the Secretary has issued an order pursuant to subsection (g) shall not be subject to chapter IV or V.

"(l) Implementing Regulations or Guidance-

"(1) SCIENTIFIC EVIDENCE- Not later than 2 years after the date of enactment of the Family Smoking Prevention and Tobacco Control Act, the Secretary shall issue regulations or guidance (or any combination thereof) on the scientific evidence required for assessment and ongoing review of modified risk tobacco products. Such regulations or guidance shall--

"(A) to the extent that adequate scientific evidence exists, establish minimum standards for scientific studies needed prior to issuing an order under subsection (g) to show that a substantial reduction in morbidity or mortality among individual tobacco users occurs for products described in subsection (g)(1) or is reasonably likely for products described in subsection (g)(2);

"(B) include validated biomarkers, intermediate clinical endpoints, and other feasible outcome measures, as appropriate;

"(C) establish minimum standards for postmarket studies, that shall include regular and long-term assessments of health outcomes and mortality, intermediate clinical endpoints, consumer perception of harm reduction, and the impact on quitting behavior and new use of tobacco products, as appropriate;

"(D) establish minimum standards for required postmarket surveillance, including ongoing assessments of consumer perception;

"(E) require that data from the required studies and surveillance be made available to the Secretary prior to the decision on renewal of a modified risk tobacco product; and

"(F) establish a reasonable timetable for the Secretary to review an application under this section.

"(2) CONSULTATION- The regulations or guidance issued under paragraph (1) shall be developed in consultation with the Institute of Medicine, and with the input of other appropriate scientific and medical experts, on the design and conduct of such studies and surveillance.

"(3) REVISION- The regulations or guidance under paragraph (1) shall be revised on a regular basis as new scientific information becomes available.

"(4) NEW TOBACCO PRODUCTS- Not later than 2 years after the date of enactment of the Family Smoking Prevention and Tobacco Control Act, the Secretary shall issue a regulation or guidance that permits the filing of a single application for any tobacco product that is a new tobacco product under section 910 and which the applicant seeks to commercially market under this section.

"(m) Distributors- Except as provided in this section, no distributor may take any action, after the date of enactment of the Family Smoking Prevention and Tobacco Control Act, with respect to a tobacco product that would reasonably be expected to result in consumers believing that the tobacco product or its smoke may present a lower risk of disease or is less harmful than one or more commercially marketed tobacco products, or presents a reduced exposure to, or does not contain or is free of, a substance or substances.

Appendix B

Chapters 1 and 2 from *Evaluation of Biomarkers and Surrogate Endpoints in Chronic Disease*[1]

[1] Institute of Medicine. 2010. *Evaluation of biomarkers and surrogate endpoints in chronic disease*. Washington, DC: The National Academies Press.

1

Introduction

Biomarkers are tools used by doctors, scientists, and other health professionals to obtain information about a patient's or research subject's health status or response to interventions. Many medical or lifestyle interventions, indispensible to modern medical care, can induce changes in biomarkers. In order for consumers, physicians, drug developers, and policy makers to make informed decisions based on biomarkers, it is important to understand the amount, strength, and quality of data supporting the use of any specific biomarker to direct decisions in clinical care, drug development, public health, and health policy decisions.

Every time a parent takes a child's temperature looking for a fever, they are using a biomarker to assess for illness. That parent may go on to monitor their child's temperature over the course of several days, both to follow the progression of an infection and to determine whether antipyretic and antimicrobial therapies are working effectively. Even this fairly simple example of a biomarker highlights some of the issues associated with their use. For example, the method used to measure body temperature matters. Using a thermometer is a more accurate approach than a hand to the forehead. Slightly different temperatures will be obtained depending on whether the measurement is an oral, ear, rectal, or axillary temperature. Although a fever is a useful piece of information about how a disease process is developing, it is only one piece of information in what could be a complex illness. To further complicate matters, some diseases present with relapsing and remitting fevers, and interpretation of temperature data in that patient population needs to be very different than an illness where a fever accompanies acute infection and resolution of the fever signals a shift to resolving the infection.

In an ideal setting, biomarkers reflect disease course and activity; many good biomarkers are useful in monitoring disease process and complications. In the diagnosis and management of prostate cancer, for example, prostate-specific antigen (PSA) can be measured in a patient's blood, and PSA levels can be followed as an indicator of whether the cancer is growing or responding to treatment. However, this example illustrates several challenges of using biomarkers. PSA may be elevated in some patients because they have prostate cancer, but it can also be elevated for other reasons. One important finding that has been reported recently is that PSA is not necessarily a good biomarker for population-wide screening for prostate cancer (Sardana et al., 2008). This illustrates the point that biomarkers are effective only to the degree that they are used in the appropriate context. It is critical to note that even a perfect biomarker cannot, with certainty, be used in place of patient outcomes in the evaluation of an intervention.

One step in supporting regulators is to institute an evidence-based, transparent process for biomarker evaluation. Biomarker evaluation is often thought of as two unlinked steps: analytical validation of biomarker tests and biomarker qualification. Biomarker qualification is the evidence-based process of linking a biomarker with one or more clinical endpoints. Decisions to use biomarkers are dependent on the intended applications. Currently, the evaluation of biomarkers is not based on uniform standards or processes, but rather on the gradual development of consensus in the scientific community. The potential value and impact of a more uniform and transparent evaluation process was noted in the 2007 Institute of Medicine (IOM) report, *Cancer Biomarkers: The Promises and Challenges of Improving Detection and Treatment* (IOM, 2007), which recommended that government agencies and non-governmental stakeholders "should work together to develop a transparent process for creating well-defined consensus standards and guidelines for biomarker development, validation, qualification, and use to reduce the uncertainty in the process of development and adoption."

The *Cancer Biomarkers* recommendation gains even more weight when considered with the emergence of pharmacogenetics, pharmacogenomics, and all of the promising medical breakthroughs of personalized medicine. Pharmacogenetics is the science of understanding how an individual's genes may interact to impact drug function and metabolism. Personalized determination of drugs that will work for given patients and dosing based on their metabolic profiles has the potential to decrease unnecessary or not helpful treatments and decrease adverse effects from treatments when they are helpful. Pharmacogenomics is the science of understanding genetic variations between populations in disease incidence, progression, and treatment. More detailed understanding of disease biology has the potential to lead to more effective prevention and treatment approaches. Biomarkers are critical to progress in these areas, and it will be important that newly discovered biomarkers be adequately studied before being adopted into routine clinical management of patients.

ORIGIN OF THE TASK

In 2008, the Food and Drug Administration's (FDA's) Center for Food Safety and Applied Nutrition (CFSAN), in conjunction with the FDA's Center for Drug Evaluation and Research, approached the IOM for advice on the topic of biomarker and surrogate endpoint evaluation, noting the limited number of surrogate endpoints available, the high cost of evaluating possible surrogate endpoints biomarkers, and the absence of an agreed-upon, systematic, transparent process for biomarker evaluation. Study developers were also interested in learning whether principles of biomarker qualification or evaluation learned in the drug development setting would also be generally applicable in other FDA-regulated product categories, such as foods and supplements. As part of its efforts within the Critical Path Initiative (CPI),[1] CFSAN requested that the IOM charge an expert committee with the following task:

> An Institute of Medicine (IOM) committee will be convened to generate recommendations on the qualification process for biomarkers, with a focus on risk biomarkers and surrogate endpoints in chronic disease. These recommendations will consider existing prototypes for qualification of biomarkers used in drug development. The committee will recommend a framework for qualification and test it using case studies of risk biomarkers and surrogate endpoints for coronary heart disease (CHD) such as low-density lipoprotein (LDL) and high-density lipoprotein (HDL) cholesterol levels. In particular, the committee will:
>
> 1. Conduct a review of current approaches to qualifying biomarkers.
> 2. Recommend a framework that can be used to rank biomarkers according to the types and quality of evidence, considering context of use for a range of product types.
> 3. Demonstrate applications through case studies.
> 4. Make ancillary recommendations for the application, enhanced development, and use of risk biomarkers and surrogate endpoints in chronic disease.[2]

[1] See http://www.fda.gov/oc/initiatives/criticalpath.

[2] The terminology in the statement of task differs in a few ways from the terminology of this report. As will be explained in Chapter 3, the committee's terminology replaces qualification with evaluation in many instances, and risk biomarker with biomarker.

CPI is a framework created by the FDA under which the challenges posed by increasing medical product development costs and lengthening time-to-market for medical products can be addressed. The need for improvement in the process for evaluation of biomarkers and surrogate endpoints was identified from the inception of the CPI at the FDA (FDA, 2004a), and formally recognized as a "Critical Path Opportunity" at CFSAN shortly thereafter (FDA, 2006). The following is an excerpt from the report published in June 2008 describing CFSAN's 2007 progress in this area (FDA, 2008):

> [The] FDA is exploring development of a framework for validating modifiable risk factors (biomarkers) for chronic diseases, such as cancer, heart disease, diabetes, and others that can be the subject of a health claim. The framework will consist of defining the level and type of evidence that is required to support a biomarker that modifies the risk of disease. The first step toward defining a framework will consist of working through the National Academy of Sciences, Institute of Medicine, to convene a panel of experts to outline the steps necessary for qualifying a biomarker for evidence-based decision making, assuming funding becomes available. The task for the panel will be to hold workshops as needed and then to issue a report that [the] FDA can use in its review of scientific evidence offered to substantiate health claims that can be used on food products, including dietary supplements. Funds from the Critical Path [I]nitiative have enabled CFSAN to develop a task order with IOM for this initiative.

Biomarkers and the FDA

With regard to biomarkers, the FDA is subject to competing forces and is expected to evaluate many factors with a limited number of resources. The desire for effective new drugs, devices, and biologics accompanied by the goal of reducing the monetary cost and time expended on development of interventions for chronic diseases serve as incentives for more aggressive use of biomarkers (IOM, 2006). The need to protect patients and consumers from undefined risks is an incentive for more conservative use of efficacy biomarkers and for the development of effective safety biomarkers.

Little consistent, reliable information is currently available regarding how consumers can know which foods might have health benefits beyond basic nutrition. Recently questions have arisen related to use of biomarkers in substantiating health claims about foods, namely whether the use of biomarkers to draw conclusions about the health benefits of nutrients, foods, and supplements should be encouraged, and how information about the uncertainty associated with using biomarkers in this way can be communicated to consumers.

Drug development costs have been estimated at $500 million to $2 billion per product depending on the size of the pharmaceutical company (Adams and Brantner, 2006). CPI began a few years after the implementation of accelerated approval regulations, and it identified a need for more biomarkers of efficacy. Public–private partnerships, such as the Critical Path Institute and the Biomarkers Consortium, were formed, in part, to foster precompetitive data sharing related to biomarker development. The Biomarkers Consortium[3] has brought together industry, academia, the FDA, the National Institutes of Health (NIH), and the Centers for Medicare & Medicaid Services to identify and address areas of greatest potential impact in the need for new qualified biomarkers. However, their focus is primarily on facilitating the discovery of new biomarkers. As a result they have not made it a priority to propose an evaluation framework for biomarkers.

At the start of CPI in 2004, it was estimated that only 8 percent of medicinal compounds reaching phase I clinical trials would eventually be approved for marketing (FDA, 2004b). One of the primary ways that CPI proposed to speed approvals was through the use of biomarkers. With accelerated approval came a greater need for postmarket studies of approved medicinal products. The FDA has faced and attempted to resolve some administrative challenges, such as manufacturers' nondisclosure and/or underreporting of adverse events that result from product usage; inadequate resources to strengthen and broaden oversight efforts; and antiquated information technology systems, in effectively requesting and enforcing these studies, as will be discussed in Chapter 5.

Nutrients, foods, and supplements are regulated under a different framework than are drugs, devices, and biologics. The FDA regulates products purchased with one out of every four consumer dollars spent. Of this amount, 75 percent is spent on products regulated by CFSAN: foods, supplements, and cosmetics. CFSAN's $470 million budget regulates the $525 billion food and cosmetics industry (FDA, 2009a). Foods do not undergo premarket evaluation. New ingredients are evaluated, but for safety only. CFSAN also regulates the labeling of foods. This includes the familiar nutrition facts panel as well as a variety of health-related claims found on food labels and promotional materials.

To a certain extent, the FDA's evaluation of health claims has been crippled by the lack of an agreed-upon, transparent process for biomarker evaluation. Authorized and qualified health claims, which describe links between a food substance and a reduction in risk for a disease, may include data based on the measurement of surrogate endpoints or risk biomarkers as justification for the claims. It is uncommon for produc-

[3] See http://www.biomarkersconsortium.org.

ers of foods or supplements to study the effects of foods and nutrients on clinical endpoints, which makes data from surrogate endpoints and biomarkers the focus of applications for health claims. These include folic acid for reducing the risk for neural tube defects and soluble oat fiber for reducing the risk of heart disease. Claims must be evaluated and authorized by the FDA in most cases. In some cases, health claims can be authorized based on a statement from an authoritative body, such as the NIH or the National Academy of Sciences.[4] The lack of an agreed-upon, transparent process for biomarker evaluation has been seen as one of the roadblocks to a broader selection of surrogate endpoints on which claims could be based.

DEFINITIONS

The committee observed a great deal of inconsistent and imprecise definition and use of terms relevant to biomarkers and biomarker evaluation. Consistent, precise definition and use of terms is critical for biomarker evaluation because it is a topic important across many disciplines and has been for several decades. The committee has attempted to be consistent with the spirit of previous efforts at standardizing the language used with reference to biomarker evaluation, and clarifies several definitions where there is overlap or potential for confusion. Several of the definitions used in the report summary (see Box 1-1 below) deserve further discussion. A definition of risk biomarker, used in the statement of task, is also defined in Box 1-1.

The definition of the term "biomarker" itself is not controversial. The definition provided by the Biomarkers Definitions Working Group is widely used, and other definitions do not differ fundamentally. The *Cancer Biomarkers* report presented two tables showing uses of biomarkers in clinical and drug development settings (see Tables 1-1 and 1-2). The committee viewed results from imaging tests as biomarkers because they are measurements that indicate normal biological processes, predict risk for disease, and monitor pathogenic processes and pharmacologic responses to therapeutic interventions. The committee also viewed genes, genetic signatures, and genetic mutations as biomarkers. While these are typically not modifiable, they do fulfill the Biomarkers Definitions Working Group definition of a biomarker, as they indicate normal biological processes, pathogenic processes, or pharmacologic responses.

The statement of task for this study cites "risk biomarkers" for chronic disease. The committee defines a risk biomarker as a biomarker that

[4] In legislation, the term National Academy of Sciences refers to the whole of the National Academies.

BOX 1-1
Important Definitions

Analytical Validation: "assessing [an] assay and its measurement performance characteristics, determining the range of conditions under which the assay will give reproducible and accurate data."[a]

Biomarker: "a characteristic that is objectively[b] measured and evaluated as an indicator of normal biological processes, pathogenic processes, or pharmacologic responses to a[n] . . . intervention."[c] Example: cholesterol level.

Chronic Disease: a culmination of a series of pathogenic processes in response to internal or external stimuli over time that results in a clinical diagnosis/ailment and health outcomes. Example: diabetes.

Clinical Endpoint: "a characteristic or variable that reflects how a patient [or consumer] feels, functions, or survives."[c] Example: death.

Fit-for-Purpose: being guided by the principle that an evaluation process is tailored to the degree of certainty required for the use proposed.

Qualification: "evidentiary process of linking a biomarker with biological processes and clinical endpoints."[d]

Surrogate Endpoint: "a biomarker that is intended to substitute for a clinical endpoint. A surrogate endpoint is expected to predict clinical benefit (or harm or lack of benefit or harm) based on epidemiologic, therapeutic, pathophysiologic, or other scientific evidence."[c] Example: blood pressure for trials of several classes of antihypertensive drugs.[e]

NOTES: [b] The committee defines "objectively" to mean "reliably and accurately." [e] Please see Chapter 2 for discussion of this biomarker.
SOURCES: [a] Wagner (2002); [c] Biomarkers Definitions Working Group (2001); and [d] Wagner (2008).

indicates a risk factor for a disease. In other words, it is a biomarker that indicates a component of an individual's level of risk for developing a disease or level of risk for developing complications of a disease. The committee viewed risk biomarkers as a subset of risk factors. Risk factors are variables that correlate with incidence of a disease or condition. Risk factors include social and environmental factors in addition to biological factors. Risk biomarkers are also to be distinguished from biomarkers of exposure used in toxicology, which were defined by the National Research Council as "the chemical or its metabolite or the product of an interaction between a chemical and some target molecule or cell that is measured in a compartment in an organism" (NRC, 2006; WHO, 2001). In its *Guidance for Industry: Evidence-Based Review System for the Scientific Evaluation of Health Claims*, CFSAN defined risk biomarkers as "biological indicators

TABLE 1-1 Use of Biomarkers in Chronic Disease Patient Care

Clinical Biomarker Use	Clinical Objective
Disease risk stratification	Assess the likelihood that the disease will develop (or recur)
Prevention	Identify and track risk factors
Screening[a]	Detect and treat early-stage disease in the asymptomatic population
Diagnosis	Definitively establish the presence of disease
Classification[b]	Classify patients by disease subset
Prognosis	Predict the probable outcome of disease to determine the aggressiveness of treatment
Prediction/treatment stratification[b]	Predict response to particular therapies and choose the drug that is mostly likely to yield a favorable response in a given patient
Therapy-related risk management[a]	Identify patients with a high probability of adverse effects of a treatment
Therapy monitoring[c]	Determine whether a therapy is having the intended effect on a disease and whether adverse effects arise
Surveillance	Early detection and treatment of advancing disease or complications

NOTES: [a] In toxicology, biomarkers of exposure help predict an individual's risk of suffering consequences from exposure to a foreign substance. Exposure biomarkers are a subset of these two categories. [b] Companion diagnostic biomarkers include features from several of these categories. These tests identify whether an individual's molecular profile associated with a disease pathophysiology is likely to respond favorably to a particular therapeutic. Examples include KRAS–cetuximab, HER-2–herceptin, and estrogen receptor status–tamoxifen. [c] Dose optimization is a subset of this category.
SOURCE: Adapted from IOM (2007).

TABLE 1-2 Use of Biomarkers in Drug Development

Biomarker Use	Drug Development Objective
Target validation	Demonstrate that a potential drug target plays a key role in the disease process
Early compound screening	Identify compounds with the most promise for efficacy and safety
Pharmacodynamic assays	Determine drug activity; select dose and schedule
Patient selection	In clinical trials, patient selection (inclusion/exclusion) by disease subset or probability of response/adverse events
Surrogate endpoint	Use of a short-term outcome measure in place of the long-term primary endpoint to determine more quickly whether the treatment is efficacious and safe in drug regulatory approval

SOURCE: IOM (2007).

that signal a changed physiological state that is associated with the risk of a disease" (CFSAN, 2009). This definition is narrower than the committee's because it would seem not to include genetic risk factors and other situations that may be present in an individual from birth. Many risk biomarkers are not modifiable in beneficial ways, even when only ones indicating changed physiological states are considered. It is important to note that while some so-called risk biomarkers have been used as surrogate endpoints, risk biomarkers are not surrogate endpoints unless they are determined to be supported for use as such for a defined context of use through use of the biomarker evaluation framework and expert panel as described in Recommendations 1 and 2.

The definition of "surrogate endpoint" is critical for clear communication and transparency in regulatory processes. Several definitions of surrogate endpoint have been used. Table 1-3 shows definitions that have appeared in regulations and other regulatory documents. Table 1-4 shows literature definitions.

TABLE 1-3 Regulatory Definitions of Surrogate Endpoint

Source	Definition
57 *FR* 13234–13242 (1992)[a]	A surrogate end point, or "marker," is a laboratory measurement or physical sign that is used in therapeutic trials as a substitute for a clinically meaningful endpoint that is a direct measure of how a patient feels, functions or survives and is expected to predict the effect of the therapy.
FDAMA (Food and Drug Administration Modernization Act) 1997 USC Section 504(b)(1)	...a surrogate endpoint that is reasonably likely to predict clinical benefit.
Title 21 – Food and Drugs 21 C.F.R. 314 Section 314.510[b]	...a surrogate endpoint that is reasonably likely, based on epidemiologic, therapeutic, pathophysiologic, or other evidence, to predict clinical benefit.
Guidance for Industry: Evidence-based review system for the scientific evaluation of health claims[c]	Surrogate endpoints are risk biomarkers that have been shown to be valid predictors of disease risk and therefore may be used in place of clinical measurements of the onset of the disease in a clinical trial.

SOURCES: [a] New drug, antibiotic and biological drug product regulations: accelerated approval. Proposed Rule. 57 *Federal Register* 13234–13242 (1992). [b] Food and Drug Modernization Act of 1997, 21 USC section 506(b)(1) (1997). Title 21—Food and Drugs, 21 CFR 314 Section 314.510 (2008) [http://frwebgate5.access.gpo.gov/cgi-bin/TEXTgate.cgi?WAISdocID=026369143256+87+1+0&WAISaction=retrieve]. [c] http://www.cfsan.fda.gov/~dms/hclmgui6.html.

TABLE 1-4 Literature Definitions of Surrogate Endpoint

Source	Definition
Biomarkers Definitions Working Group (2001)	A biomarker that is intended to substitute for a clinical endpoint. A surrogate endpoint is expected to predict clinical benefit (or harm or lack of benefit or harm) based on epidemiologic, therapeutic, pathophysiologic, or other scientific evidence.
Guide to Clinical Trials (Spilker, 1991)	The ideal surrogate endpoint is a disease marker that reflects what is happening with the underlying disease. The relationship between the marker and the true endpoint is important to establish. After this is done, the validity of data based on how the marker is affected by a medicine or other treatment can be translated into a valid statement about the disease and true endpoint.
Prentice (1989)[a]	A response variable for which a test of the null hypothesis of no relationship to the treatment groups under comparison is also a valid test of the corresponding null hypothesis based on the true endpoint.
Temple (1995)[a]	A surrogate endpoint of a clinical trial is a laboratory measurement or a physical sign used as a substitute for a clinically meaningful endpoint that measures directly how a patient feels, functions or survives. Changes induced by a therapy on a surrogate endpoint are expected to reflect changes in a clinically meaningful endpoint.
Johnston (1999)[a]	A surrogate outcome measure is simply one that is used in place of a clinical endpoint ... an adequate surrogate measure must not only correlate with the clinical endpoint, but it must be predictive of the clinical endpoint in the presence of the intervention under study.
Baker et al. (2005)[a]	A surrogate endpoint is defined as a measure or indicator of a biological process that is obtained sooner, at less cost or less invasively than a true endpoint of health outcome, and is used to make conclusions about the effect of an intervention on the true endpoint.
Grimes and Schulz (2005)[a]	A valid surrogate endpoint must both correlate with and accurately predict the outcome of interest.
Gluud et al. (2007)[a]	A surrogate outcome measure is a laboratory measurement, a physical sign, or any other intermediate substitute that is able to predict a treatment response on a clinically meaningful outcome measure.
Pryseley et al. (2007)[a]	A surrogate for a true endpoint is an endpoint that can be used in lieu of the true endpoint to assess treatment benefits. That is, the effect of the treatment on the surrogate endpoint should reliably predict the effect of the treatment on the true endpoint.
Gobburu (2009); Lathia et al. (2009)	A biomarker that is intended to substitute for a clinical endpoint.

NOTE: [a] See also the Shi and Sargent (2009) compilation of these definitions.

There are a few common features in the overwhelming majority of the surrogate endpoint definitions. First, a surrogate endpoint is meant to substitute for a clinically meaningful endpoint. Second, the surrogate endpoint needs to predict change in those clinical outcomes given an intervention. The last definition in Table 1-4 appears to be for a proposed surrogate endpoint, not for one that has already been determined to satisfy the requirements of a true surrogate endpoint. The committee views this definition as being too inclusive to be accurate. This definition is not consistent with consensus and most regulatory definitions of surrogate endpoint. The last definition in Table 1-3 is the definition used by CFSAN for review of health claims that industry submits for inclusion in food labeling. The citation given in the guidance document is for Spilker's *Guide to Clinical Trials* (shown in Table 1-4; 1991); however, the definition in the guidance document is not consistent with the one it cites. In Dr. Spilker's more recent book, *Guide to Drug Development: A Comprehensive Review and Assessment* (2009), the Biomarkers Definitions Working Group definition is used. The CFSAN definition does not include a critical component of the definition of surrogate endpoints: the ability to predict clinical benefit or harm of an intervention based on a change in the surrogate endpoint. The use of the word "valid" in this definition is also ambiguous, as will be discussed below. Finally, the CFSAN definition accounts only for use of surrogate endpoints in clinical trials and does not allow for use in observational studies. The Biomarkers Definitions Working Group's definition takes into account uses of surrogate endpoints in observational studies.

There are a number of other important concepts to understand when considering surrogate endpoints. The Prentice criteria are succinctly summarized in two parts: correlation and capture. Under correlation, the surrogate endpoint must be statistically correlated to the clinical endpoint. In other words, the surrogate endpoint should have prognostic value relative to the clinical endpoint. Under capture, an intervention's entire effect on the clinical endpoint should be explained by the intervention's effect on the surrogate endpoint. In other words, the surrogate endpoint should account for all of an intervention's effects; the surrogate endpoints should be a perfect proxy for the effect of an intervention on the recipient's risk of important clinical outcomes (Desai et al., 2006; Prentice, 1989).

The terms "clinical endpoint" and "true endpoint" are sometimes used interchangeably. The definition of clinical endpoint given in Box 1-1 is widely accepted and consistently used, while the term true endpoint is broader and ill defined. To some, only all-cause mortality is a true endpoint. In practice, however, a trial's true endpoint is defined by the experimenters. It can be mortality due to the disease being studied, failure of the treatment (which can be defined in several ways), time to progression, or something else. Sometimes, a surrogate endpoint in one study can be the clinical endpoint in another study. Practically, the true endpoint is the endpoint for which a surrogate endpoint is sought. Myocardial infarction (MI) is an example. Because an MI in a person outside the hospital is detected from symptoms, it is a plausible clinical endpoint. However, it should be acknowledged that a significant element of the importance of MI derives from both the fact that it is a biomarker for risk of future events (death, heart failure) and that it requires objective biomarker measurements for the diagnosis.

The term "validation" encompasses many different aspects of biomarker development. In the statistics literature, validation means what other fields term "qualification." Validation and analytical validation are often used interchangeably, as are clinical validation and qualification. Clinical utility is often used interchangeably with utilization. In this report, the committee uses validation and analytical validation interchangeably, qualification but not clinical validation, and utilization but not clinical utility.

Correct definition of the terms food, substance, disease, and drug are important for understanding FDA regulations. Food is defined as (1) articles used for food or drink for humans or other animals, (2) chewing gum, and (3) articles used for components of any such article.[5] As was noted in the summary of this report, however, the committee has been more explicit in its definition: the term "food" is inclusive of foods consumed as part of meals and snacks, dietary supplements, and components contained in them (nutrients, other bioactive substances). A substance is "a specific food (tomato) or component of food (lycopene), whether in conventional food or dietary supplement form"[6] (Trumbo and Ellwood, 2009). A disease or health-related condition is "Damage to an organ, part, structure, or system of the body such that it does not function properly (e.g., CHD), or a state of health leading to such dysfunctioning (e.g., hypertension)"[7] (Trumbo and Ellwood, 2009). A drug is defined as "articles intended for use in the diagnosis, cure, mitigation, treatment, or prevention of disease" and "articles (other than food) intended to affect the structure or any function of the body of man or other animals"[8] (FDA, 2002). The term "intervention" refers to any drug, device, biologic, behavioral modification, nutritional modification, lifestyle modification, or other treatment intended to improve health.

[5] FDCA, Sec. 201(II)(f).

[6] 21 C.F.R. 101.14(a)(2).

[7] 21 C.F.R. 101.14(a)(5).

[8] FDCA, Sec. 201(g)(1).

RELATED IOM WORK

The committee views this report as building on and supporting the recommendations of several previous committees. In particular, the committee would like to reemphasize the recommendations of the report on *Cancer Biomarkers* (Box B-1) and the report on *The Future of Drug Safety* (Box B-2). The recommendations from both of these reports are included in Appendix B. *Cancer Biomarkers* grouped its recommendations into three categories: (1) methods, tools, and resources needed to discover and develop tools for cancer; (2) guidelines, standards, oversight, and incentives needed for biomarker development; and (3) methods and processes needed for clinical evaluation and adoption. Government agencies, academics, healthcare practitioners, industrial stakeholders, and the Institute of Medicine have been working to explore and implement changes that reflect the needs identified in the recommendations. As mentioned earlier, the current report was requested by the FDA as a path forward on recommendation 6 from the *Cancer Biomarkers* report.

The recommendations from *The Future of Drug Safety* were grouped into categories: organizational culture, science and expertise, regulation, communication, and resources. Following the release of the report in 2007, the Food and Drug Administration Amendments Act was passed. It reauthorized a number of key pieces of legislation important for increasing drug safety and expanded FDA responsibilities and capabilities to respond to a number of *The Future of Drug Safety* report's recommendations (FDA, 2009b). In 2009, the FDA published a table describing the significant progress made on implementation of the IOM recommendations (FDA, 2009c).

FRAMEWORK OF THE REPORT

The framework of the report follows the statement of task and the committee's recommendations. Chapter 2 reviews previous biomarker and surrogate endpoint evaluation processes. Chapter 3 presents the committee's recommended biomarker evaluation framework. Chapter 4 contains the case studies that exemplify use of the biomarker evaluation process. Finally, Chapter 5 describes data collection and data infrastructure needs to support the FDA's work.

REFERENCES

Adams, C. P., and V. V. Brantner. 2006. Estimating the cost of new drug development: Is it really $802 million? *Health Affairs* 25(2):420–428.

Baker, S. G., G. Izmirlian, and V. Kipnis. 2005. Resolving paradoxes involving surrogate end points. *Journal of the Royal Statistical Society Series A* 168(4):753–762.

Biomarkers Definitions Working Group. 2001. Biomarkers and surrogate endpoints: Preferred definitions and conceptual framework. *Clinical Pharmacology and Therapeutics* 69(3):89–95.

CFSAN (Center for Food Safety and Applied Nutrition). 2009. *Guidance for industry: Evidence-based review system for the scientific evaluation of health claims*. http://www.cfsan.fda.gov/~dms/hclmgui6.html (accessed February 25, 2009).

Desai, M., N. Stockbridge, and R. Temple. 2006. Blood pressure as an example of a biomarker that functions as a surrogate. *AAPS Journal* 8(1):E146.

FDA (Food and Drug Administration). 2002. *Is it a cosmetic, a drug, or both? (Or is it soap?)* http://www.fda.gov/Cosmetics/GuidanceComplianceRegulatoryInformation/ucm074201.htm (accessed October 2, 2009).

FDA. 2004a. *Challenge and opportunity on the critical path to new medical products*. http://www.fda.gov/oc/initiatives/criticalpath/whitepaper.html (accessed December 10, 2008).

FDA. 2004b. *Challenge and opportunity on the critical path to new medical products*. Washington, DC: FDA.

FDA. 2006. *Critical Path opportunities list*. http://www.fda.gov/oc/initiatives/criticalpath/reports/opp_list.pdf (accessed December 10, 2008).

FDA. 2008. *Key FDA Critical Path activities under way in 2007*. http://www.fda.gov/oc/initiatives/criticalpath/report2007.html (accessed December 10, 2008).

FDA. 2009a. *CFSAN – What we do*. http://www.fda.gov/AboutFDA/CentersOffices/CFSAN/WhatWeDo/default.htm (accessed October 1, 2009).

FDA. 2009b. *Regulatory information: Food and Drug Administration Amendments Act (FDAAA) of 2007*. http://www.fda.gov/RegulatoryInformation/Legislation/FederalFoodDrugandCosmeticActFDCAct/SignificantAmendmentstotheFDCAct/FoodandDrugAdministrationAmendmentsActof2007/default.htm (accessed November 20, 2009).

FDA. 2009c. *Report to Congress: Changing the future of drug safety: FDA initiatives to strengthen and transform the drug safety system*. http://www.fda.gov/downloads/Safety/SafetyofSpecificProducts/UCM184046.pdf (accessed November 20, 2009).

Gluud, C., J. Brok, Y. Gong, and R. L. Koretz. 2007. Hepatology may have problems with putative surrogate outcome measures. *Journal of Hepatology* 46(4):734–742.

Gobburu, J. V. S. 2009. Biomarkers in clinical drug development. *Clinical Pharmacology and Therapeutics* 86(1):26–27.

Grimes, D. A., and K. F. Schulz. 2005. Surrogate end points in clinical research: Hazardous to your health. *Obstetrics and Gynecology* 105(5 Pt 1):1114–1118.

IOM (Institute of Medicine). 2006. *Developing biomarker-based tools for cancer screening, diagnosis, and treatment: The state of the science, evaluation, implementation, and economics*. Washington, DC: The National Academies Press.

IOM. 2007. *Cancer biomarkers: The promises and challenges of improving detection and treatment*. Washington, DC: The National Academies Press.

Johnston, K. 1999. What are surrogate outcome measures and why do they fail in clinical research? *Neuroepidemiology* 18(4):167–173.

Lathia, C. D., D. Amakye, W. Dai, C. Girman, S. Madani, J. Mayne, P. MacCarthy, P. Pertel, L. Seman, A. Stoch, P. Tarantino, C. Webster, S. Williams, and J. A. Wagner. 2009. The value, qualification, and regulatory use of surrogate end points in drug development. *Clinical Pharmacology and Therapeutics* 86(1):32–43.

NRC (National Research Council). 2006. *Human biomonitoring for environmental chemicals*. Washington, DC: The National Academies Press.

Prentice, R. L. 1989. Surrogate endpoints in clinical trials: Definition and operational criteria. *Statistics in Medicine* 8(4):431–440.

Pryseley, A., A. Tilahun, A. Alonso, and G. Molenberghs. 2007. Information-theory based surrogate marker evaluation from several randomized clinical trials with continuous true and binary surrogate endpoints. *Clinical Trials* 4(6):587–597.

Sardana, G., B. Dowell, and E.P. Diamandis. 2008. Emerging biomarkers for the diagnosis and prognosis of prostate cancer. *Clinical Chemistry* 54(12):1951–1960.

Shi, Q., and D. J. Sargent. 2009. Meta-analysis for the evaluation of surrogate endpoints in cancer clinical trials. *International Journal of Clinical Oncology* 14(2):102–111.

Spilker, B. 1991. Surrogate endpoints. In *Guide to clinical trials*. New York: Raven Press. Pp. 608–609.

Spilker, B. 2009. Surrogate endpoints and biomarkers. In *Guide to drug development: A comprehensive review and assessment*. Philadelphia, PA: Lippincott Williams and Wilkins.

Temple, R. J. 1995. A regulatory authority's opinion about surrogate endpoints. In *Clinical measurement in drug evaluation*, edited by W. S. Nimmo and G. T. Tucker. New York: John Wiley and Sons. Pp. 3–22.

Trumbo, P., and K. Ellwood. 2009. *Developing a framework for biomarker qualification for chronic disease*. Paper presented at Institute of Medicine Committee on Qualification of Biomarkers as Surrogate Endpoints for Chronic Disease Risk, Washington, DC, January 12.

Wagner, J. A. 2002. Overview of biomarkers and surrogate endpoints in drug development. *Disease Markers* 18(2):41–46.

Wagner, J. A. 2008. Strategic approach to fit-for-purpose biomarkers in drug development. *Annual Review of Pharmacology and Toxicology* 48:631–651.

WHO (World Health Organization). 2001. Biomarkers in risk assessment: Validity and validation. Environmental health criteria 222. http://www.inchem.org/documents/ehc/ehc/ech222.htm (accessed March 16, 2010).

2

Review: Evaluating and Regulating Biomarker Use

INTRODUCTION

The context within which this study is set has developed from the contributions of various scientific fields, industries, and government bodies. From toxicology to cardiology, from the food industry to the drug industry, and from the Food and Drug Administration (FDA) to the federal courts, biomarkers and the scientific evidence needed to substantiate their use have been topics of discussion for several decades. Along with a brief review of biomarker evaluation methods and their uses, this chapter seeks to describe critical areas of background information so that readers from different fields can gain a more comprehensive understanding of the policy and regulatory issues with respect to biomarkers.

Methods for evaluation of biomarkers and surrogate endpoints have been reviewed successfully and systematically in the recent past (Lassere, 2008; Shi and Sargent, 2009). This chapter will direct the readers toward appropriate reviews, and it will discuss the evolution of thinking at the FDA—focusing on the Center for Food Safety and Applied Nutrition (CFSAN), in particular—regarding surrogate endpoints. It will also discuss the evolution in thinking in academic and industry communities, to a lesser extent. The contents of this chapter are as follows:

- Use of biomarkers in areas as diverse as scientific research, medical practice, product development, and public health policy
- Use of biomarkers as surrogate endpoints
- Evaluation frameworks proposed from academia and industry

- The broader context of biomarker and surrogate endpoint evaluation by the FDA, including the legal and regulatory basis for claims made on CFSAN-regulated products

Examples are included on blood pressure as a surrogate endpoint, HIV/AIDS drug development, arrhythmia suppression interventions, exercise tolerance in congestive heart failure, and kidney toxicity biomarkers.

SURVEY OF BIOMARKER USES

Biomarkers have a wide array of uses in a variety of fields. These fields include medicine, oral health, mental health, nutrition, environmental health, toxicology, developmental biology, and basic scientific research. They are used to study the safety and efficacy of interventions, develop understanding of the mechanisms of disease, make good decisions in clinical care, and guide the policies that impact public health. Table 2-1 gives a list of several categories of biomarker use.

For the uses in Table 2-1, any biomarker would need to be evaluated to ensure that data supporting the biomarker's association with the disease or condition of interest and the analytical validation of the test are adequate for the proposed use. In situations, however, where biomarker data will not or is not yet anticipated to be submitted to the FDA for a regulatory purpose or used by professional societies or other groups for clinical practice guidelines or other decision-making processes impacting public health or the practice of medicine, this may be an informal process. Ideally, evaluations are already done by clinicians, product developers, government regulators, professional societies, and scientists; this report's contribution is to propose a systematic process for biomarker evaluation.

Use of Biomarkers and Surrogate Endpoints for Clinical Efficacy Studies and Formation of Clinical Practice Guidelines

Surrogate endpoints were defined in Chapter 1 and can be found in several locations in Table 2-1. First, they have been used in approvals of products or claims for drugs, biologics, devices, foods, and supplements. This will be discussed further in several subsections of this chapter's section on evolution of regulatory perspectives on surrogate endpoints and in Chapter 5. Second, they have been used in the formulation of clinical practice guidelines. As defined by an Institute of Medicine (IOM) committee in 1990, "practice guidelines are systematically developed statements to assist practitioner and patient decisions about appropriate health care

TABLE 2-1 Categories of Biomarker Use

Use	Description
Discovery	Identification of biochemical, image, or other biomarkers associated with a disease, condition, or behavior of interest; biomarkers identified may be screened for many potential uses, including as a target for intervention to prevent, treat, or mitigate a disease or condition
Early product development	Biomarkers used for target validation, compound screening, pharmacodynamic assays, safety assessments, and subject selection for clinical trials, and as endpoints in early clinical screening (i.e., phase I and II trials)
Surrogate endpoints for claim and product approvals	Biomarkers used for phase III clinical testing and biomarkers used to substantiate claims for product marketing
Clinical endpoints	Biomarkers used as endpoints for clinical trials that measure how a patient feels, functions, or survives; for example, measures of depression, blindness, and muscle weakness are biomarkers that may be used as clinical endpoints
Clinical practice	Biomarkers used by clinicians for uses such as risk stratification, disease prevention, screening, diagnosis, prognosis, therapeutic monitoring, and posttreatment surveillance
Clinical practice guidelines	Biomarkers used to make generalized recommendations for healthcare practitioners in the areas of risk stratification, disease prevention, treatment, behavior/lifestyle modifications, and more
Comparative efficacy and safety	Biomarkers used in clinical studies looking at the relative efficacy, safety, and cost effectiveness of any or all interventions used for a particular disease or condition, including changes in behavior, nutrition, or lifestyle; these studies are a component of comparative effectiveness research
Public health practice	Biomarkers used to track public health status and make recommendations for prevention, mitigation, and treatment of diseases and conditions at the population level

for specific clinical circumstances" (IOM, 1990). Clinical practice guidelines and the systematic reviews that inform them are the subjects for two current IOM studies;[1] the reports are expected in 2011. A guideline regarding treatment of a particular disease may identify target levels for specific biomarkers. In order to arrive at a recommendation for a particular biomarker level, clinical trial and observational data must be evaluated. It is possible that more trials will measure a particular surrogate endpoint in addition to or rather than the clinical endpoint of interest. In these cases, it may be desirable to include data from trials that did not measure the clinical endpoints of interest in the systematic reviews.

It is useful to mention that professional societies play an essential role in helping stakeholders understand the best ways to use biomarker-related information in clinical practice. One way in which professional societies assist in the understanding and use of biomarker data is through the promulgation of clinical practice guidelines. The committee recognized that clinical practice guidelines could use the committee's proposed biomarker evaluation framework in reaching decisions. Other methods of rigorous, systematic review, including the Cochrane Collaboration, may also be valuable in assessing the evidence associated with clinical practice guidelines. One consideration that bodies involved in the work of determining the best clinical practice guideline may need to make is that of cost effectiveness. The committee viewed this topic as being beyond the statement of task for this study and well studied elsewhere, but the committee recognizes that comparisons of interventions looking at the number of quality-adjusted life-years gained through use of an intervention or relative to no intervention are useful.

The IOM recently released a report, *Initial National Priorities for Comparative Effectiveness Research* (IOM, 2009c), which identified six characteristics of comparative effectiveness research, or CER (Box 2-1). In general, use of surrogate endpoints in CER would not fulfill the fourth characteristic of comparative effectiveness research, as identified in the report (IOM, 2009c). Quoted below is the report's description of this characteristic of CER:

> **CER measures outcomes—both benefits and harms—that are important to patients.**
>
> The committee is using the term "effectiveness" in reference to the extent to which a specific intervention, procedure, regimen, or service does what it is intended to do when used under *real-world* circumstances.

---

[1] Standards for Developing Trustworthy Clinical Practice Guidelines (http://www8.nationalacademies.org/cp/projectview.aspx?key=49125) and Standards for Systematic Reviews of Clinical Effectiveness Research (http://www8.nationalacademies.org/cp/projectview.aspx?key=49124).

BOX 2-1
Characteristics of Comparative Effectiveness Research (CER)

1. CER has the objective of directly informing a specific clinical decision from the patient perspective or a health policy decision from the population perspective.
2. CER compares at least two alternative interventions, each with the potential to be "best practice."
3. CER describes results at the population and subgroup levels.
4. CER measures outcomes—both benefits and harms—that are important to patients.
5. CER employs methods and data sources appropriate for the decision of interest.
6. CER is conducted in settings that are similar to those in which the intervention will be used in practice.
7. CER has the objective of directly informing a specific clinical decision from the patient perspective or a health policy decision from the population perspective.
8. CER compares at least two alternative interventions, each with the potential to be "best practice."
9. CER describes results at the population and subgroup levels.
10. CER measures outcomes—both benefits and harms—that are important to patients.
11. CER employs methods and data sources appropriate for the decision of interest.
12. CER is conducted in settings that are similar to those in which the intervention will be used in practice.

SOURCE: IOM (2009c).

This can be contrasted with "efficacy," which is the extent to which an intervention produces a beneficial result under controlled conditions (Cochrane, 1971; Higgins and Green, 2008). This implies an important distinction between much clinical research and CER, in that CER places high value on external validity, or the ability to generalize results to real-world decision making. Harms or risks of unintended consequences are also outcomes of interest, because they influence the net benefits of an intervention. Including and giving weight to patient-reported outcomes is particularly important for CER studies in which patient ratings of effectiveness or adverse events may differ from clinical measures. Finally, resource utilization may be highly relevant to net benefits when comparing the full clinical course of interventions over time. Cost-effectiveness analysis is a useful tool of CER, allowing evaluation of the full range

> of treatment outcomes in relationship to the difference in costs. Robust evidence of comparative clinical effectiveness is a building block necessary for resource allocation decisions. Moreover, just as clinical effects may vary in different settings, costs vary as well, so a given set of cost-effectiveness results is often not generalizable. (IOM, 2009c)

Comparative effectiveness research is meant to fill gaps in evidence that prevent comparison of available treatments (IOM, 2009c) with a focus on outcome measurements that are tangible to the person rather than biomarkers or putative surrogate endpoints. Occasionally, it may be impractical for many of these studies to examine clinical endpoints; careful selection of surrogate endpoints after significant interaction with patient groups and expert investigators would be necessary. Finally, surrogate endpoints can be found in public health practice when there is a need to estimate the health of populations or short-term impacts of longer-term programs for prevention, treatment, or mitigation of infectious or chronic diseases when health outcomes important to patients cannot be measured. For example, reporting to stakeholders about interventions to decrease diseases and conditions of importance in the population, such as stroke or heart attack, may be done by measuring and reporting blood pressure as a surrogate for the desired improvement in health status, although measuring health outcomes important to patients such as stroke or quality of life would be preferable as guidance to public health interventions unless such measures were deemed impractical.

Surrogate Endpoints: Successes

The most widely discussed use of surrogate endpoints is in phase III clinical studies used to support applications for new drugs, biologics, and devices and to support claims on foods and supplements. In his presentation to the committee during its April public workshop, Dr. Robert Temple of the Center for Drug Evaluation and Research (CDER) at the FDA outlined the reasons why researchers and clinicians use surrogate endpoints (Temple, 2009).

These reasons include when the clinical endpoint is rare or takes years to develop; when the surrogate endpoints seem to be obviously linked to the clinical endpoint of interest (e.g., tumor size in cancer or maintenance of regular heart rhythm in arrhythmia patients); and when other treatments exist, to alleviate the difficulties of conducting trials when a new intervention must be proven as non-inferior to existing treatments. In addition, although it may be possible to use a clinical endpoint in a population at high risk for the disease or condition, studying a population at relatively lower risk using the clinical endpoint may be too burdensome since the number of subjects required would be very large. Dr. Temple noted that the idea of a surrogate endpoint is to enable faster, smaller, more efficient clinical trials that can address urgent needs and facilitate the advancement of medicine.

Two notable successes of the use of surrogate endpoints are discussed in the next sections: blood pressure and HIV-1 RNA. The first example details the history of the evaluation of blood pressure as a surrogate endpoint. It may be surprising to readers that blood pressure as a surrogate endpoint for cardiovascular disease endpoints was hotly debated for decades before reaching its current status. Still, there is no broad agreement that blood pressure is a universal surrogate endpoint (Carter, 2002; Psaty et al., 1996). Even though these examples describe successful use of surrogate endpoints, important caveats are also described. Dr. Temple and others have noted surprises and mistakes in the selection and use of surrogate endpoints, and so several examples of these are discussed after the sections on blood pressure and HIV-1 RNA.

Blood Pressure

Blood pressure is often looked to as an exemplar surrogate endpoint for cardiovascular mortality and morbidity due to the levels and types of evidence that support its use. More than 75 antihypertensive agents in more than 9 therapeutic classes demonstrate the wide availability of agents to treat hypertension (Israili et al., 2007). Although new antihypertensive drugs are approved on the basis of blood pressure reductions, blood pressure's history as a surrogate endpoint is unusual in that many drugs used to treat hypertension (thiazides, methyldopa, reserpine, hydralazine, guanethidine) were approved prior to the FDA's effectiveness requirement or the availability of clinical trial data supporting the impact of blood pressure control on cardiovascular outcomes (Desai et al., 2006).

The status of blood pressure as a surrogate endpoint for cardiovascular disease endpoints was debated for decades (Perry et al., 1978). Even as one of the most well-established surrogate endpoints, an effect on blood pressure may not fully capture the benefit—or risk—of an intervention.

Although some issues are still outstanding, the benefits of blood pressure control are mostly well understood due to comprehensive epidemiologic and clinical trial evidence. Hypertension has been identified as the most common risk biomarker for cardiovascular morbidity and mortality, with a World Health Organization report suggesting that hypertension is the single most important preventable cause of premature death in developed countries (Ezzati et al., 2002). Data suggest that in the United States, hypertension is responsible for 35 percent of myocardial infarctions and strokes, 49 percent of episodes of heart failure, and 24 percent of premature deaths (Wolff and Miller, 2007). Hypertension affects one in four U.S. adults, but the majority of those affected remain either untreated or undertreated in spite of the substantial health benefits gained from modest blood pressure reductions (Wang and Vasan, 2005).

Epidemiological, clinical trial data Williams (2005) suggested that the blood pressure–cardiovascular outcomes relationship is substantiated by one of the strongest evidence bases in clinical medicine. Epidemiologic studies consistently demonstrate the relationship between blood pressure and cardiovascular mortality and morbidity, including one meta-analysis of nine studies that demonstrated an association between diastolic blood pressure and coronary heart disease and stroke in 420,000 subjects (MacMahon et al., 1990). Observational studies have also demonstrated the robustness of blood pressure's relationship to heart disease in adults; despite different assessment parameters (systolic alone, diastolic alone, or systolic and diastolic), the relationship is maintained (Desai et al., 2006). This relationship has also been confirmed in diverse populations, including different genders, adult age groups, and race/ethnicities. In children, this relationship does not hold (Brady and Feld, 2009).

Both placebo- and active-controlled clinical trials conducted in the past three to four decades have demonstrated that pharmacologic reductions in blood pressure reduce cardiovascular mortality and morbidity (Desai et al., 2006). While earlier trials compared hypertension agents against placebo, the growing evidence base supporting the benefit of hypertension therapy necessitated head-to-head trials comparing two or more agents, which reduced power of the studies and required much larger numbers of patients to see an effect (Williams, 2005). Many different therapeutic agents—including diuretics, beta blockers, angiotension converting enzyme (ACE) inhibitors, calcium channel blockers, and angiotensin receptor blockers—are approved to lower blood pressure.

Effects of blood pressure-lowering drugs Impact on blood pressure may or may not capture an intervention's entire risk–benefit balance. Different classes of agents, or even agents within a specific class, may have multiple effects, one of which is lowering blood pressure (NHLBI Working Group, 2005). For example, ACE inhibitors are known to have at least 10 pharmacologic effects (Borer, 2004). This notion has generated trials testing whether agents have beneficial effects that go beyond blood pressure lowering. ALLHAT (Antihypertensive and Lipid Lowering Treatment to Prevent Heart Attack Trial) compared the efficacy of four different drug classes (a calcium channel blocker, an ACE inhibitor, an alpha adrenergic blocker, and a diuretic) for initial therapy of hypertension. Study results demonstrated that three classes of drugs (calcium channel blocker, ACE inhibitor, and diuretic) could not be distinguished for the primary endpoint, coronary heart disease (CHD) mortality and non-fatal myocardial infarction, but the lower cost diuretics were superior in regard to secondary outcomes and should be the preferred first step therapy (ALLHAT Officers and Coordinators, 2002). The alpha adrenergic blocker arm of the trial was dropped because of the significantly higher incidence of combined cardiovascular events in the alpha adrenergic blocker arm compared to the diuretic, including a two-fold relative risk of congestive heart failure compared to the diuretic (ALLHAT Officers and Coordinators, 2000).

Other conclusions have also been drawn from these large, prospective head-to-head comparison trials; some investigators suggest that it is the blood pressure reduction, rather than the specific drug used, that confers cardiovascular benefit (Williams, 2005). In an analysis of 147 randomized trials, investigators found that all classes of blood pressure-lowering drugs have similar effects in reducing coronary heart disease events and strokes for a given level of blood pressure reduction, with the exception of an extra protective effect of beta blockers administered shortly after myocardial infarction and minor protective effect of calcium channel blockers in stroke (Law and Morris, 2009). Although there is still some ambiguity about the use of differing blood pressure agents, the fact that pharmacologically distinct agents have directionally similar effects on cardiovascular outcomes has provided more support for the use of blood pressure as a surrogate endpoint for coronary heart disease and stroke.

Regulatory use of blood pressure as a surrogate endpoint The consistent demonstration that diverse blood pressure-lowering agents confer cardiovascular benefits, as well as the substantial epidemiological data linking hypertension to cardiovascular events, provides the basis for the FDA's use of blood pressure as a surrogate endpoint (Desai et al., 2006; Temple, 1999). However, clear guidance on the use of surrogate endpoints within the FDA is lacking because the Food, Drug, and Cosmetic Act does not specifically state which endpoints—or criteria—can be used for drug approval. Through case law, the FDA has the authority to deny approval of a drug on the basis of its effect on the surrogate endpoint if the surrogate endpoint's clinical value is unknown.[2] In 1992, FDA regulation provided a new method for drug approval on the basis of effects on a surrogate endpoint, called accelerated approval, for serious or life-threatening conditions without available therapy. The regulation stated that drugs could be approved on the basis of surrogate endpoint data if it "is reasonably

[2] *Warner-Lambert* v. *Heckler*, 787 F.2d 147 (3rd Cir. 1986).

likely, based on epidemiologic, therapeutic, pathophysiologic, or other evidence, to predict clinical benefit"[3] and required confirmatory clinical evidence. The regulation also referenced "well-established" surrogates on which drug approval had been based, but did not define well-established endpoints. Temple (1999) noted that "well-established" surrogates would need to be more than "reasonably likely" to predict benefit.

Despite the lack of clarity in the regulations concerning surrogate endpoints, the FDA accepts surrogate endpoints for drug approval and as the basis for authorized health claims. However, different divisions and centers within the FDA accept different surrogate endpoints. For example, the Cardio-Renal Division within the CDER accepts blood pressure reduction as a surrogate endpoint for cardiovascular event reduction, but requires direct clinical benefit measurement for other endpoints, while the Metabolic-Endocrine Division also accepts LDL-C lowering as a surrogate endpoint for cardiovascular events (Borer, 2004). The Metabolic-Endocrine Division also accepts use of glycosylated hemoglobin level and blood glucose control as surrogate endpoints for diabetes control (Borer, 2004). Even so, the FDA has recognized the inadequacy of small six-month trials that address effects of type 2 diabetes mellitus treatments on HbA1c, and now the FDA requires large-scale randomized cardiovascular safety clinical endpoint trials be conducted pre- and post-approval.

Within CFSAN, blood pressure is recognized as a surrogate endpoint for hypertension (FDA, 1999). Hypertension is considered a disease-related health condition. As discussed earlier, hypertension—high blood pressure—is recognized as a strong risk factor for cardiovascular disease. CFSAN has authorized a health claim for low-sodium foods based on the surrogate endpoint–disease-related condition relationship, stating either "diets low in sodium may reduce the risk of high blood pressure, a disease associated with many factors" or "development of hypertension or high blood pressure depends on many factors. [This product] can be part of a low sodium, low salt diet that might reduce the risk of hypertension or high blood pressure."[4]

HIV Drug Development

One of the motivations for the earliest efforts at surrogate endpoint evaluation arose from the acute need for effective therapeutics early in the HIV/AIDS epidemic. The early trials of anti-HIV therapies used progression to AIDS or death as the clinical outcome measures. These studies could be short in some settings, like those in which the effects of the

---

[3] 21 C.F.R. § 601 (2008).
[4] 21 C.F.R. § 101.74 (2009).

intervention were large and participants had advanced disease (Fischl et al., 1987; Hammer et al., 1997). Studies could also be short when they were large enough so that only a small percentage of patients who progress to advanced disease drove the principal finding (Volberding et al., 1994). However, the latter type of study could produce misleading results in that a small number of patients destined to progress quickly might benefit from an intervention, like AZT monotherapy, while an even larger number might experience no benefit and even positive harm following the conclusion of the study, because of factors like the development of resistance to the drug under study and others with similar mechanisms of action. Such concerns underscored the need for a more rapid means of evaluating the benefit of antiviral therapy that might reflect risk or benefit to a larger proportion of the study population more rapidly.

Early in the AIDS epidemic, it was observed that clinical disease progression was associated with a decline of $CD4^+$ T-lymphocytes (CD4 cells); in the 1990s, a virologic measure that both responded to therapy and predicted outcomes was developed (HIV-1 RNA). The earliest approval of a drug based on a biomarker—didanosine was approved in 1991—used CD4 cell count; however, the development of measurement of plasma HIV-1 RNA by polymerase chain reaction (PCR), which made a direct measurement of viral replication possible, rapidly became the standard endpoint in HIV clinical trials. In the mid-1990s, representatives from industry, drug regulatory agencies, and academia sought to formally evaluate CD4 cell count and HIV-1 RNA as surrogate endpoints for disease progression in clinical trials and in patient management (Hughes et al., 1998).

To evaluate HIV-1 RNA and CD4 cell count as surrogate endpoints, the HIV Surrogate Marker Collaborative Group, a group involving statisticians and clinicians from pharmaceutical companies and government-funded cooperative clinical trials groups, was formed. The HIV Surrogate Marker Collaborative Group undertook a meta-analysis of clinical trials to evaluate treatment-mediated changes in HIV-1 RNA and CD4 cell count as surrogate endpoints (HIV Surrogate Marker Collaborative Group, 2000). The meta-analysis found that HIV-1 RNA and CD4 cell count have independent value as prognostic biomarkers. However, the meta-analysis also found that short-term changes in the values of these biomarkers were not adequate surrogate endpoints for determining the impact of an intervention on long-term clinical endpoints such as progression to AIDS and death (HIV Surrogate Marker Collaborative Group, 2000). Their analysis also showed that changes in HIV-1 RNA explained only about half of the benefit of treatment. However, these results mostly reflected the experience of patients on drug regimens that were not capable of suppressing most patients' viral loads below levels of assay detection.

In 2002, the FDA issued a guidance for industry that advocated the use of HIV-1 RNA in plasma as the primary basis for assessing efficacy of antiretroviral drugs for accelerated and traditional approval, although it had begun approving drugs based on evidence of lower levels of plasma HIV-1 RNA a few years earlier (Behrman, 1999). Additionally, it recommended that "changes in CD4 cell counts be consistent with observed HIV-1 RNA changes when considering approval of an antiretroviral drug" (FDA, 2002). In most cases, approval was based on demonstrations that new drugs, used in combination with existing drugs, were able to suppress virus among patients who had not been previously exposed to therapy and had virus that was sensitive to at least one other agent in the regimen. An important distinction must be made between using HIV-1 RNA as a surrogate for a clinical endpoint in a setting where virus can be fully suppressed and a setting where virus is only partly, and often therefore temporarily, suppressed. Complete viral suppression often leads to durable suppression, perhaps because of the lower risk of development of viral resistance mutations in patients without replicating virus. Tolerable drugs that produce durable suppression are likely to benefit patients because such suppression is associated with steady improvements in CD4 and reduced risk of clinical events associated with HIV infection.

The value of HIV-1 RNA as a surrogate in settings where suppression of HIV-1 RNA is partial is much more problematic and contingent on context, because partial HIV suppression invites development of new drug resistance mutations that limit the future usefulness of the drugs under study and similar drugs. Therefore a drug that induces a temporary reduction in HIV-1 RNA, while perhaps valuable in reducing risk of clinical disease over a short interval, may reduce the possibility of later construction of a durable three-drug regimen. Such loss of future drug options is an important consequence of drug treatment that is not captured by plasma HIV-1 RNA levels (Jiang et al., 2003). Another important factor is viral fitness, which is affected by treatment and may also be relevant for long-term outcomes (Deeks and Martin, 2007).

As a consequence the use of HIV-1 RNA as a surrogate for clinical endpoint in settings where viral suppression is not complete has not been supported with evidence and probably cannot be. As mentioned above, the relative benefit of different degrees of partial HIV suppression are highly context specific and dependent on the availability of other drugs. De Gruttola et al. (2006), in a discussion of the approval of tipranavir in exactly such a context, recommended that only complete suppression of plasma HIV-1 RNA be used in such studies, and that partial suppression endpoints not be used in clinical trials.

Historically, it is important to note that the FDA's guidance to industry occurred prior to the approval of newer types of antiretroviral drugs that use different mechanisms than those formally evaluated in the meta-analysis (Hughes, 2005). More potent antiretroviral drugs, which can fully suppress HIV-1 viral load, have since become standard of care. This suggests that although HIV-1 RNA has become the primary endpoint to determine efficacy in many antiretroviral trials, collection of additional and longer term information that relates to both risk and benefit—especially in studies of newer types of antiretroviral drugs—is warranted.

In conclusion, the rapid development of HIV drugs in the 1990s was enabled through the use of surrogate endpoints. While this use of surrogate endpoints inspired the creation of the Critical Path Initiative, the process of biomarker evaluation used was not systematic and so was not easily translated into other disease areas. Nonetheless, the success of this effort to speed approvals of HIV drugs highlighted the value that a systematic biomarker evaluation process could have for drug regulation in general.

Cautionary Statements Regarding the Use of Surrogate Endpoints

Remarkably, the cautionary voices speaking about the risks of using surrogate endpoints have been repeating the same messages for 20 years. What has been changing is the continually increasing amount of data supporting their arguments. In 1989, Ross Prentice initiated the conversation about surrogate endpoints with his influential paper, which provided a statistical definition of a surrogate endpoint. In this paper, he wrote, "I am somewhat pessimistic concerning the potential of the surrogate endpoint concept" (Prentice, 1989). This statement was made in acknowledgment of the hope, already palpable, that a surrogate endpoint, once shown useful for one intervention, would be extensible to other interventions and that relative reductions in one risk factor would be comparable to others for a given clinical endpoint.

Editorials in the early 1990s looked at the rapid advances—and mistakes—enabled through use of surrogate endpoints at the beginning of the HIV/AIDS epidemic (Cotton, 1991; De Gruttola et al., 1997; Holden, 1993; Lagakos and Hoth, 1992). The potential benefits and hazards of the use of surrogate endpoints have been understood since the beginning of this discussion. In 1991, Cotton noted several standing questions in relation to use of surrogate endpoints in the treatment of HIV/AIDS. Due to contemporaneous failures of surrogate endpoints in cardiology trials, researchers were wary when they did not understand the role a surrogate played in disease pathogenesis and progression. They noted that the role and importance of a biomarker may change over the course of a disease, such that extension of results in a population with more advanced disease may not translate to a population with less advanced disease and vice versa. Finally, researchers were not confident in the analytical validation of the tests being used to measure the surrogate endpoints (Cotton, 1991). In 1992, Lagakos and Hoth noted that experience from use of CD4 cell count as a surrogate endpoint in HIV/AIDS trials led to the idea that "it seems unrealistic to expect that any single marker can fully explain all of a drug's clinical effects." Furthermore, they recommended that "we cannot confidently abandon clinical endpoints as the basis for judging efficacy in these large trials. . . . It is therefore important that we continue to conduct comparative efficacy trials that collect data on both clinical outcomes and surrogate markers to establish CD4 count or other markers as valid surrogates for clinical effect" (Lagakos and Hoth, 1992). In 1993, Holden noted the desire of some to obtain a list of preapproved surrogate endpoints has been worrying to regulators because of the relevance of a biomarker's context of use in every application. In the article, Holden summarized a statement of Sidney Wolfe of the Public Citizen Health Research Group, saying that "drug companies could abuse [approvals of surrogate endpoints by the FDA] by failing to do careful clinical trials once they get a marker approved. . . . If clinical trials don't pan out, it might be very hard to ban the unapproved drug" that had been provisionally approved on the basis of the proposed surrogate endpoint (Holden, 1993).

Several of these warnings have been repeated since the early 1990s. Psaty et al. (1996) pointed out that different blood pressure-lowering interventions do not result in the same effects on clinical outcomes for a given reduction in blood pressure. De Gruttola et al. (1997) noted that unless disease mechanism of action is understood, uncertainty is inherent in the assumption that the surrogate can predict all of an intervention's effect. Schatzkin and Gail (2002) discussed use of surrogate endpoints in cancer research in 2002; they again noted the difficult balance between strong evidence that a surrogate endpoint has predictive value for the clinical endpoint and use of surrogates to achieve new drug approvals before full clinical trials using clinical endpoints can be completed. In the same year, DeMets and Califf (2002) reviewed principles of cardiovascular research and focused on the important distinctions between putative surrogate endpoints and clinical endpoints, reviewing multiple cases in which naïve use of putative surrogates had endangered patients with cardiovascular disease. In these cases, therapies, including antiarrhythmic, heart failure, and antiatherosclerosis treatments that had been assumed to be beneficial based on putative surrogate endpoints were indeed detrimental to health when confirmatory trials were done, usually because of off-target effects of systemically administered drugs. Manns et al. (2006) cited problems with the use of surrogate endpoints in a 2006 editorial. They discussed the opportunity cost of making decisions about allocation of healthcare resources (monetary, professional, and tangible), treatment decisions to use one treatment and forgo others, and allocation of research funding. The authors suggest that "it would seem prudent for [clinical practice guideline] developers to refrain from recommending the use of new agents until they have been proved to improve clinically meaningful outcomes" (Manns et al., 2006). Krumholz and Lee (2008) wrote in the *New England Journal of Medicine* that although use of surrogate endpoints can simplify the practice of medicine, it can do so at the cost of quality and outcomes. In 2009, Colatsky noted that surrogate endpoint biomarkers, low-density lipoprotein cholesterol (LDL-C) levels and carotid intima-media thickness (IMT) in this example, do not always correlate well with one another, making interpretation of trial results difficult (Colatsky, 2009).

These cautionary statements have gathered strength as some surrogate endpoints have failed. Examples of these failures and the reasons for their occurrence are discussed in the next section.

Failure of Surrogate Endpoints: Reasons and Examples

Putative surrogate endpoints often fail to predict clinical outcomes. In 1996, Fleming and DeMets published a paper explaining the failures in surrogate endpoints that had occurred mostly during the late 1980s and early 1990s (Fleming and DeMets, 1996). As described in Figure 2-1, according to Fleming and DeMets (1996), several factors explain the failure of surrogate endpoints: (1) the surrogate endpoint does not involve the same pathophysiologic process that results in the clinical outcome; (2) the intervention affects only one pathway mediated through the surrogate, of several possible causal pathways of the disease; (3) the surrogate is not part of the causal pathway of the intervention's effect, or is insensitive to its effect; and (4) the intervention has mechanisms of action independent of the disease process. As noted in Figure 2-2, the most promising setting in which to qualify a surrogate endpoint occurs when the surrogate is on the only causal pathway of the disease process, and the intervention's entire effect on the clinical outcome is mediated through its effect on the surrogate (Fleming and DeMets, 1996). However, even in the best of circumstances, it is possible for surrogate endpoints to be misleading by either overestimating or underestimating an intervention's effect on clinical outcomes.

A number of biomarkers have been proposed as rational surrogate endpoints, but have failed to demonstrate usefulness for that purpose upon further scrutiny in clinical trials. One example was the use of beta-carotene and retinol as biomarkers for cancer, cardiovascular disease, and (later) cataract risk, and as interventions for chemoprevention of these diseases. Observational studies indicated that lower dietary intakes of

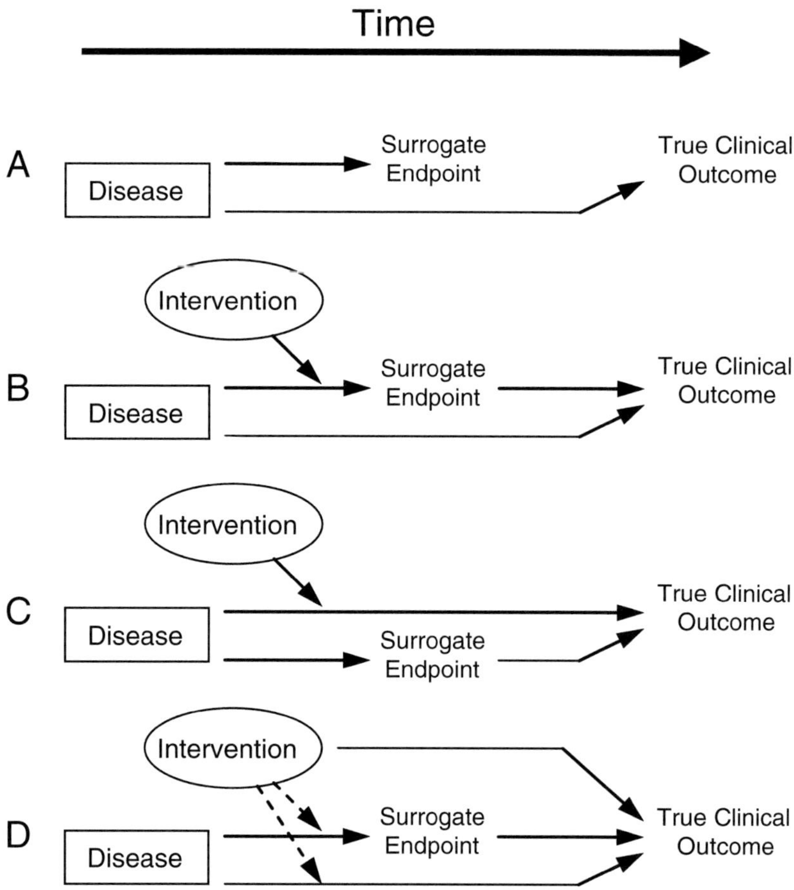

FIGURE 2-1 Reasons for failure of surrogate endpoints. (A) The surrogate is not in the causal pathway of the disease process. (B) Of several causal pathways of disease, the intervention affects only the pathway mediated through the surrogate. (C) The surrogate is not in the pathway of the intervention's effect or is insensitive to its effect. (D) The intervention has mechanisms of action independent of the disease process. Dotted lines = mechanisms of action that might exist.
SOURCE: Fleming and DeMets (1996). Reprinted, with permission, from the *Annals of Internal Medicine*. Copyright 1996 by American College of Physicians.

beta-carotene and lower serum levels of beta-carotene were associated with greater risk of cancer. It is useful to note that while serum level of beta-carotene is a biomarker for adequate intake of the nutrient and a proposed surrogate endpoint for prevention of cancer and atherosclerotic disease, supplementation of the diet with beta-carotene is an intervention to either address deficiencies or conditions for which it is used as a sur-

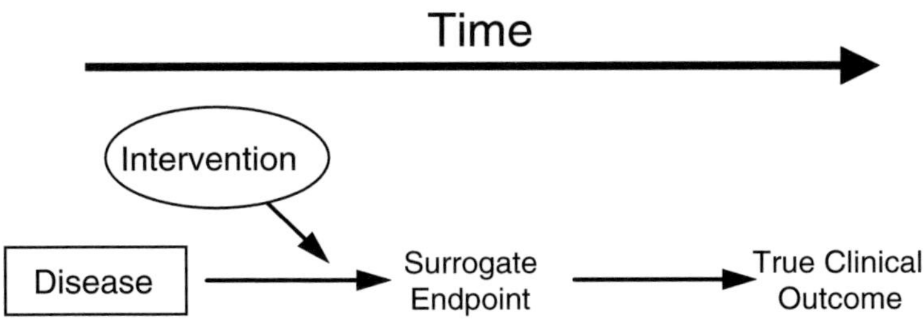

FIGURE 2-2 The setting that provides the greatest potential for the surrogate endpoint to be valid.
SOURCE: Fleming and DeMets (1996). Reprinted, with permission, from the *Annals of Internal Medicine*. Copyright 1996 by American College of Physicians.

rogate. Beta-carotene was shown to have in vitro antioxidant effects, and supplementing the diet with beta-carotene as a dietary supplement was expected to lower risk for atherosclerotic disease and cancer. However, its use in large population studies with mortality as the endpoint was not shown to lower risk for atherosclerosis or cancer; instead, it was shown to increase cancer incidence (Omenn et al., 1996; Peto et al., 1981). Beta-carotene will be discussed further in Chapter 4.

In another example, elevated serum levels of homocysteine were found to be associated with greater risk for atherosclerotic disease in observational associations and serum homocysteine was thought to be a surrogate endpoint. Homocysteine can exacerbate endothelial dysfunction, thrombosis, and other risk mechanisms for atherosclerosis. Folic acid was shown to decrease levels of circulating homocysteine. Researchers were confident that cardiovascular endpoints of death and vascular morbidity would be reduced with the administration of folic acid supplements. During this period, the use of folic acid supplements was found to decrease fetal development of neural tube defects when administered to pregnant women, and grain products were fortified with folic acid in the United States and other countries. The incidence of neural tube defects decreased following fortification. However, atherosclerotic disease, either coronary heart disease or peripheral vascular disease, did not decrease following folic acid fortification or with the administration of folic acid supplements in several large clinical trials despite important decreases in serum homocysteine levels with both interventions (Clarke et al., 2007).

From these examples, it is apparent that without a detailed understanding of a biomarker's role in the disease or treatment mechanism, biomarker evaluation can be difficult. The recent failure of some surrogate endpoints to predict clinical outcomes has elicited concern over guidelines and performance measures used in clinical decision making. Traditionally, clinicians focus on reducing risk factors below certain levels to prevent disease; for example, clinical guidelines and performance measures "encourage treatment geared toward achieving ambitious goals for levels of glycated hemoglobin, lipids, and blood pressure" (Krumholz and Lee, 2008). In light of recent trials that demonstrate a reduction in a risk biomarker without a corresponding reduction in risk, Krumholz and Lee suggest a rethinking of risk factor reduction. Instead of focusing on just the amount a risk biomarker is reduced, clinicians should also be aware of the strategy involved in risk reduction. According to Krumholz and Lee (2008), "We are now beginning to appreciate that a strategy's effect on a risk biomarker may not predict its effect on patient outcomes." Since it is recognized that "[s]ome strategies are known to improve patient outcomes, whereas others are known to affect only risk-factor levels or other intermediate outcomes," Krumholz and Lee believe that guidelines and performance measures should not specify targets without strategies used to achieve them. Additionally, practice guidelines and performance measurement should discuss risks of disease and adverse events in a more sophisticated and explicit way so that an assessment of net clinical benefit can be made (Krumholz and Lee, 2008).

As Krumholz and Lee (2008) pointed out, changes in surrogate endpoints do not always correspond with changes in clinical outcomes. Data from additional clinical trials have supplemented the notion that effects on proposed surrogate endpoints may fail to predict clinical outcomes. Nambi and Ballantyne (2007) emphasized that "we must use a great deal of caution before substituting a surrogate for a clinical endpoint" because the scientific community has been misled by biomarkers in the past. Patients and the credibility of science in the eyes of the public can be negatively impacted when the scientific community is misled by a biomarker. Fleming and DeMets (1996) further noted that "a review of recent experiences with surrogates is sobering, revealing many cases for which biological markers were correlates of clinical outcomes but failed to predict the effect of treatment on the clinical outcome." The following examples related to cardiovascular disease (CVD)—arrhythmia suppression, exercise tolerance in congestive heart failure, and lowering lipids—were outlined by Fleming and DeMets as telling examples of failed surrogate endpoints.

Arrhythmia Suppression

As described by Fleming and DeMets (1996), an example of the failure of a surrogate endpoint to predict clinical outcomes is the reduction of ventricular ectopic contractions for decreased cardiovascular mortality. When drugs were being developed and clinically tested, it was well known that compared to patients without ventricular arrhythmia, ventricular arrhythmia was independently associated with a significant increase in the risk of death related to cardiac complications, including sudden death (Bigger et al., 1984; Cardiac Arrhythmia Suppression Trial [CAST] Investigators, 1989; Echt et al., 1991; Mukharji et al., 1984; Ruberman et al., 1977). Researchers hypothesized that suppression of ventricular arrhythmias after myocardial infarction would reduce the rate of death. Scientists were so confident in this hypothesis that three drugs were approved by the FDA—encainide, flecainide, and moricizine—using arrhythmia suppression as the surrogate endpoint in phase III clinical trials. To illustrate the confidence scientists had in arrhythmia suppression as a surrogate endpoint, many of them believed that randomizing patients to either one of the study drugs or a placebo would be unethical. After approvals based on positive echocardiogram data, a feasibility trial was first conducted to determine whether a placebo-controlled trial would be safe enough to undertake (Cardiac Arrhythmia Pilot Study [CAPS] Investigators, 1986, 1988; CAST Investigators, 1989; Emanuel and Miller, 2001; Ruskin, 1989). After approval, more than 200,000 people eventually took these drugs each year, despite the lack of data evaluating the reduction of arrhythmias on mortality rates. The Cardiac Arrhythmia Suppression Trial (CAST) was designed to assess the drugs' impact on survival for patients who had had myocardial infarction and at least 10 premature beats per hour. Both the encainide and flecainide arms of the trial were terminated early when 33 sudden deaths occurred, as compared to only 9 in the matching placebo group. In total, 56 patients in the encainide and flecainide groups died, compared to 22 patients in the placebo group. Later data confirmed that patients taking moricizine were also at increased risk for death (Fleming and DeMets, 1996).

In addition to the CAST study, two other examples of failed surrogate endpoints have occurred with arrhythmia treatment. Quinidine had been used for many years to restore and maintain sinus rhythm in patients with atrial fibrillation. However, a meta-analysis indicated that quinidine increased the mortality rate from 0.8 percent to 2.9 percent, which outweighed the benefit of maintaining sinus rhythm (Fleming and DeMets, 1996). According to Lesko and Atkinson (2001), "unanticipated adverse consequences of drug therapy are a frequent confounding factor when biomarkers [such as maintaining normal sinus rhythm] are relied on as surrogates for definitive endpoints." Ventricular tachycardia, in the case of lidocaine drug therapy, was also shown to be an inadequate surrogate endpoint. Although a meta-analysis indicated lidocaine therapy produced a one-third reduction in the risk of ventricular tachycardia, it was also accompanied by a one-third increase in death rate (Fleming and DeMets, 1996). The failure of surrogate endpoints (e.g., maintenance of normal sinus rhythm and reduction of risk of ventricular tachycardia) to predict clinical endpoints "underlies much of the controversy surrounding the use of surrogate endpoints as the basis for regulatory evaluation of new therapeutic entities" (Lesko and Atkinson, 2001).

Exercise Tolerance in Congestive Heart Failure

Decreased cardiac output, decreased exercise capacity, and high risk of death are conditions associated with congestive heart failure, noted Fleming and DeMets (1996). Heart failure is a leading problem in cardiology; for example, 12 percent of a cohort of individuals age 65 or over were found to have symptomatic heart failure (Afzal et al., 2007). Heart failure patients may experience shortness of breath, congestion in the lungs, difficulty exercising, swelling in the legs, and quality-of-life–reducing effects. During the time leading up to the Prospective Milrinone Survival Evaluation (PROMISE) trial, cardiac output and ejection fraction had been used as surrogate endpoints, while exercise tolerance and symptomatic improvement had been used as intermediate endpoints. The PROMISE trial was requested by the FDA, which was concerned about long-term adverse effects of milrinone (Fleming and DeMets, 1996). Milrinone, a drug that was used to treat congestive heart failure, was shown to increase total mortality in the PROMISE trial, even though earlier studies demonstrated milrinone's effectiveness in improving cardiac output and increasing exercise tolerance. The drug flosequinan, a vasodilator that reduces cardiac workload, was also conditionally approved by the FDA to treat congestive heart failure in patients who did not respond to or tolerate other drugs. However, the Prospective Flosequinan Longevity Evaluation (PROFILE) trial demonstrated that flosequinan increased total mortality, even though it improved exercise tolerance. According to Fleming and DeMets (1996), "[a]lthough cardiac output, ejection fraction, and exercise tolerance are correlated with longer survival of patients with congestive heart failure, a treatment-induced improvement in those measurements is not a reliable predictor of the effect of treatment on mortality rates."

EVALUATION FRAMEWORKS

Biomarkers differ in their contexts of use and thus in the types of evidence needed for evaluation. Furthermore, use of surrogate endpoints for collection of evidence in support of policy or regulatory decisions is subject to the challenges and risks discussed in the previous sections (see Figures 2-1 and 2-2 and associated discussion). For additional detail, see Figure 2 in the paper by Boissel et al. (1992), outlining an approach for selection of surrogate endpoints. As each of these figures illustrate, the evaluation of a biomarker as a surrogate endpoint is particularly challenging because of the biological complexity of human disease and response to drugs and nutrients. Neither correlation of the biomarker with clinical outcome nor biological plausibility is sufficient to establish the usefulness of a biomarker as a surrogate endpoint. Moreover, qualification of a biomarker for a particular disease or treatment does not necessarily translate to qualification for related uses or even for an essentially identical use at a different point of time (and thus a different context of use).

Several frameworks for biomarker qualification and several for biomarker assay validation have been published. Appendix A presents a time line of critical developments in the discussion about biomarker and surrogate endpoints evaluation, republished with permission from the 2008 review in *Statistical Methods in Medicine* by Lassere. Terminology is presented as it was by Lassere, which was consistent with the original publications. Since 2007, there have been a few important publications, which have also been tabulated in Appendix A.

The next section discusses the evolution of thought on association and causation between exposure to a pathogenic agent, biomarkers, and incidence and mortality from disease. Several examples of the evaluation and use of surrogate endpoints in drug development are then discussed. The last two sections address the two main directions in the discussion of biomarker evaluation: those focusing on statistical methods and those focusing on qualitative methods. The reason is that while it is straightforward to establish a statistical association, it is difficult to definitively establish causality. Qualitative criteria have been used to fill this gap in the quantitative methods. Furthermore, decisions sometimes must be made when sufficient data are not available to make a quantitative analysis, and so qualitative methods are used.

Biomarker–Clinical Endpoint Relationships: Association Versus Causation

Many students of biology and epidemiology are familiar with Koch's postulates for determining the cause of infectious diseases. These postulates state that in order to conclude that a particular infectious agent is the cause of a disease, the following conditions must be fulfilled:

1. The agent must be associated with all cases of the disease;
2. The agent must be isolable and cultured from a diseased organism;
3. The cultured agent must be able to infect a new host; and
4. The agent must be reisolable from the host in postulate 3.

These postulates were developed in the 1880s, and in the 1900s, scientists sought to establish causality in diseases that were not infectious, such as cancer. In a report outlining the evidence supporting a causal link between smoking and lung cancer, an advisory committee to the Surgeon General of the Public Health Service outlined five criteria for the case of non-infectious or chronic diseases: the strength, specificity, temporality, and consistency of the association (Advisory Committee to the Surgeon General, 1964). These criteria were refined when, in 1965, Sir Austin Bradford Hill discussed these criteria in a famous lecture to the section of occupational medicine of the UK's Royal Society of Medicine (Hill, 1965). The criteria are now known as Hill's criteria and are outlined in Box 2-2. Since the 1960s, these criteria have been used in environmental health, toxicology, pharmacology, epidemiology, and medicine.

Surrogate endpoints have been discussed for a little over 20 years. In 1989, Ross Prentice defined the term "surrogate endpoint" in his paper entitled "Surrogate endpoints in clinical trials: Definition and operational criteria" (Prentice, 1989). This paper was accompanied by three other papers in an issue of *Statistics in Medicine* exploring the possible use of biomarkers as surrogate endpoints, using examples from cancer (Ellenberg and Hamilton, 1989), cardiovascular disease (Wittes et al., 1989), and ophthalmologic disorders (Hillis and Seigel, 1989). As discussed briefly in the previous chapter, the Prentice criteria specify that a biomarker under consideration as a potential surrogate endpoint must correlate with the clinical outcome it is meant to replace and that the biomarker must capture the entire effect of the intervention on the clinical endpoint (Prentice, 1989). Further development of statistical methods has occurred since 1989, as statisticians search for methods to ease the burden of the second criterion (Fleming, 2005). These approaches include meta-analysis of data from multiple trials (Alonso et al., 2006; Burzykowski et al., 2004; Buyse and Molenberghs, 1998; Buyse et al., 2000; Hughes, 2002; Hughes et al., 1995) as well as addressing the following: (1) the proportion of treatment effect described by the surrogate endpoint; (2) the relative effect and adjusted association; and (3) the surrogate threshold effect. These methods are summarized in Lassere's (2008) review, and several of them are discussed in detail in this chapter's section on statistical approaches to biomarker evaluation.

Nonetheless, surrogate endpoints were used before these conversations began. One of the best examples of this is blood pressure, which is used as a surrogate endpoint for CVD clinical outcomes. Blood pressure represents the historical course of biomarker evaluation, gradual accumulation of data, and agreement among stakeholders on the utility of a biomarker, as described in the earlier section on the history of the evaluation of blood pressure as a surrogate endpoint.

BOX 2-2
Hill's Criteria

1. **Strength**—Causation is supported if the relative risk due to the exposure is very large.
2. **Consistency**—Causation is supported if the relationship is seen in different populations at different times and in different circumstances.
3. **Specificity**—Causation is supported if an exposure appears to cause only a specific effect.
4. **Temporality**—Causation is supported if the exposure precedes the effect.
5. **Biological Gradient**—Causation is supported when the magnitude of the exposure is proportional to the magnitude of the effect.
6. **Plausibility**—Data elucidating the biological pathways leading from exposure to effect are useful.
7. **Coherence**—"The cause-and-effect interpretation of [the] data should not seriously conflict with the generally known facts of the natural history and biology of the disease."
8. **Experiment**—In some circumstances, evidence that removing the exposure lessens or removes the effect can be used to draw conclusions about causality.
9. **Analogy**—In some circumstances, comparison between weaker evidence of causation between an exposure and its effect and strong evidence of causality between another exposure and its similar effect is appropriate.

SOURCE: Hill (1965).

HIV/AIDS drug development provides another historical example of the use of surrogate endpoints. On October 11, 1988, frustrated with the length of time-to-approval for new therapies to treat HIV infection, ACT-UP, an AIDS patient advocacy group, staged a demonstration in front of FDA headquarters. Eight days later, on October 19, Frank Young, then commissioner of the FDA, announced regulations by which review times would be shortened for drugs designed to treat "life-threatening or severely debilitating" diseases (Arno and Feiden, 1988; AVERT, 2009; FDA, 1988). For that reason, HIV/AIDS drugs were some of the first to be approved explicitly on the basis of surrogate endpoints, and served as the foundation for the laws on accelerated approval of drugs and biologics. HIV/AIDS was also the first example of a more systematic, prospective approach to biomarker evaluation, although its precedent was not easily translatable into general guidance.

Finally, after the early 1990s, much of the literature has focused on the use of surrogate endpoints to approve oncology drugs. There is a substantial literature in this area, which is discussed in relation to use of tumor size as a surrogate endpoint for cancer treatment interventions in Chapter 4. Research and development in oncology has been working to implement broader use of biomarkers, but this effort continues with lack of a standard approach.

Statistical Approaches to Biomarker Evaluation

Although randomized clinical trials with clinically meaningful endpoints provide the most rigorous means of assessing benefit of an intervention, such trials may be lengthy and expensive, and not always feasible. Therefore considerable interest has been shown in development of a framework for "statistical validation" of surrogates for clinical endpoints that can reliably provide information more quickly and cheaply about medical interventions. While much work has been done in this area, there remains no widely accepted research paradigm for statistical validation, in the way that, for example, randomized clinical trials provide such a paradigm for comparing new to existing therapies. Below we describe why no single paradigm is likely to arise soon, or perhaps ever. We also show, however, that existing frameworks and methods are useful for investigating the properties of surrogate endpoints.

It is useful to restate Prentice's influential definition of a statistically valid surrogate, which required that a test of the null hypothesis of no relationship of the surrogate endpoint to the treatment assignment must also be a statistically valid test of the corresponding null hypothesis based on the true endpoint (Prentice, 1989). Statistical validation was based on two conditions: (1) correlation of the surrogate with the true clinical endpoint; and (2) the ability of the surrogate to fully capture the treatment's "net effect" on the clinical endpoint. As described by Fleming and DeMets (1996), the net effect is the aggregate effect accounting for all mechanisms of action of the intervention. Considerable effort has been made to assess the degree to which this second condition holds in a variety of settings, but such analyses are complicated by difficulty in reliably estimating the quantities of interest and in the need for extensive assumptions (see below).

An alternative approach is based on meta-analyses across studies. Daniels and Hughes (1997) used Bayesian methods to construct prediction intervals for the true difference in clinical outcome associated with a given estimated treatment effect on the potential surrogate. By "borrowing" information regarding estimates of the effects of treatment on the clinical endpoint, and on the relationships between the surrogate and the clinical endpoint given treatment from previous studies, one predicts effects of a new treatment from data on the surrogate.

An important recent paper by Joffe and Greene (2009) attempts to provide a broader intellectual framework, using ideas from causal inference, that subsumes several different approaches (including those described above) and also provides insight into why this research is so challenging. They describe four different frameworks for statistical validation of surrogacy, and show connections among them. The first is based on the Prentice criteria described above. A second considers the estimation of direct and indirect effects of treatment; the latter are those mediated through a biomarker. Joffe and Greene describe these two approaches as belonging to a category of causal effects frameworks, in which knowledge of the effects of the treatment on the surrogate and of the surrogate on the clinical outcome is used to predict the effect of the treatment on the clinical outcome.

The use of causal graphs modeling shows the challenge of basing a statistical validation procedure on the Prentice criteria. For true surrogate markers, there should be no direct effect of treatment independent of the marker, but instead all of the effect should be mediated by the surrogate. If there were no other causes of the clinical endpoint besides the treatment and the surrogate, analyses would be straightforward; in reality many other factors are likely to be involved. While randomization assures that treatment is not associated causally with any confounding variable, there is no reason to believe this to be true for the surrogate. In fact, the relationship of surrogate to clinical endpoint may well be confounded by other variables, each of which may or may not be measured. Joffe and Greene point out that even if the surrogate mediates the entire effect of treatment on the outcome (a most unlikely situation), the presence of confounding factors would imply that the treatment is not independent of the endpoint given the surrogate—in other words the Prentice criteria will not be met.

Model-based estimation of direct and indirect effects, possibly making use of the causal modeling approaches of Robins and Greenland (1992, 1994), offer some hope of addressing this issue, but such methods still require strong assumptions. One such assumption is that the intervention directly affects the surrogate, which in turn affects the clinical endpoint. Another is that one can control for confounding of the effect of the surrogate on the clinical outcome by proper inclusion of baseline covariates in a regression. In reality, baseline covariates may not be sufficient—an occasion that arises when a postrandomization covariate, influenced by treatment, affects the surrogate and is independently associated with outcome. For example, suppose that a blood pressure medication induced fatigue and therefore caused a reduction in the amount of exercise patients undertook; such an adverse consequence of treatment could affect both blood pressure and clinical events, such as time to myocardial infarction. Procedures are available to permit assessment of surrogacy in this situation, but they require that the confounding be controllable, through measurement and appropriate modeling. Unfortunately, there can be no way to test such that confounding can be appropriately controlled.

The third framework mentioned by Joffe and Greene is that of meta-analysis. As described above, meta-analysis investigates the relationship of the effects of treatment on surrogates with its effects on clinical outcomes over a series of trials. The fourth framework is defined in terms of the ideas of principal stratification, developed by Frangakis and Rubin (2002). These approaches belong to the causal-association paradigm, in which the effect of treatment on the surrogate is associated, across studies or population groups, with its effect on the clinical outcome, thereby allowing prediction of the effect on the clinical outcome from the effect on the surrogate.

For the meta-analysis approach, the average value of the surrogate measured in each trial should be able to predict the outcome for that trial. Of course, such an approach requires variability in the effect of treatment on the surrogate across studies. This approach may be the most promising because of its avoidance of the need for strong assumptions regarding confounding; nonetheless, even in this case, interpretation must be made with care. For example, Daniels and Hughes (1997) demonstrated that the change in CD4 count was associated with clinical endpoints (time to new AIDS definition or death). But in their example, all of the studies with large treatment and surrogate effects compared active treatments to placebo, whereas all of the studies with small treatment or surrogate effects had active controls. Therefore, extension of the results to a setting where a trial with an active control had a strong surrogate effect may not be warranted, as the biological processes might be quite different in this case than among those that were studied.

In contrast to the meta-analytic approaches, the principal surrogacy approach focuses on the association of the individual-level effects on surrogate and outcome. As is true in general of principal stratification, the group for whom the causal effects of treatment are defined is not observable, because for each individual, the surrogate can be observed only on one treatment and not the other(s). Full description of this approach is beyond the scope of this chapter, but such analyses are most likely to be useful in settings whether there is a strong effect of treatment on both surrogate and endpoint.

In conclusion, no simple paradigm for evaluation of surrogates is possible; consistency of findings across all of the approaches described by Joffe and Greene would probably provide the most convincing evidence. But the statistical methods do not in themselves provide the type of compelling evidence that a randomized trial with nearly complete follow-up can provide. Both a deep understanding of biological context combined with a thorough knowledge of causal research are necessary for any attempt at statistical validation of markers.

Decision Analysis Approaches to Biomarker Evaluation

Decision theory allows for logical and reproducible decision-making based on both quantitative and qualitative inputs. For biomarker evaluation, decision theory may be useful for the utilization step, and many principles from decision theory can be found throughout the report. Dr. Rebecca Miksad from Harvard University gave a presentation to the committee on decision theory as it could be applied to biomarker evaluation at the committee's April 2009 workshop. In the presentation, Miksad defined decision science as a "field of science which rigorously and quantitatively evaluates the short and long term outcomes of complex clinical situations through analysis of clinical decisions" (Miksad, 2009). Decision analysis formalizes complex decision-making processes involving ambiguity in data, variation in data interpretation, competing benefits and risks, gaps in information, and personal preferences when applicable. Decision analysis requires that decision makers break down decisions into their component parts and make any assumptions explicit. Miksad identified five unique features of decision analysis in her presentation (Box 2-3).

While analytical sensitivity and specificity of biomarker tests are important aspects of analytical validation, it is also important to take variability between individual interpreters of data. Receiver operating characteristic (ROC) graphs are a common decision analysis tool for accomplishing this goal. An ROC graph plots the impact of data interpretation variability on use of a given decision threshold, such as a cutoff value for a diagnostic test, for example (IOM, 2005). The x-axis of an ROC curve is the likelihood of a false positive result, or 1-specificity, while the y-axis of an ROC curve is the likelihood of a false negative result, or the sensitivity (IOM, 2005). ROC curves are described in Figure 2-3.

During decision analysis, all possible choices are mapped onto a decision tree. Then, mathematical models are used to compare possible outcomes of each choice. From these models, decision makers can then choose the most appropriate course of action or identify areas where more information is needed.

Miksad outlined important questions that can be addressed using decision analysis for biomarker evaluation (Miksad, 2009):

- What are the optimal characteristics and analytical thresholds for the biomarker assays themselves?
- What are the positive and negative predictive values of the biomarker assays?

BOX 2-3
Five Unique Features of Decision Analysis for Surrogate Endpoint Evaluation

- Directly addresses clinical complexity:
  - Multiple and potentially contradictory data
  - Multiple treatment options
  - Multiple potential interactions
  - Competing risks from patient comorbidities
- Explicitly incorporates uncertainty:
  - Data errors
  - Ambiguity and variations in data interpretation
  - Discordance between data and true disease state
  - Variable treatment effects, side effects and disease courses
- Identifies and compares trade-offs between competing objectives and risks:
  - Benefit of diagnosis versus risks of procedure
  - Therapeutic effects versus side effects
- Extends existing trial data to project outcomes across long time periods, including estimations of uncertainty
- Component parts of clinical decisions are broken down and data is recombined systematically

SOURCE: Miksad Presentation (2009). Reprinted, with permission, from Rebecca Miksad. Copyright 2009 by Rebecca Miksad.

- Does use of the biomarker assay lead to improved clinical outcomes?
- What are the areas of uncertainty that lead to the largest differences in predicted affects on clinical outcomes?
- Is additional data needed before use of the biomarker can be adopted?

Decision theory can be useful as a way to formalize the biomarker evaluation framework. While each biomarker evaluation would require a unique decision analysis, these analyses would provide stakeholders with a transparent accounting of the assumptions and subjective judgments that were needed for making specific decisions. In addition, these analyses would provide details on where biomarkers may benefit from the collection of additional data.

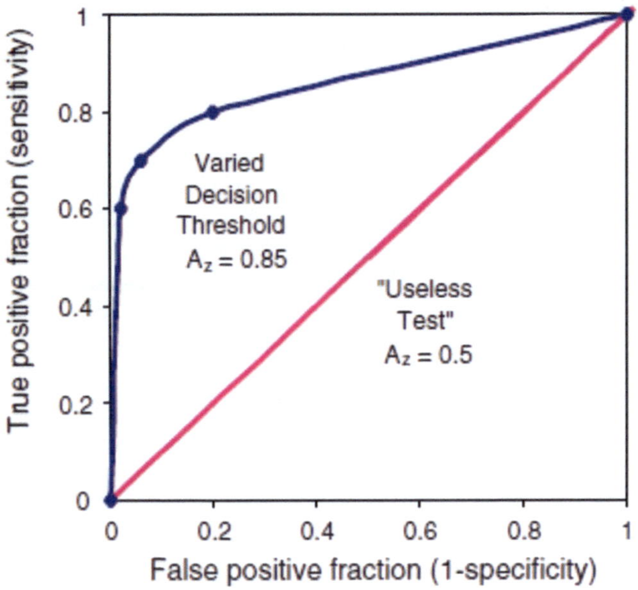

FIGURE 2-3 Receiver operating characteristic (ROC) graph of a varying decision threshold compared with a "useless test." The best-fit curve drawn through these points is the ROC curve, which represents the overall performance of the diagnostic test across all possible interpretations (decision thresholds). The overall accuracy of this test under varying conditions is determined by the area under the complete curve, 0.85. The leftmost point shows low sensitivity and high specificity. The middle point shows moderate sensitivity and specificity. The rightmost point shows high sensitivity and low specificity. Yet because they all lie on the same curve they have the same overall statistical accuracy, which is quantified by A_Z. The 45-degree-angle line represents a series of guesses between two choices, as in a coin toss. This would be considered a "useless test" if the outcome of the test was dichotomous (for example cancer vs. no cancer) for diagnostic purposes. For instance, radiologists reading mammograms with their eyes closed would tend to fall on this line. The number of true positives would approach the number of false negatives.The area under such a curve, 0.5, represents 50 percent accuracy of the test. In contrast, the ROC curve for a test with 100 percent accuracy will trace the y-axis up at a false-positive fraction of zero and follow along the top of the graph at a true-positive fraction of one. The area under such a curve would be 1.0 and represent a perfect test.
SOURCE: IOM (2005).

Qualitative Approaches to Biomarker Evaluation—Drug Development

This section describes one of the biomarker evaluation frameworks presented in the tables in Appendix A. In particular, this section discusses efforts made through public–private partnerships to develop a standardized, fit-for-purpose biomarker evaluation process. Beginning in the late 1990s and early 2000s, drug developers began participating in the development of biomarker evaluation processes (Colburn, 1997, 2000; Wagner, 2002). This effort was further strengthened by the formation of public–private partnerships such as the Biomarker Consortium and other Foundation for National Institutes of Health (NIH) efforts, as well as the Critical Path Institute (C-Path). The frameworks proposed in collaborations with pharmaceutical industry representatives strive for several characteristics: reproducibility, clear process, risk management, and incremental or fit-for-purpose evaluations (Altar, 2008; Altar et al., 2008; Lathia et al., 2009; Wagner, 2002, 2008; Wagner et al., 2007; Williams et al., 2006). In addition, several also consider cost effectiveness in frameworks to make decisions on biomarker evaluation (Altar et al., 2008).

A 2008 paper proposed use of an "evidence map" for use in biomarker evaluations (Table 2-2) (Altar et al., 2008). This map was developed as a collaboration between pharmaceutical industry representatives, a representative from the Foundation for NIH, and an FDA representative. The paper subsequently received attention from FDA staff at a conference entitled "2008 Cardiovascular Biomarkers and Surrogate Endpoints Symposium: Building a Framework for Biomarker Application." Briefly, the evaluation method proposed involves use of a committee to make decisions based on data and non-quantitative factors, such as public tolerability of the proposed decision. The first step in the process is for the committee to define and agree on a purpose and context of use for the biomarker. The next step is to assess the potential benefits and harms of the future success or failure of the biomarker in its proposed use. The third step is to come to an agreement about the tolerability for risk for the particular biomarker, given its proposed purpose and context of use. The fourth step is to assess the evidentiary status of the biomarker through use of the evidence map. During this step, the purpose and context-of-use combination is given a grade the biomarker needs to achieve in order to be deemed qualified. The final step is to summarize the committee's proceedings for the stakeholders.

The authors of the paper tested this framework with a panel of experts at a workshop, and found it to be useful; they also suggested next steps to improve the framework (Altar et al., 2008). This framework provided some of the basis for Recommendations 1 and 2.

In 2009, many industry authors of the Altar et al. (2008) paper published a paper commenting on the use of surrogate endpoints for drug approvals. They described characteristics of successful surrogate endpoints: biologic plausibility, prognostic value, and a positive correlation between an intervention's effect on the surrogate endpoint and the clinical endpoint (Lathia et al., 2009). A representative from CDER commented on their paper in the same issue, providing important examples of how biomarkers can be used to speed drug development without being used as surrogate endpoints (Gobburu, 2009).

Inclusion of Cost-Effectiveness Analysis in Biomarker Evaluation

A controversial issue in the drug development community is whether or not cost–benefit analysis should be part of a biomarker evaluation process. In 2006, Williams and colleagues outlined principles for biomarker evaluation that were the basis for the 2008 evidence map discussed above (Williams et al., 2006). Principle 8 was that "post hoc review of cost effectiveness should be performed at regular intervals as new information is available and conclusions recorded systematically as to how this should modify the qualification and use" (Williams et al., 2006). In 2008, this idea was discussed again: "some individuals from industry expressed great concern about the use and potential misuse of cost–benefit analyses and principles and did not wish to see them used here" (Altar et al., 2008).

Some additional considerations of the committee considered during its deliberations included the following:

- The FDA does not include analysis of cost in decisions to approve drugs or in other regulatory decisions.
- In their 2009 study entitled *The use of surrogate outcomes in model-based cost-effectiveness analyses: a survey of UK Health Technology Assessment reports*, Taylor and Elston stated that their "literature searches found no empirical studies examining the use of surrogate outcomes in [health technology assessments] and [cost-effectiveness models] therein."
- Conclusions regarding the cost-effectiveness analysis on drug development processes cannot be definitively drawn until evidence relating the use of a new intervention with clinical outcomes is available.

An explanation of why the committee did not include cost-effectiveness analysis as part of its biomarker evaluation process is included in Chapter 3.

TABLE 2-2 Altar et al. (2008) Proposed Evidence Map for Biomarker Qualification

Evidence Type	Grade D	Grade D+/C-	Grade C
Theory on biological plausibility	Observed association only	Theory, indirect evidence of relevance of the biomarker from animals	As for lower grade but evidence is direct
Interaction with pharmacologic target	Biomarker identifies target in in vitro binding		
Pharmacologic mechanistic response	In vitro evidence that the drug affects the biomarker	In vitro evidence that multiple members of this drug class affects the biomarker	In vivo evidence that this drug affects biomarker in animals
Linkage to clinical outcome of a disease or toxicity		Biomarker epidemiologically associated with outcome without any intervention	Biomarker associated with change in outcome from intervention in another drug class
Mathematics replication, confirmation		An algorithm is required to interpret the biomarker and was developed from the dataset	
Accuracy and precision (analytic validation)			
Relative performance		Does not meet performance benchmark	

SOURCE: Altar et al. (2008). Adapted, with permission, from Macmillian Publishers Ltd: Clinical Pharmacology and Therapeutics. Altar, C. A., D. Amakye, D. Buonos, J. Bloom, G. Clack, R. Dean, V. Devanarayan, D. Fu, S. Furlong, C. Girman, L. Hinman, C. Lathia,

Grade C+/B-	Grade B	Grade B+/A-	Grade A
Theory, indirect evidence of relevance in humans	Theory, direct evidence in humans, non-causal pathway possible	As for lower grade, but biomarker on causal path	Human evidence based on mathematical model of biology showing biomarker is on causal pathway
Biomarker identifies target in in vivo binding in animals	Biomarker identifies target in in vivo studies or from human tissue, no truth standard		Biomarker identifies target in in vivo studies or from tissues in humans, with accepted truth standard
As for lower grade but effect shown across drug class	Human evidence that this drug affects the biomarker OR animal evidence of specificity	Human evidence across this mechanistic drug class	Human evidence that multiple members of this drug class affect the biomarker and the effect is specific to this class/mechanism
As for lower grade but in this drug class	As for lower grade but multiple drug classes albeit inconsistent or a minority of disease effect		As for lower grade but consistent linkage and explains majority of disease effect
Algorithm was developed from a different dataset and applied here prospectively			Algorithm developed from different dataset, replicated prospectively in other sets and applied prospectively here
Sources of technical variation are unknown but steps are taken to ensure consistent test application	Major sources of variation known and controlled to be less than biological signal; standardization methods applied		All major sources of technical imprecision are known, and controlled test/assay accuracy is defined against standards
Similar performance to benchmark			Exceed performance of benchmark or best alternative biomarker

L. Lesko, S. Madani, J. Mayne, J. Meyer, D. Raunig, P. Sager, S. A. Williams, P. Wong, and K. Zerba. 2008. A prototypical process for creating evidentiary standards for biomarkers and diagnostics. *Clinical Pharmacology and Therapeutics* 83(2):368–371, Copyright 2007.

EVOLUTION OF REGULATORY PERSPECTIVES ON SURROGATE ENDPOINTS

Table 2-3 outlines the regulations and guidances pertaining to surrogate endpoints and the FDA. FDA regulatory authority for drugs, biologics, devices, foods, and supplements is discussed in detail in Chapter 5. While not discussed in detail, the NIH has historically played a vital role in the discovery, development, and regulatory perspective toward biomarkers; this is discussed briefly in Box 2-4.

2006–2008: FDA Pilot Process for Biomarker Qualification

Federico Goodsaid and Felix Frueh developed a biomarker qualification pilot process at the FDA, in collaboration with C-Path (Goodsaid, 2008a, 2008b; Goodsaid and Frueh, 2006, 2007a, 2007b; Goodsaid et al., 2008). The FDA pilot process for biomarker qualification was designed to qualify biomarkers incrementally, based on the data that are available for drug development or clinical applications. A biomarker would first be qualified in a narrow context of use, and then the context of use would be expanded as additional information became available. The

TABLE 2-3 List of Regulations and Guidances Pertaining to Surrogate Endpoints

Regulation or Guidance	Significance
21 C.F.R. 314.510	Accelerated approval: drugs. "Surrogate - Approval based on a surrogate endpoint or on an effect on a clinical endpoint other than survival or irreversible morbidity."[a]
21 C.F.R. 601.41	Accelerated approval: biologics. "Surrogate - Approval based on a surrogate endpoint or on an effect on a clinical endpoint other than survival or irreversible morbidity."[a]
Guidance for industry: Available therapy (FDA, 2004)	This guidance states that "the approval of one therapy under the accelerated approval regulations (either on the basis of a surrogate endpoint or with restricted distribution) should not preclude the approval under the accelerated approval regulations of additional therapies."

TABLE 2-3 Continued

Regulation or Guidance	Significance
21 C.F.R. 314.520	Postmarket authority of Food and Drug Administration (FDA) for drug accelerated approvals: "Restricted - Approval with restrictions to assure safe use."[a]
21 C.F.R. 601.42	Postmarket authority of FDA for biologic accelerated approvals: "Restricted - Approval with restrictions to assure safe use."[a]
21 C.F.R. parts 862–872, among others	The C.F.R. mentions surrogate endpoints in exceptions to the exemption of class I and II medical devices from premarket review: devices measuring analytes that are to serve as surrogate endpoints must undergo premarket review.
Guidance for industry and FDA staff: Postmarket surveillance under section 522 of the Federal Food, Drug, and Cosmetic Act (CDRH, 2006)	Postmarket surveillance may be requested when "premarket evaluation of the device may have been based on surrogate markers. Once the device is actually marketed, postmarket surveillance may be appropriate to assess the effectiveness of the device in detecting or treating the disease or condition, rather than the surrogate."
Guidance for industry: Clinical studies section of labeling for human prescription drug and biological products—Content and format (FDA, 2006a)	This guidance document recommends that manufacturers include more information in the Clinical Studies section of the label when "The study uses an unfamiliar endpoint (e.g., a novel surrogate endpoint), or there are important limitations and uncertainties associated with an endpoint."
Guidance for industry: Clinical data needed to support the licensure of seasonal inactivated influenza vaccines (CBER, 2007)	The document states that "For influenza vaccines, the immune response elicited following receipt of the vaccine may serve as a surrogate endpoint that is likely to predict clinical benefit, that is, prevention of influenza illness and its complications."
Guidance for industry: Clinical trial endpoints for the approval of cancer drugs and biologics (FDA, 2007)	The document describes current and past thought on use of non-survival endpoints in oncology approvals. A table comparing important cancer endpoints is presented.

continued

TABLE 2-3 Continued

Regulation or Guidance	Significance
Guidance for industry and FDA staff: Clinical study designs for catheter ablation devices for treatment of atrial flutter (CDRH, 2008)	The document states that "acute procedural success may be appropriate to serve as a surrogate effectiveness endpoint for catheters provided all of the following device characteristics are present: • Creates endocardial lesions • Manipulated in the endovascular space • A single ablation electrode • The energy source is radiofrequency (RF) • Temperature sensing capability • 'Steerable' (i.e., catheter has a tip which is manually-deflectable via a thumb-wheel or similar mechanism residing on the handle of the catheter) • Percutaneous placement."
Guidance for industry: Evidence-based review system for the scientific evaluation of health claims (CFSAN, 2009a)	Includes the definition of surrogate endpoint discussed in Chapter 1. The document lists the four currently accepted surrogate endpoints for health claims: "(1) serum low-density lipoprotein (LDL) cholesterol concentration, total serum cholesterol concentration, and blood pressure for cardiovascular disease; (2) bone mineral density for osteoporosis; (3) adenomatous colon polyps for colon cancer; and (4) elevated blood sugar concentrations and insulin resistance for type 2 diabetes." However, it also stipulates that biomarkers not on the biological pathway of a particular nutrient–disease risk link may not be used as surrogate endpoints for development of health claims.

NOTE: [a] http://www.fda.gov/Drugs/DevelopmentApprovalProcess/HowDrugsareDevelopedandApproved/DrugandBiologicApprovalReports/ucm121606.htm.

qualification process, as outlined in Figure 2-4, involves FDA reviewers, outside experts, and advisory committees. The process started with a two-page letter submitted to the FDA. The letter includes a description of the biomarker, an accurate definition of the context of use that the biomarker is being proposed for, and a list of the data supporting the request. Submissions are made by companies, consortia, and academics.

BOX 2-4
FDA's Risk Communication Advisory Committee

The FDA's Risk Communication Advisory Committee was created in 2008 with the following purpose:

> The Committee advises the Commissioner of the Food and Drugs or designee on methods to effectively communicate risk associated with products regulated by the Food and Drug Administration and in discharging responsibilities as they relate to helping to ensure safe and effective drugs for human use and any other product for which the Food and Drug Administration has regulatory responsibility. The Committee reviews and evaluates strategies and programs designed to communicate with the public about the risks and benefits of FDA-regulated products so as to facilitate optimal use of these products. It also reviews and evaluates research relevant to such communication to the public by both FDA and other entities, and facilities interactively sharing risk and benefit information with the public to enable people to make informed independent judgments about use of FDA-regulated products. (FDA, 2010)

The committee is currently chaired by Dr. Baruch Fischoff, professor in the Departments of Social & Decision Sciences and Engineering & Public Policy at Carnegie Mellon University. The committee has ten additional members. The committee meets four times a year. In 2009, the committee discussed topics such as

- Risk communication research needs,
- Quality of consumer drug information,
- Communicating about food recalls and food-borne illness,
- Communicating about tobacco and health,
- Clinical trials database, and
- Use of social media as surveillance tools.

SOURCE: FDA (2010).

The next step is the recruitment of a biomarker qualification review team. A briefing document is requested from the group submitting the request, and then a face-to-face meeting is held between the review team and the group submitting the request. The gaps in evidence are evaluated, revised data packages are requested, and the process goes back and forth until the package is as complete as possible. Then, the review team writes a document, and a regulatory briefing is submitted (Goodsaid et al., 2008). Goodsaid emphasized in his presentation at the Cardiovascular Markers of Disease (CMOD) conference that "*biomarker qualification* is the process by which data are provided to show that exploratory biomarkers are qualified for application in a specific context of use," and that "the *context of use* for a biomarker is the general area of biomarker application, specific

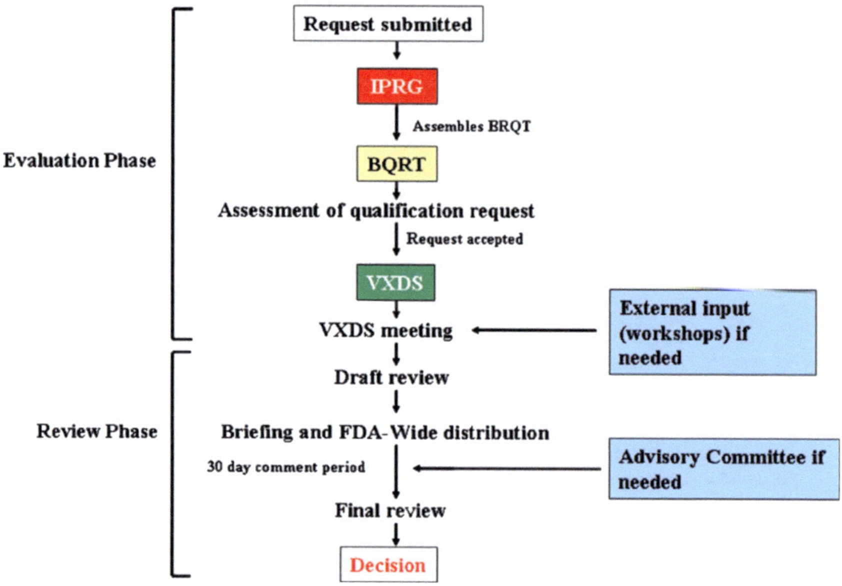

FIGURE 2-4 Outline of the Food and Drug Administration's (FDA's) biomarker qualification pilot process.
NOTE: BQRT = Biomarker Qualification Review Team; IPRG = Interdisciplinary Pharmacogenomic Review Group; VXDS = voluntary data submission.
SOURCE: Goodsaid et al. (2008). Reprinted with permission from Elsevier. Copyright 2008, Elsevier.

applications/implementations and critical factors which define where a biomarker is to be used and how the information from measurement of this biomarker is to be integrated in drug development and regulatory review" (Goodsaid, 2008a).

Melanie Blank, medical officer in the Division of Cardiovascular and Renal Products in CDER, has also discussed the FDA pilot process for biomarker qualification and how the evidentiary standards would be higher when the consequences of false results are graver (Blank, 2008): the qualification process as it would be applied to several problems such as how efficacy biomarkers can help in large, expensive drug trials where the clinical endpoint is rare and delayed, how safety biomarkers contribute when there is late discovery of toxicity resulting in late abandonment of the drug development program, and how safety biomarkers contribute when there are no sensitive methods to detect observed preclinical toxicities.

Example: Biomarkers of Kidney Toxicity

Dr. Joseph Bonventre of Harvard University spoke to the committee members at their first meeting. He has been involved on FDA committees as well as the only academic participant in the Critical Path Institute's biomarker qualification effort, which was done in collaboration with Federico Goodsaid at the FDA and industry partners. The Predictive Safety Testing Consortium (PSTC), as part of the Critical Path Institute's efforts in the area of biomarker evaluation, assembled a panel of scientists to evaluate potential safety biomarkers of acute kidney injury. These biomarkers are needed for use in "early diagnosis, to monitor severity and progression of disease, predict an outcome without an intervention, better stratify patients for clinical trials, predict who will respond to an intervention, [determine whether] the intervention [is] working ([through use of a] surrogate [endpoint]), and to identify therapeutic targets for an intervention" (Bonventre, 2009).

The most commonly used biomarkers for kidney injury are functional biomarkers rather than biomarkers of injury: serum creatinine and blood urea nitrogen. As in many organ systems, there are different stages of injury: risk, damage, reduction in function, organ failure, and death. Complications are associated with each stage. Elevations of serum creatinine and blood urea nitrogen above established normal ranges occur only after significant renal damage is present. Biomarkers of injury were the target of the preclinical studies.

Preclinical studies were conducted under the context of the PSTC, mostly internally at the FDA or in industry. Conferences were held early in the process with the European Medicines Agency (EMEA) and the Japanese drug regulatory agency. Following the process outlined in the previous section, seven biomarkers were validated and qualified: KIM-1, albumin, total protein, 2-microglobulin, cystatin C, urinary clusterin, and urinary trefoil factor 3.

As a result of the new biomarkers and validation information obtained in these studies, creatinine is no longer sufficient for showing safety at the FDA. The final step in the process occurred in June 2008, when the FDA and EMEA released a statement: "In the first use of a framework allowing submission of a single application to the two agencies, the FDA and the EMEA worked together to allow drug companies to submit the results of seven new tests that evaluate kidney damage during animal studies of new drugs" (FDA, 2008a). The need for better safety biomarkers relating to kidney toxicity and efforts to address this issue are also described in the IOM's recent workshop summary *Accelerating the Development of Biomarkers for Drug Safety* (IOM, 2009a).

Surrogate Endpoints in Nutrition: Foods, Supplements, and Public Health

The following sections describe the types of claims found on food packaging in the United States and how biomarkers play a role in their evidentiary substantiation.

Health Claim Definition

Health claims for foods and dietary supplements are "voluntary statements that characterize the relation between a substance and its ability to reduce the risk of disease or health-related condition" (Schneeman, 2007). Third-party references, written statements, symbols, or vignettes (e.g., brand names including the word heart or heart symbols) that relate a food substance to reduced risk of disease are considered health claims. Implied health claims are statements, symbols, vignettes, and other forms of communication that suggest a relationship between a substance and a disease or health-related condition.[5]

Health claims consist of two parts, a substance (specific food or component of food, including a dietary supplement) and a disease or health-related condition (damage to an organ, part, structure, or system of the body such that it does not function properly or a state of health leading to such dysfunction).[6] In addition, health claims are directed to the general population or population subgroups (e.g., the elderly, women) with the intent to assist the consumer in maintaining healthful dietary practices (CFSAN, 2009a).

As a point of history, prior to the 1990 legislation authorizing health claims, a claim on a food label that referred to a disease condition resulted in the product being classified as a drug and subject to drug regulations. However, emerging science of the 1970s and 1980s had begun to demonstrate a relationship between dietary substances and reduced risk of disease. Taylor and Wilkening (2008) note that "it seemed untenable that only drug products could mention diseases on their labels and even less tenable that food substances with the potential to reduce risk be regulated as drugs." To avoid drug status,[7] health claims cannot assert or imply that they prevent, treat, or mitigate disease, but instead only to reduce the risk of disease.

---

[5] 21 C.F.R. § 101.14(a)(1) (2008).

[6] 21 C.F.R. § 101.14(a)(2) (2008) and 21 C.F.R. § 101.14(a)(5) (2008).

[7] A drug is defined as an article intended for use in the diagnosis, cure, mitigation, treatment, or prevention of disease. 21 U.S.C. § 321(g)(1)(b).

Legal Basis for Health Claims and Review of Evidence for Health Claims

The Federal Food, Drug, and Cosmetic Act (FDCA) authorizes the Food and Drug Administration to regulate food and dietary supplement labels. In respect to health claims, the FDCA has been amended over time by the 1990 Nutrition Labeling and Education Act (NLEA), the 1994 Dietary Supplement and Health Education Act (DSHEA), and the Food and Drug Administration Modernization Act of 1997. A 1999 court decision (*Pearson* v. *Shalala)* further influenced the FDA's process of evaluating health claims by allowing claims of lesser evidence, accompanied with qualifying language.

The NLEA made nutrition labeling on most foods mandatory and allowed health claims that are based on significant scientific agreement (SSA), or:

> based on the totality of publicly available scientific evidence (including evidence from well-designed studies conducted in a manner which is consistent with generally recognized scientific procedures and principles), that there is significant scientific agreement among experts qualified by scientific training and experience to evaluate such claims, that the claim is supported by such evidence.[8]

DSHEA further amended the FDCA to provide for the use of health claims and nutrient content claims on eligible supplement products, and to provide for the use of structure/function claims. The FDA Modernization Act amended the FDCA to allow health claims based on an authoritative statement of a scientific body of the U.S. government or the National Academy of Sciences. In 1999, the U.S. Court of Appeals found that the SSA standard was overly stringent and violated First Amendment rights by constricting commercial free speech.[9] The court found that claims that did not meet the SSA standard were legal if accompanied by appropriate qualifying language.

In 2009, CFSAN completed a guidance for industry that outlined the agency's current thinking on the process for evaluating scientific evidence for a health claim, the meaning of the SSA standard, and credible scientific evidence to support qualified health claims (CFSAN, 2009a). In the evidence-based review system for the scientific evaluation of health claims, CFSAN has outlined a process to evaluate the strength of the scientific evidence to support a claim about a substance/disease relationship. First, the agency conducts a literature search to identify studies that evaluate the substance/disease relationship, primarily in humans. Studies

[8] 21 C.F.R. § 101.14(c) (1998).

[9] *Pearson* v. *Shalala*, 164 F.3d 650 (D.C. Cir. 1999).

are categorized into intervention studies, observational studies, research synthesis studies, and animal and in vitro studies, and are evaluated and assessed for methodological quality. The agency then sets out to evaluate the totality of scientific evidence about a substance/disease relationship by considering study type, methodological quality rating, number of the various types of studies and sample sizes, relevance of the body of scientific evidence to the U.S. population or target subgroup, replications of findings, and overall consistency of the scientific evidence. Assessing whether the SSA standard is met and specifying the approved claim language are also part of this evidence-based review system.

According to Kathy Ellwood and Paula Trumbo's presentation at the first committee meeting, there is no difference in how the scientific evidence is reviewed for an SSA-level claim or qualified health claim: "Health claims represent a continuum of scientific evidence that extends from very limited or inconclusive evidence to consensus, with evidence supporting SSA health claims lying closer to consensus" (Trumbo and Ellwood, 2009). In the scientific review of evidence for health claims, "Surrogate endpoints of disease risk" considered valid by the FDA's Center for Food Safety and Applied Nutrition include serum LDL cholesterol, total serum cholesterol, and blood pressure for cardiovascular disease; bone mineral density for osteoporosis; adenomatous colon polyps for colon cancer; and elevated blood sugar concentrations and insulin resistance for type 2 diabetes (CFSAN, 2009a). Health claims based on surrogate endpoints include both authorized and qualified claims (Table 2-4). It is important to note that structure/function claims, nutrient content claims, and dietary guidance statements are not based on this scientific evidence review. Because most of these claims do not make reference to disease or health-related conditions, surrogate endpoints are generally not relevant to these types of claims.

Types of Health Claims

Health claims based on significant scientific agreement (authorized health claims) According to Schneeman, the SSA standard "is based on a high level of confidence in the validity of the relation between the substance and the disease or health-related condition" (Schneeman, 2007) and considers the totality of publicly available evidence. When the NLEA was implemented, it required the FDA to consider health claims for 10 specific relationships, of which 8 were approved (Taylor and Wilkening, 2008):

- Calcium and osteoporosis;
- Sodium and hypertension;
- Dietary fat and cancer;

- Dietary saturated fat and cholesterol and CHD;
- Fiber-containing grain products, fruits, and vegetables and cancer;
- Fruits, vegetables, and grain products containing fiber, especially soluble fiber, and CHD;
- Fruits and vegetables and cancer; and
- Folic acid and neural tube defects.

In addition to these initial approved health claims, the NLEA provided a petition process for the consideration of future health claims, involving the petitioner submitting all relevant scientific findings to the FDA. Through this process, an additional seven claims have been approved. Approved health claims that were based on surrogate endpoint data are shown in Table 2-5.

Health claims approved under the SSA standard require specific claim language to be followed. For example, the model health claim language approved for sodium and high blood pressure includes: "Development of hypertension or high blood pressure depends on many factors. [This

TABLE 2-4 Health Claims Based on Surrogate Endpoints

Nutrient	Disease	Surrogate Endpoint	Type of Claim
Phytosterols, soy protein, corn oil, canola oil, and olive oil	Coronary heart disease	LDL and total cholesterol	Phytosterols: Authorized Soy protein: Authorized Corn oil: Qualified Canola oil: Qualified Olive oil: Qualified
Chromium picolinate	Type 2 diabetes	Insulin resistance	Qualified
Calcium and sodium	Hypertension	Systolic and diastolic blood pressure	Calcium: Qualified Sodium: Authorized
Calcium and vitamin D	Osteoporosis	Bone mineral density	Authorized
Calcium	Colorectal cancer	Colorectal polyps	Qualified

NOTE: LDL = low-density lipoprotein.
SOURCE: Trumbo and Ellwood (2009).

TABLE 2-5 Qualified Health Claims Approved by the Food and Drug Administration

Category of Disease	Approved Qualified Health Claims
Cancer	Tomatoes and prostate, ovarian, gastric, and pancreatic cancers
	Calcium and colon/rectal cancer and calcium and colon/rectal polyps
	Green tea and risk of breast, prostate cancer
	Selenium and site-specific cancers
	Antioxidant vitamins C and E and risk of certain cancers
Cardiovascular disease	Folic acid, vitamin B6, vitamin B12 and vascular disease
	Walnuts and coronary heart disease
	Nuts and coronary heart disease
	Omega-3 fatty acids and reduced risk of coronary heart disease
	Corn oil and corn oil-containing products and a reduced risk of heart disease
	Unsaturated fatty acids from canola oil and reduced risk of coronary heart disease
	Monounsaturated fatty acids from olive oil and coronary heart disease
Cognitive function	Phosphatidylserine and cognitive function and dementia
Diabetes	Chromium picolinate and a reduced risk of insulin resistance, type 2 diabetes
Hypertension	Calcium and hypertension, pregnancy-induced hypertension, and preeclampsia
Neural tube defects	Folic acid and neural tube defects

SOURCE: CFSAN (2009b).

product] can be part of a low sodium, low salt diet that might reduce the risk of hypertension or high blood pressure."[10]

Health claims based on authoritative statements The Food and Drug Administration Modernization Act of 1997 specified that the FDA's scientific review process could be circumvented if other scientific bodies of the U.S. government or the National Academy of Sciences[11] had issued authoritative statements about the substance/disease relationship. Authoritative statements from the National Academy of Sciences were

[10] 21 C.F.R. § 101.74(e) (2009).

[11] In legislation, the term National Academy of Sciences refers to the whole of the National Academies.

used to approve three additional health claims—the relationship between whole grains and heart disease, the relationship between certain cancers and potassium, and the relationship between high blood pressure and stroke (Taylor and Wilkening, 2008).

Qualified health claims Litigation over the SSA standard for dietary supplements resulted in an FDA process to approve claims with lesser evidence, given additional qualifying language (qualified health claims). In *Pearson* v. *Shalala*, appellants argued that the high SSA standard impeded First Amendment commercial free speech. According to Schneeman (2007), "courts indicated that the FDA had not presented any data that potentially misleading claim language would not be cured by qualifying language enabling consumers to understand the nature of the evidence supporting a claim." The FDA used a mechanism known as enforcement discretion to allow for the use of qualified health claims (rather than through authorization and publication in the *Federal Register*, as required in the NLEA for SSA health claims) (Taylor and Wilkening, 2008).

As part of a guidance on interim procedures for health claims, FDA proposed a scientific ranking system for health claims, where A-level evidence refers to SSA-level health claims and B-, C-, and D-level evidence refers to the differing levels of evidence for qualified health claims (see Figure 2-5). This ranking system is not used. The FDA approved a B-level qualified health claim for the relationship between walnuts and coronary heart disease. The qualifying language approved was: "supportive but not conclusive research shows that eating 1.5 ounces per day of walnuts, as part of a low saturated fat and low cholesterol diet and not resulting in increased caloric intake, may reduce the risk of coronary heart disease. See nutrition information for fat [and calorie] content" (CFSAN, 2004). The relationship between selenium and cancer was approved as a C-level health claim with the associated qualifying language: "Selenium may reduce the risk of certain cancers. Some scientific evidence suggests that consumption of selenium may reduce the risk of certain forms of cancer. However, [the] FDA has determined that this evidence is limited and not conclusive" (CFSAN, 2003).

An example of qualifying language for a D-level qualified health claim is the relationship between tomatoes/tomato sauce and prostate cancer. The disclaimer language the FDA approved included "very limited and preliminary scientific research suggests that eating one-half to one cup of tomatoes and/or tomato sauce a week may reduce the risk of prostate cancer. [The] FDA concludes that there is little scientific evidence supporting this claim" (CFSAN, 2005). Likewise, the relationship between tomatoes and pancreatic cancer was also approved as a D-level qualified health claim with the associated disclaimer: "one study suggests

FIGURE 2-5 2003 Food and Drug Administration ranking system for health claims. Claims that met the significant scientific agreement standard were considered A-level claims and were unqualified (requiring no disclaimer). Qualified claims (levels B through D) required disclaimers, such as "evidence is not conclusive."
SOURCES: FDA (2003). See also Mitka (2003).

that consuming tomatoes does not reduce the risk of pancreatic cancer, but one weaker, more limited study suggests that consuming tomatoes may reduce this risk. Based on these studies, [the] FDA concludes that it is highly unlikely that tomatoes reduce the risk of pancreatic cancer" (CFSAN, 2005).

To date, dozens of qualified health claim petitions have been submitted to the FDA. Qualified health claim petitions have been approved for several categories of disease, including cancer, cardiovascular disease, cognitive function, diabetes, hypertension, and neural tube defects (see Table 2-5). On the FDA's website, the denied petitions for qualified health claims are also listed, and include lycopene and cancer, green tea and reduced risk of cardiovascular disease, vitamin E and heart disease, among others (a total of 15 letters of denial have been produced, with one petition—soy protein and cancer—withdrawn) (CFSAN, 2009b).

Other Types of Claims

Nutrient content claims Nutrient content claims expressly or implicitly characterize a level of a nutrient (e.g., "low in fat," "high in vitamin C") in a product (IFT, 2005). Nutrient content claims were established to provide consistent usage throughout the food supply. Prior to the NLEA, nutrient content claims were not standardized, enabling manufacturers to claim "rich in oat bran," "extremely low in saturated fat," with "no assurance that the levels in the food were in fact high or low relative to other similar foods or to an overall diet" (Taylor and Wilkening, 2008).

The FDA currently accepts a number of content claims including free, low, lean, extra lean, high, good source, reduced, less, light, fewer, and more. In addition, the FDA has allowable synonyms for each of the core terms (Taylor and Wilkening, 2008). Nutrient content claims have been authorized for substances that have established Daily Reference Values (DRVs) or Reference Daily Intakes (RDIs), collectively referred to as Daily Values (DVs). For example, a label may claim that the food is "high in," "rich in," or an "excellent source" of a nutrient if the food provides 20 percent or more of the DVs per RACC (Reference Amount Customarily Consumed) (IFT, 2005). Although foods without established DVs cannot have core content claims, manufacturers can make labeling statements, such as "contains *x* mg lycopene per serving," because it does not imply whether the amount of the nutrient is high or low based on DVs, as long as the statement is not misleading (IFT, 2005).

Structure/function claims Claims about the dietary impact of a nutrient on the structure or function of the human body are generally allowed. However, these types of claims cannot suggest that the food or nutrient will cure, mitigate, prevent, or treat disease because that makes it a drug claim. Several structure/function claim examples include "calcium helps build strong bones" or "protein helps build strong muscles." The Institute of Food Technologists note that there is "considerable uncertainty about how far this type of structure/function claim can be 'pushed' before [the] FDA will assert either drug status or health claim status" (IFT, 2005).

Dietary guidance statements Although not considered claims, dietary guidance statements also appear on food labeling. As compared to health claims, dietary guidance statements make reference to either a food substance or a disease, but do not relate these two components in the claim. For example, a dietary guidance statement may say "carrots are good for your health" or "calcium is good for you." Unlike health claims, truthful, non-misleading dietary guidance statements may be used on food labels without premarket review by the FDA (CFSAN, 2008).

BIOMARKERS AND COMMUNICATION STRATEGIES AT THE FDA

Effective use of biomarkers for many purposes depends on the ability of regulators, health-care practitioners, and even advertisers to clearly communicate information about the biomarkers as well as the risks and benefits related to their use. Biomarker use also depends on the ability of the public and others to understand this information. In this and the next section, communication strategies as well as numeracy are discussed, with attention to topics most relevant to public understanding and acceptance of biomarker use.

Research on effective communications in the clinical setting and with respect to prescription and over-the-counter drugs has shown the dramatic effects that good communication strategies can have on patient outcomes. In the clinical setting, studies have pointed to the need for clinicians to receive training on how to communicate with their patients about potential risks of medical treatment (IOM, 2007b; NCI, 2007; Nicholson, 1999). In a review of effective risk communication strategies for cancer genetic counseling, Julian-Reynier and colleagues (2003) emphasized the importance and challenges of providing standardized information about risks of testing to relevant populations as well as individually tailored information based on the patient's immediate concerns. Berry explained many issues of risk communication from a psychology perspective in the book *Risk, Communication and Health Psychology* (Berry, 2004); the understanding and approaches suggested in this book are generally applicable across different health-related settings. The Cochrane Collaboration has reviewed strategies and decision aids for helping patients make decisions about screening tests or health treatments (Edwards et al., 2006; O'Connor et al., 2009). In general, research has found that symbolic representations of probabilistic information, when presented well, are the most effective at enhancing patient–provider communication (Akl et al., 2007; Kim et al., 2009; Lipkus, 2007).

As the primary agency in charge of the safety of foods and drugs, the FDA uses and provides access to a great deal of information on the safety of food, supplements, drugs, biologics, and devices, and on the strength of evidence supporting certain types of health claims on foods and supplements. However, this information can be difficult to access or interpret. Therefore, the main sources of information for clinicians and consumers about the safety, efficacy, and accuracy of product claims that are subject to regulatory review are (1) the labels and package inserts of drugs, biologics, and devices, (2) the drug facts panels found on over-the-counter medication packaging, and (3) the nutrition facts panels and health claims on food packaging.

A recent perspective by Schwartz and Woloshin (2009) in the *New England Journal of Medicine* highlighted some of the problems with drug labels:

- Drug labels are written by drug companies, and not the FDA. As a result, the FDA may overlook omissions, exaggerations, or inconsistencies in the drug labels.
- For this reason, important information about drug risks may not appear in the final drug label.
- For the same reason, information about the possible benefits of the drugs also may not appear on the drug label.
- A reflection of the reviewers' confidence in the approval decision is rarely reflected in the drug label.

Schwartz and Woloshin noted that the FDA has recognized these problems and has begun to address them. The Risk Communication Advisory Committee was initiated at the FDA in 2008 (see Box 2-4) (FDA, 2008b). A draft guidance not yet finalized was issued in 2006 recommending the use of a prescription drug information highlights panel to "provide immediate access to the information that practitioners most commonly refer to and view as most important" (FDA, 2006b). Inclusion of summaries of the following information was suggested: date of initial U.S. approval, boxed warnings, recent major changes in the label, indications and usage, dosage and administration, dosage forms and strengths, contraindications, warnings and precautions, adverse reactions, drug interactions, and use in special populations.

Effective drug labels have been studied, and the data show that concise, balanced information with symbolic communication aids are useful (Davis et al., 2009; Dowse and Ehlers, 2005; Mansoor and Dowse, 2003; Schwartz et al., 2009). These findings have been discussed at several IOM workshops (IOM, 2007c, 2008), where speakers have suggested that a standardized drug label would improve patient understanding and adherence (IOM, 2008). The challenges of accomplishing this goal were highlighted by Shrank and colleagues (2009) after the conclusion of a study on the ability of a new drug label design to improve patient outcomes in several chronic diseases.

In 2006, the FDA began requiring companies to submit drug label information in an electronic format to enable public access to this information on the FDA website (FDA, 2005). To enhance the usefulness of this information to the public, the committee identified a need to improve the description of the balance of risks and benefits and to expand the product categories included in the database. The website, Drugs@FDA, is not readily found (FDA, 2009). It does not appear on the first 10 pages of results in a Google search on "FDA electronic drug label," for example.[12] Improvement and expansion of this database and the accessibility of the website would be beneficial.

FURTHER ISSUES WITH USE OF BIOMARKERS

The need for effective communication is important for foods and supplements in addition to drugs, biologics, and devices. A recent report to the FDA Science Board recommended interfacing with universities to improve risk communication (Subcommittee on Science and Technology, 2007). Recommendations 6.1 and 6.2 of the IOM's *The Future of Drug Safety* report focused on ways that the FDA Center for Drug Evaluation and Research could improve risk communication with stakeholders (IOM, 2007a). As a result of these recommendations, the Risk Communication Advisory Committee was created at the FDA. To build on these recommendations, this biomarker evaluation report seeks to extend the intent of these recommendations across regulated product categories and a broader range of stakeholders.

Healthcare providers face a challenging task in conveying health-related information to the public. Professional societies can help healthcare providers obtain skills in how to communicate with their patients about the probabilistic nature of health-related evidence and decisions. Professional societies have an important role to play in helping physicians, consumers, dietitians, other healthcare workers, and individuals in the pharmaceuticals, biologics, medical devices, supplements, and food industries to understand the consequences of innumeracy, evidence gaps, and the insufficiency of evidence to predict all outcomes when evidence is based on surrogate endpoints, other biomarkers, short-term clinical trials, or observational studies alone rather than clinical endpoints.

Numeracy

The need to improve health literacy has been widely recognized. The IOM made recommendations for addressing the issue in a 2004 report in which health literacy was defined as "the degree to which individuals have the capacity to obtain, process, and understand basic health information and services needed to make appropriate health decisions" (IOM, 2004). That definition had been in use previously by several other groups (HHS, 2000; IOM, 2004; Selden et al., 2000).

---

[12] Date of the Google search: November 11, 2009. As of March 3, 2010, Drugs@FDA is the second entry on the first page of results.

One important component of health literacy is numeracy, the ability to understand and interpret the integers, decimals, percentages, and fractions encountered in daily life and to perform related arithmetic (Peters et al., 2007). Its importance actually goes far beyond the ability to understand and make health-related decisions for one's self and family; it is needed for financial transactions, cooking, sewing, building, navigating, and making health-related decisions. Golbeck et al. (2005) define health numeracy as "the degree to which individuals have the capacity to access, process, interpret, communicate, and act on numerical, quantitative, graphical, biostatistical, and probabilistic health information needed to make effective health decisions."

Lower numeracy is associated with less consumer comprehension of drug labels (Davis et al., 2006; Nelson et al., 2008) and food labels (Levy and Fein, 1998; Rothman et al., 2006). Lower numeracy is also associated with poorer health outcomes (Ancker and Kaufman, 2007; Nelson et al., 2008). A great deal of research focuses on strategies for communication between healthcare providers and patients about risks and probabilities (Akl et al., 2007; Apter et al., 2008; Fagerlin et al., 2005; Montori and Rothman, 2005; Peters et al., 2007).

Innumeracy is a problem that goes beyond the general public, however. Researchers have found that numeracy does not necessarily correlate as closely with education as literacy (Jacobson, 2007; Nelson et al., 2008). Nelson et al. (2008) and others recommend the use of short assessments by practitioners so they can better tailor their communication to their patients (Keller and Siergrist, 2009). Furthermore, healthcare practitioners themselves must deal with innumeracy. The adoption and practice of evidence-based medicine depends on physicians' ability to understand and communicate risk and other probabilistic information (Jacobson, 2007; Nusbaum, 2006; Rao, 2008). Innumeracy among other health professionals also needs to be addressed. For example, researchers have examined this issue in nursing (Jukes and Gilchrist, 2006) and psychology (Mulhern and Wylie, 2004).

Numeracy is important to the successful adoption of the biomarker evaluation framework recommended in this report. Understanding biomarker use and the probabilities involved requires comfort with mathematical reasoning. Without adequate numeracy, individuals will have difficulty making decisions under conditions of uncertainty, such as when there are multiple possible outcomes. Without numeracy, regulators will have difficulty explaining to industry the reasoning behind biomarker evaluation, healthcare practitioners will experience difficulty communicating with patients about the probabilities involved with predictions based on biomarkers, and the media will have difficulty in communicating about these topics with the public in general. More work is needed to determine the best ways to communicate probabilistic information and address innumeracy. The National Research Council has made recommendations on ways to improve numeracy (NRC, 1990, 2005), and the Institute of Medicine has taken several looks at the impact of numeracy on health (IOM, 2001, 2004, 2007b, 2009b). Public support and understanding are important for successful adoption of new policies; informed consumers can help to drive change with respect to careful biomarker evaluation and use.

Cognitive Biases and Impacts of Evidence Gaps

Every day individuals make decisions on the basis of incomplete information on a variety of issues, such as education, safety, diet, health, and more. Although any decision an individual makes may be important in the course of one's life, arguably the decisions related to health are the most likely to affect the length and quality of one's life. For this reason, the stakes are high for these decisions, which are often guided by physicians. But just because the stakes are high does not mean more information is available to use to make an informed decision. Health-related decisions have the same uncertainties as other life decisions. In addition, decisions that policy makers and regulators must make to maximize and protect public health also have these uncertainties. To manage both risks and benefits, all stakeholders—including patients, physicians, and regulatory bodies—need access to reliable information about the uncertainties involved in health decisions.

The goal of access to information can be undermined by the strained resources of government agencies, the overload of information presented to consumers, the profit motivation of companies, and the desire by physicians to reassure their patients. The FDA has a unique relationship with all of these stakeholders and the authority to take actions to protect and promote public health. With better risk communication and access to reliable and complete information about the benefits and risks involved in health decisions, agencies like the FDA will be better able to respond and adjust to the most accurate and current data available for its regulatory decisions.

The committee identified two types of evidence gaps observed when surrogate and other types of biomarkers are used to make decisions about the efficacy of a drug or health benefits of a food. First, they do not explain the entire effect of the food or drug on a person. Second, changes in a biomarker caused by a particular drug, food, or other health intervention do not always predict changes in the clinical outcome of interest. Use of surrogate biomarkers, short-term clinical trials, or observational studies alone cannot adequately predict clinical benefit or harm, and in some cases they do not predict clinical benefit or harm at all. This caution is even more relevant to decisions based solely on biomarkers whose data do not support use as surrogate endpoints. Without information about an intervention's effect on clinical endpoints, it is impossible to have complete information about the efficacy and safety of the intervention.

Humans tend to oversimplify or ignore evidence gaps in order to make decisions, and are often unaware of evidence gaps. In situations of insufficient or overly complex information, humans use cognitive biases to make decisions; in other words, the types of mistakes people make when making decisions in the absence of complete information are predictable. Tversky and Kahneman explored this area in a famous 1974 paper entitled "Judgment under uncertainty: Heuristics and biases." In this article, Tversky and Kahneman explored the heuristics of representativeness, availability, and anchoring and the biases in judgment that arise from them. Tversky and Kahneman outlined the following heuristics and related cognitive biases in their important 1974 paper:

- The representativeness heuristic (the tendency to make judgments based on how well an element matches to preconceptions of a larger group) leads to the following biases:
    - Insensitivity to prior probability of outcomes (this is also known as neglect of probability bias, or ignoring available probabilistic information when making decisions)
    - Insensitivity to sample size
    - Misconceptions of chance
    - Insentivitiy to predictability (also known as neglect of probability bias, or ignoring available probabilistic information when making decisions)
    - The illusion of validity
    - Misconceptions of regression
- The availability heuristic (making decisions based on the most readily available memories or examples) leads to the following biases:
    - Biases due to the retrievability of instances
    - Biases of imaginability
    - Illusory correlation
- Adjustment and anchoring heuristic (anchoring is the tendency to allow some factor to weigh too heavily in a decision)
    - Insufficient adjustment
    - Biases in the evaluation of conjunctive and disjunctive events
    - Anchoring in the assessment of subjective probability distributions

Each of these heuristics and biases are explained in the referenced paper (Tversky and Kahneman, 1974). An example of insensitivity to probability bias, also known as neglect of probability bias, is when a person chooses to eat a nutrient or other substance that has been shown in observational studies to be associated with a reduced risk of disease, while ignoring the fact that this research alone does not confirm a substance's causal connection to a reduced risk of disease. Because these biases are well known, some may try to take advantage of them to mislead consumers.

Cognitive biases of healthcare professionals in health-related decision making have been studied in the context of emergent (Pines, 2006), acute (Aberegg et al., 2005; Freshwater-Turner et al., 2007), and chronic healthcare settings (Gruppen et al., 1994; Lutfey and McKinlay, 2009; Redelmeier and Shafir, 1995; Roswarski and Murray, 2006), while cognitive biases of patients have been evaluated in regard to illnesses such as myocardial infarction (Khraim and Carey, 2009) and cancer (Han et al., 2006).

Efforts by professional societies can help physicians, dietitians, and other healthcare practitioners be aware of information gaps and common cognitive biases when helping their patients or clients make decisions about their health care. With this knowledge, strategies can be developed and disseminated. In situations where the public and health professionals need to make decisions in the absence of complete, definitive evidence, decision makers need to be able to access balanced, non-misleading data, or they will be likely to make systematic errors in their thinking.

REFERENCES

Aberegg, S. K., E. F. Haponik, and P. B. Terry. 2005. Omission bias and decision making in pulmonary and critical care medicine. *Chest* 128(3):1497–1505.

Advisory Committee to the Surgeon General. 1964. *Report of the Advisory Committee to the Surgeon General*. Washington, DC: U.S. Department of Health, Education, and Welfare.

Afzal, A. K S. J. Jacobsen, D. W. Mahoney, J. A. Kors, M. M. Redfield, J. C. Burnett, and R. J. Rodeheffer. 2007. Prevalence and prognostic significance of heart failure stages: Application of the American College of Cardiology/American Heart Association heart failure staging criteria in the community. *Circulation* 115(12):1563–1570.

Akl, E. A., N. Maroun, G. Guyatt, A. D. Oxman, P. Alonso-Coello, G. E. Vist, P. J. Devereaux, V. M. Montori, and H. J. Schunemann. 2007. Symbols were superior to numbers for presenting strength of recommendations to health care consumers: A randomized trial. *Journal of Clinical Epidemiology* 60(12):1298–1305.

ALLHAT Officers and Coordinators for the ALLHAT Collaborative Research Group. 2000. Major cardiovascular events in hypertensive patients randomized to doxazosin vs chlorthalidone. *Journal of the American Medical Association* 283(15):1967–1975.

ALLHAT Officers and Coordinators for the ALLHAT Collaborative Research Group. 2002. Major outcomes in high-risk hypertensive patients randomized to angiotensin-converting enzyme inhibitor or calcium channel blocker vs. diuretic: The Antihypertensive and Lipid-Lowering Treatment to Prevent Heart Attack Trial (ALLHAT). *Journal of the American Medical Association* 288(23):2981–2997.

Alonso, A., G. Molenberghs, H. Geys, M. Buyse, and T. Vangeneugden. 2006. A unifying approach for surrogate marker validation based on Prentice's criteria. *Statistics in Medicine* 25(2):205–221.

Altar, C. A. 2008. The Biomarkers Consortium: On the critical path of drug discovery. *Clinical Pharmacology and Therapeutics* 83(2):361–364.

Altar, C. A., D. Amakye, D. Buonos, J. Bloom, G. Clack, R. Dean, V. Devanarayan, D. Fu, S. Furlong, L. Hinman, C. Girman, C. Lathia, L. Lesko, S. Madani, J. Mayne, J. Meyer, D. Raunig, P. Sager, S. A. Williams, P. Wong, and K. Zerba. 2008. A prototypical process for creating evidentiary standards for biomarkers and diagnostics. *Clinical Pharmacology and Therapeutics* 83(2):368–371.

Ancker, J. S., and D. Kaufman. 2007. Rethinking health numeracy: A multidisciplinary literature review. *Journal of the American Medical Informatics Association* 14(6):713–721.

Apter, A. A., M. K. Paasche-Orlow, J. T. Remillard, I. M. Bennett, E. P. Ben-Joseph, R. M. Batista, H. Hyde, and R. E. Rudd. 2008. Numeracy and communication with patients: They are counting on us. *Journal of General Internal Medicine* 23(12):2117–2124.

Arno, P. S., and K. L. Feiden. 1988. *Against the odds, the story of AIDS drug development, politics and profits*. New York: Harper Collins Publishers.

AVERT. 2009. *History of AIDS: 1987–1992*. http://www.avert.org/aids-history87-92.htm# (accessed October 3, 2009).

Behrman, R. E. 1999. FDA approval of antiretroviral agents: An evolving paradigm. *Drug Information Journal* 33(2):337–341.

Berry, D. C. 2004. *Risk, communication and health psychology*. New York: Open University Press.

Bigger, J. T., J. L. Fleiss, R. Kleiger, J. P. Miller, and L. M. Rolnitzky. 1984. The relationships among ventricular arrhythmias, left ventricular dysfunction, and mortality in the 2 years after myocardial infarction. *Circulation* 69(2):250-258.

Blank, M. 2008. *Clinical Evidentiary Standards for Biomarker Qualification*. Paper presented at 2008 Cardiovascular Biomarkers and Surrogate Endpoints Symposium: Building a Framework for Biomarker Application, Bethesda, MD, September 12.

Boissel, J. P., J. P. Collet, P. Moleur, and M. Haugh. 1992. Surrogate endpoints: A basis for a rational approach. *European Journal of Clinical Pharmacology* 43(3):235–244.

Bonventre, J. V. 2009. *Biomarkers of Kidney Injury: Dawn of a New Era*. Paper read at Institute of Medicine Committee on Qualification of Biomarkers and Surrogate Endpoints in Chronic Disease, Meeting 2, Washington, DC, April 6.

Borer, J. S. 2004. Development of cardiovascular drugs: The U.S. regulatory milieu from the perspective of a participating nonregulator. *Journal of the American College of Cardiology* 44(12):2285–2292.

Brady, T. M., and L. G. Feld. 2009. Pediatric approach to hypertension. *Seminars in Nephrology* 29(4):379–388.

Burzykowski, T., G. Molenberghs, and M. Buyse. 2004. The validation of surrogate end points by using data from randomized clinical trials: A case-study in advanced colorectal cancer. *Journal of the Royal Statistical Society A* 167(Pt 1):103–124.

Buyse, M., and G. Molenberghs. 1998. Criteria for the validation of surrogate endpoints in randomized experiments. *Biometrics* 54(3):1014–1029.

Buyse, M., G. Molenberghs, T. Burzykowski, D. Renard, and H. Geys. 2000. The validation of surrogate endpoints in meta-analyses of randomized experiments. *Biostatistics* 1(1):49–67.

CAPS (Cardiac Arrhythmia Pilot Study) Investigators. 1986. The Cardiac Arrhythmia Pilot Study. *American Journal of Cardiology* 57(1):91–95.

CAPS Investigators. 1988. Effects of encainide, flecainide, imipramine and moricizine on ventricular arrrhythmias during the year after acute myocardial infarction: The CAPS. *American Journal of Cardiology* 61(8):501–509.

CAST (Cardiac Arrhythmia Suppression Trial) Investigators. 1989. Preliminary report: Effect of encainide and flecainide on mortality in a randomized trial of arrhythmia suppression after myocardial infarction. *New England Journal of Medicine* 321(6):406–412.

Carter, B. L. 2002. Blood pressure as a surrogate end point for hypertension. *Annals of Pharmacotherapy* 36(1):87–92.

CBER (Center for Biologics Evaluation and Research). 2007. *Guidance for industry: Clinical data needed to support the licensure of seasonal inactivated influenza vaccines.* http://www.fda.gov/BiologicsBloodVaccines/GuidanceComplianceRegulatoryInformation/Guidances/Vaccines/ucm074794.htm (accessed September 24, 2009).

CDRH (Center for Devices and Radiological Health). 2006. *Guidance for industry and FDA staff: Postmarket surveillance under section 522 of the Federal Food, Drug, and Cosmetic Act.* http://www.fda.gov/MedicalDevices/DeviceRegulationandGuidance/GuidanceDocuments/ucm072517.htm (accessed September 24, 2009).

CDRH. 2008. *Guidance for industry and FDA staff: Clinical study designs for catheter ablation devices for treatment of atrial flutter.* http://www.fda.gov/MedicalDevices/DeviceRegulationandGuidance/GuidanceDocuments/ucm070919.htm (accessed November 10, 2009).

CFSAN (Center for Food Safety and Applied Nutrition). 2003. *Selenium and certain cancers (Qualified health claim: Final decision letter).* http://www.cfsan.fda.gov/~dms/ds-ltr35.html (accessed March 17, 2009).

CFSAN. 2004. *Qualified health claims: Letter of enforcement discretion—walnuts and coronary heart disease* (accessed March 17, 2009).

CFSAN. 2005. *Qualified health claims: Letter regarding "Tomatoes and prostate, ovarian, gastric and pancreatic cancers (American Longevity Petition)."* http://www.cfsan.fda.gov/~dms/qhclyco.html (accessed March 17, 2009).

CFSAN. 2008. *A food labeling guide.* http://www.cfsan.fda.gov/~dms/2lg-8.html#health (accessed March 26, 2009).

CFSAN. 2009a. *Guidance for industry: Evidence-based review system for the scientific evaluation of health claims.* http://www.cfsan.fda.gov/~dms/hclmgui6.html (accessed February 25, 2009).

CFSAN. 2009b. *Qualified health claims.* http://www.foodsafety.gov/~dms/lab-qhc.html (accessed March 17, 2009).

Clarke, R., S. Lewington, P. Sherliker, and J. Armitage. 2007. Effects of B-vitamins on plasma homocysteine concentrations and on risk of cardiovascular disease and dementia. *Current Opinion in Clinical Nutrition and Metabolic Care* 10(1):32–39.

Cochrane, A. L. 1971. *Effectiveness and efficiency: Random reflections on health services.* London, England: Nuffield Provincial Hospitals Trust.

Colatsky, T. J. 2009. Reassessing the validity of surrogate markers of drug efficacy in the treatment of coronary artery disease. *Current Opinion in Investigational Drugs* 10(3):239–244.

Colburn, W. A. 1997. Selecting and validating biologic markers for drug development. *Journal of Clinical Pharmacology* 37(5):355–362.

Colburn, W. A. 2000. Optimizing the use of biomarkers, surrogate endpoints, and clinical endpoints for more efficient drug development. *Journal of Clinical Pharmacology* 40(12 Pt 2):1419–1427.

Cotton, P. 1991. HIV surrogate markers weighed. *Journal of the American Medical Association* 265(11):1357, 1361–1362.

Daniels, M. J., and M. D. Hughes. 1997. Meta-analysis for the evaluation of potential surrogate markers. *Statistics in Medicine* 16(17):1965–1982.

Davis, T. C., M. S. Wolf, P. F. Bass, J. A. Thompson, H. H. Tilson, M. Neuberger, and R. M. Parker. 2006. Literacy and misunderstanding prescription drug labels. *Annals of Internal Medicine* 145(12):887–894.

Davis, T. C., A. D. Federman, P. F. Bass, R. H. Jackson, M. Middlebrooks, R. M. Parker, and M. S. Wolf. 2009. Improving patient understanding of prescription drug label instructions. *Journal of General Internal Medicine* 24(1):57–62.

De Gruttola, V., T. Fleming, D. Y. Lin, and R. Coombs. 1997. Perspective: Validating surrogate markers—Are we being naive? *Journal of Infectious Diseases* 175(2):237–246.

De Gruttola, V., C. Flexner, J. Schapiro, M. Hughes, M. van der Laan, and D. R. Kuritzkes. 2006. Drug development strategies for salvage therapy: Conflicts and solutions. *AIDS Research and Human Retroviruses* 22(11):1106–1109.

Deeks, S. G., and J. N. Martin. 2007. Partial treatment interruptions. *Current Opinion in HIV and AIDS* 2(1):46–55.

DeMets, D. L., and R. M. Califf. 2002. Lessons learned from recent cardiovascular clinical trials: Part I. *Circulation* 106(6):746–751.

Desai, M., N. Stockbridge, and R. Temple. 2006. Blood pressure as an example of a biomarker that functions as a surrogate. *AAPS Journal* 8(1):E146.

Dowse, R., and M. Ehlers. 2005. Medicine labels incorporating pictograms: Do they influence understanding and adherence? *Patient Education and Counseling* 58(1):63–70.

Echt, D. S., P. R. Liebson, L. B. Mitchell, R. W. Peters, D. Obias-Manno, A. H. Barker, D. Arensberg, A. Baker, L. Friedman, H. L. Greene, M. L. Huther, D. W. Richardson, and the CAST Investigators. 1991. Mortality and morbidity in patients receiving encainide, flecainide, or placebo: The Cardiac Arrhythmia Suppression Trial. *New England Journal of Medicine* 324(12):781-788.

Edwards, A. G. K., R. Evans, J. Dundon, S. Haigh, K. Hood, and G. J. Elwyn. 2006. Personalised risk communication for informed decision making about taking screening tests. *Cochrane Database of Systematic Reviews* 2006(4):Art. No. CD001865.

Ellenberg, S. S., and J. M. Hamilton. 1989. Surrogate endpoints in clinical trials: Cancer. *Statistics in Medicine* 8(4):405–413.

Emanuel, E. J., and F. G. Miller. 2001. The ethics of placebo-controlled trials—A middle ground. *New England Journal of Medicine* 345(12):915–919.

Ezzati, M., A. D. Lopez, A. Rodgers, S. V. Hoorn, C. J. L. Murray, and C. R. A. C. Group. 2002. Selected major risk factors and global and regional burden of disease. *The Lancet* 360(9343):1347–1360.

Fagerlin, A., C. Wang, and P. A. Ubel. 2005. Reducing the influence of anecdotal reasoning on people's health care decisions: Is a picture worth a thousand statistics? *Medical Decision Making* 25(4):398–405.

FDA (Food and Drug Administration). 1988. *Making drugs available for life-threatening diseases.* http://www.aegis.com/news/FDA/1988/Fd881001.html (accessed October 3, 2009).

FDA. 1999. *Significant scientific agreement in the review of health claims for conventional foods and dietary supplements.* http://www.fda.gov/Food/GuidanceComplianceRegulatoryInformation/GuidanceDocuments/FoodLabelingNutrition/ucm059132.htm (accessed October 3, 2009).

FDA. 2002. *Guidance for industry: Antiretroviral drugs using plasma HIV RNA measurements—Clinical considerations for accelerated and traditional approval.* http://www.fda.gov/cder/guidance/3647fnl.pdf (accessed December 29, 2008).

FDA. 2003. *FDA to encourage science-based labeling and competition for healthier dietary choices.* http://www.fda.gov/bbs/topics/NEWS/2003/NEW00923.html (accessed March 17, 2009).

FDA. 2004. *Guidance for industry: Available therapy.* http://www.fda.gov/RegulatoryInformation/Guidances/ucm126586.htm (accessed September 24, 2009).

FDA. 2005a. *FDA announces the use of new electronic drug labels to help better inform the public and improve patient safety.* http://www.fda.gov/NewsEvents/Newsroom/PressAnnouncements/2005/ucm108509.htm (accessed November 22, 2009).

FDA. 2006a. *Guidance for industry: Clinical studies section of labeling for human prescription drug and biological products—Content and format*. http://www.fda.gov/RegulatoryInformation/Guidances/ucm127509.htm (accessed September 24, 2009).

FDA. 2006b. *Draft guidance for industry: Labeling for human prescription drug and biological products—Implementing the new content and format requirements*. http://www.fda.gov/downloads/Drugs/GuidanceComplianceRegulatoryInformation/Guidances/ucm075082.pdf (accessed November 22, 2009).

FDA. 2007. *Guidance for industry: Clinical trial endpoints for the approval of cancer drugs and biologics*. http://www.fda.gov/downloads/Drugs/GuidanceComplianceRegulatoryInformation/Guidances/UCM071590.pdf (accessed September 24, 2009).

FDA. 2008a. *FDA, European Medicines Agency to consider additional test results when assessing new drug safety*. http://www.fda.gov/NewsEvents/Newsroom/PressAnnouncements/2008/ucm116911.htm (accessed November 23, 2009).

FDA. 2008b. *Risk Communication Advisory Committee*. http://www.fda.gov/AdvisoryCommittees/CommitteesMeetingMaterials/RiskCommunicationAdvisoryCommittee/default.htm (accessed November 21, 2009).

FDA. 2009. *Drugs@FDA*. http://www.accessdata.fda.gov/Scripts/cder/DrugsatFDA/ (accessed November 11, 2009).

FDA. 2010. *Risk Communication Advisory Committee*. http://www.fda.gov/AdvisoryCommittees/CommitteesMeetingMaterials/RiskCommunicationAdvisoryCommittee/default.htm (accessed March 14, 2010).

Fischl, M. A., D. D. Richman, M. H. Grieco, M. S. Gottlieb, P. A. Volberding, O. L. Laskin, J. M. Leedom, J. E. Groopman, D. Mildvan, R. T. Schooley et al. 1987. The efficacy of azidothymidine (AZT) in the treatment of patients with AIDS and AIDS-related complex. A double-blind, placebo-controlled trial. *New England Journal of Medicine* 317(4):185–191.

Fleming, T. R. 2005. Surrogate endpoints and FDA's accelerated approval process. *Health Affairs* 24(1):67–78.

Fleming, T. R., and D. L. DeMets. 1996. Surrogate end points in clinical trials: Are we being misled? *Annals of Internal Medicine* 125(7):605–613.

Frangakis, C. E., and D. B. Rubin. 2002. Principal stratification in causal inference. *Biometrics* 58(1):21–29.

Freshwater-Turner, D. A., R. J. Boots, R. N. Bowman, H. G. Healy, and A. C. Klestov. 2007. Difficult decisions in the intensive care unit: An illustrative case. *Anaesthesia and Intensive Care* 35(5):748–759.

Gobburu, J. V. S. 2009. Biomarkers in clinical drug development. *Clinical Pharmacology and Therapeutics* 86(1):26–27.

Golbeck, A. L., C. R. Ahlers-Schmidt, A. M. Paschal, and S. E. Dismuke. 2005. A definition and operational framework for health numeracy. *American Journal of Preventive Medicine* 29(4):375–376.

Goodsaid, F. 2008a. *Impact of the Biomarker Qualification Project on Drug Development*. Paper presented at 2008 Cardiovascular Biomarkers and Surrogate Endpoints Symposium: Building a Framework for Biomarker Application, Bethesda, MD, September 12.

Goodsaid, F. 2008b. *Markers of Renal Toxicity*. Paper presented at 2008 Cardiovascular Biomarkers and Surrogate Endpoints Symposium: Building a Framework for Biomarker Application, Bethesda, MD, September 11.

Goodsaid, F., and F. Frueh. 2006. Process map proposal for the validation of genomic biomarkers. *Pharmacogenomics* 7(5):773–782.

Goodsaid, F., and F. Frueh. 2007a. Biomarker qualification pilot process at the U.S. Food and Drug Administration. *AAPS Journal* 9(1):E105–E108.

Goodsaid, F., and F. Frueh. 2007b. Questions and answers about the pilot process for biomarker qualification at the FDA. *Drug Discovery Today: Technologies* 4(1):9–11.

Goodsaid, F., F. W. Frueh, and W. Mattes. 2008. Strategic paths for biomarker qualification. *Toxicology* 245:219–223.

Gruppen, L. D., J. Margolin, K. Wisdom, and C. M. Grum. 1994. Outcome bias and cognitive dissonance in evaluating treatment decisions. *Academic Medicine* 69(10 Suppl):S57–S59.

Hammer, S. M., K. E. Squires, M. D. Hughes, J. M. Grimes, L. M. Demeter, J. S. Currier, J. J. Eron Jr., J. E. Feinberg, H. H. Balfour Jr., L. R. Deyton, J. A. Chodakewitz, and M. A. Fischl. 1997. A controlled trial of two nucleoside analogues plus indinavir in persons with human immunodeficiency virus infection and CD4 cell counts of 200 per cubic millimeter or less. AIDS Clinical Trials Group 320 Study Team. *New England Journal of Medicine* 337(11):725–733.

Han, P. K., R. P. Moser, and W. M. Klein. 2006. Perceived ambiguity about cancer prevention recommendations: Relationship to perceptions of cancer preventability, risk, and worry. *Journal of Health Communication* 11(Suppl 1):51–69.

HHS (Department of Health and Human Services). 2000. *Healthy People 2010: Understanding and improving health*. Washington, DC: HHS.

Higgins, J., and S. Green. 2008. *Cochrane handbook for systematic reviews of interventions: Version 5.0.1.* http://www.cochrane.org/resources/handbook/index.htm (accessed April 2, 2009).

Hill, A. B. 1965. The environment and disease: Association or causation? *Proceedings of the Royal Society of Medicine* 58:295–300.

Hillis, A., and D. Seigel. 1989. Surrogate endpoints in clinical trials: Ophthalmologic disorders. *Statistics in Medicine* 8(4):427–430.

HIV Surrogate Marker Collaborative Group. 2000. Human immunodeficiency virus type 1 RNA level and CD4 count as prognostic markers and surrogate end points: A meta-analysis. *AIDS Research and Human Retroviruses* 16(12):1123–1133.

Holden, C. 1993. FDA okays surrogate markers. *Science* 259(5091):32–33.

Hughes, M. D. 2002. Evaluating surrogate endpoints. *Controlled Clinical Trials* 23(6): 703–707.

Hughes, M. D. 2005. The evaluation of surrogate endpoints in practice: Experience in HIV. In *The evaluation of surrogate endpoints*, edited by T. Burzykowski, G. Molenberghs, and M. Buyse. New York: Springer. Pp. 295–321.

Hughes, M. D., V. De Gruttola, and S. L. Welles. 1995. Evaluating surrogate markers. *Journal of Acquired Immune Deficiency Syndromes and Human Retrovirology* 10(Suppl 2):S1–S8.

Hughes, M. D., M. J. Daniels, M. A. Fischl, S. Kim, and R. T. Schooley. 1998. CD4 cell count as a surrogate endpoint in HIV clinical trials: A meta-analysis of studies of the AIDS Clinical Trials Group. *AIDS* 12(14):1823–1832.

IFT (Institute of Food Technologists). 2005. *Functional foods: Opportunities and challenges*. Chicago, IL: IFT.

IOM (Institute of Medicine). 1990. *Clinical practice guidelines: Directions for a new program*. Washington, DC: National Academy Press.

IOM. 2001. *Health and behavior: The interplay of biological, behavioral, and societal influences*. Washington, DC: National Academy Press.

IOM. 2004. *Health literacy: A prescription to end confusion*. Washington, DC: The National Academies Press.

IOM. 2005. *Saving women's lives: Strategies for improving breast cancer detection and diagnosis*. Washington, DC: The National Academies Press.

IOM. 2007a. *The future of drug safety: Promoting and protecting the health of the public*. Washington, DC: The National Academies Press.

IOM. 2007b. *Understanding the benefits and risks of pharmaceuticals: Workshop Summary*. Washington, DC: The National Academies Press.

IOM. 2008. *Standardizing medication labels: Confusing patients less: Workshop summary*. Washington, DC: National Academies Press.

IOM. 2009a. *Accelerating the development of biomarkers for drug safety*. Washington, DC: The National Academies Press.

IOM. 2009b. *Health literacy, eHealth, and communication: Putting the consumer first: Workshop Summary*. Washington, DC: The National Academies Press.

IOM. 2009c. *Initial national priorities for comparative effectiveness research*. Washington, DC: The National Academies Press.

Israili, Z. H., R. Hernandez-Hernandez, and M. Valasco. 2007. The future of antihypertensive treatment. *American Journal of Therapeutics* 14(2):121–134.

Jacobson, R. M. 2007. Teaching numeracy to physicians-in-training: Quantitative analysis for evidence-based medicine. *Minnesota Medicine* 90(11):37–38, 46.

Jiang, H., S. G. Deeks, D. R. Kuritzkes, M. Lallemant, D. Katzenstein, M. Albrecht, and V. De Gruttola. 2003. Assessing resistance costs of antiretroviral therapies via measures of future drug options. *Journal of Infectious Diseases* 188(7):1001–1008.

Joffe, M. M., and T. Greene. 2009. Related causal frameworks for surrogate outcomes. *Biometrics* 65(2):530–538.

Jukes, L., and M. Gilchrist. 2006. Concerns about numeracy skills of nursing students. *Nurse Education in Practice* 6(4):192–198.

Julian-Reynier, C., M. Welkenhuysen, L. Hagoel, M. Decruyenaere, P. Hopwood, and on behalf of the CRISCOM Working Group. 2003. Risk communication strategies: State of the art and effectiveness in the context of cancer genetic services. *European Journal of Human Genetics* 11(10):725–736.

Keller, C., and M. Siergrist. 2009. Effect of risk communication formats on risk perception depending on numeracy. *Medical Decision Making* 29(4):483–490.

Khraim, F. M., and M. G. Carey. 2009. Predictors of pre-hospital delay among patients with acute myocardial infarction. *Patient Education and Counseling* 75(2):155–161.

Kim, H., C. Nakamura, and Q. Zeng-Treitler. 2009. Assessment of pictographs developed through a participatory design process using an online survey tool. *Journal of Medical Internet Research* 11(1):e5.

Krumholz, H. M., and T. H. Lee. 2008. Redefining quality—Implications of recent clinical trials. *New England Journal of Medicine* 358(24):2537–2539.

Lagakos, S. W., and D. F. Hoth. 1992. Surrogate markers in AIDS: Where are we? Where are we going? *Annals of Internal Medicine* 116(7):599–601.

Lassere, M. N. 2008. The biomarker-surrogacy evaluation schema: A review of the biomarker-surrogate literature and a proposal for a criterion-based, quantitative, multidimensional hierarchical levels of evidence schema for evaluating the status of biomarkers as surrogate endpoints. *Statistical Methods in Medical Research* 17(3):303–340.

Lathia, C. D., D. Amakye, W. Dai, C. Girman, S. Madani, J. Mayne, P. MacCarthy, P. Pertel, L. Seman, A. Stoch, P. Tarantino, C. Webster, S. Williams, and J. A. Wagner. 2009. The value, qualification, and regulatory use of surrogate end points in drug development. *Clinical Pharmacology and Therapeutics* 86(1):32–43.

Law, M. R., and J. K. Morris. 2009. Use of blood pressure lowering drugs in the prevention of cardiovascular disease: Meta-analysis of 147 randomized trials in the context of expectations from prospective epidemiological studies. *British Medicine Journal* 338:b1665.

Lesko, L. J., and A. J. Atkinson Jr. 2001. Use of biomarkers and surrogate endpoints in drug development and regulatory decision making: Criteria, validation, strategies. *Annual Review of Pharmacology and Toxicology* 41:347–366.

Levy, A. S., and S. B. Fein. 1998. Consumers' ability to perform tasks using nutrition labels. *Journal of Nutrition Education* 30(4):210–217.

Lipkus, I. M. 2007. Numeric, verbal, and visual formats of conveying health risks: Suggested best practices and future recommendations. *Medical Decision Making* 27(5):696–713.

Lutfey, K. E., and J. B. McKinlay. 2009. What happens along the diagnostic pathway to CHD treatment? Qualitative results concerning cognitive processes. *Sociology of Health and Illness* 31(7):1077–1092.

MacMahon, S., R. Peto, J. Cutler, R. Collins, P. Sorlie, J. Neaton, R. Abbott, J. Godwin, A. Dyer, and J. Stamler. 1990. Blood pressure, stroke, and coronary heart disease. Part 1, prolonged differences in blood pressure: Prospective observational studies corrected for the regression dilution bias. *Lancet* 335(8692):765–774.

Manns, B., W. F. Owen, W. C. Winkelmayer, P. J. Deveraux, and M. Tonelli. 2006. Surrogate markers in clinical studies: Problems solved or created? *American Journal of Kidney Diseases* 48(1):159–166.

Mansoor, L. E., and R. Dowse. 2003. Effect of pictograms on readability of patient information materials. *Annals of Pharmacotherapy* 37(7–8):1003–1009.

Miksad, R. 2009. *Decision science for qualification of biomarker surrogate endpoints*. Paper read at Institute of Medicine Committee on Qualification of Biomarkers and Surrogate Endpoints in Chronic Disease, Meeting 2, Washington, DC, April 6.

Mitka, M. 2003. Food fight over product label claims: Critics say proposed changes will confuse consumers. *Journal of the American Medical Association* 290(7):871–875.

Montori, V. M., and R. L. Rothman. 2005. Weakness in numbers: The challenge of numeracy in health care. *Journal of General Internal Medicine* 20(11):1071–1072.

Mukharji, J., R. E. Rude, W. K. Poole, N. Gustafson, L. J. Thomas Jr., H. W. Strauss, A. S. Jaffe, J. E. Muller, R. Roberts, D. S. Raabe, C. H. Croft, E. Passamani, E. Braunwald, J. T. Willerson, and the MILIS Study Group. 1984. Risk factors for sudden death after acute myocardial infarction: Two-year follow-up. *American Journal of Cardiology* 54(1):31–36.

Mulhern, G., and J. Wylie. 2004. Changing levels of numeracy and other core mathematical skills among psychology undergraduates between 1992 and 2002. *British Journal of Psychology* 95(Pt 3):355–370.

Nambi, V., and C. M. Ballantyne. 2007. Role of biomarkers in developing new therapies for vascular disease. *World Journal of Surgery* 31(4):676–681.

NCI (National Cancer Institute). 2007. *Patient-centered communication in cancer care*. Washington, DC: NCI.

Nelson, W., V. F. Reyna, A. Fagerlin, I. Lipkus, and E. Peters. 2008. Clinical implications of numeracy: Theory and practice. *Annals of Behavioral Medicine* 35(3):261–274.

NHLBI Working Group on Future Directions in Hypertension Treatment Trials. 2005. Major clinical trials of hypertension: What should be done next? *Hypertension* 46(1):1–6.

Nicholson, P. J. 1999. Communicating health risk. *Occupational Medicine* 49(4):253–256.

NRC (National Research Council). 1990. *On the shoulders of giants: New approaches to numeracy*. Washington, DC: National Academy Press.

NRC. 2005. *Measuring literacy: Performance levels for adults*. Washington, DC: The National Academies Press.

Nusbaum, N. J. 2006. Mathematics preparation for medical school: Do all premedical students need calculus? *Teaching and Learning in Medicine* 18(2):165–168.

O'Connor, A. M., C. L. Bennett, D. Stacey, M. Barry, N. F. Col, K. B. Eden, V. A. Entwistle, V. Fiset, M. Holmes-Rovner, S. Khangura, H. Llewellyn-Thomas, and D. Rovner. 2009. Decision aids for people facing health treatment or screening decisions. *Cochrane Database of Systematic Reviews* 2009(3): Art. No. CD001431.

Omenn, G. S., G. E. Goodman, M. D. Thornquist, J. Balmes, M. R. Cullen, A. Glass, J. P. Keogh, F. L. Meyskens Jr., B. Valanis, J. H. Williams Jr., S. Barnhart, and S. Hammar. 1996. Effects of a combination of beta-carotene and vitamin A on lung cancer and cardiovascular disease. *New England Journal of Medicine* 334:1150–1155.

Perry, H. M., A. I. Goldman, M. A. Lavin, H. W. Schnaper, A. E. Fitz, E. D. Frohlich, B. Steele, and H. G. Richman. 1978. Evaluation of drug treatment in mild hypertension: VA-NHLBI feasibility trial. Plan and preliminary results of a two-year feasibility trial for a multicenter intervention study to evaluate the benefits versus the disadvantages of treating mild hypertension. Prepared for the Veterans Administration–National Heart, Lung, and Blood Institute Study Group for Evaluating Treatment in Mild Hypertension. *Annals of the New York Academy of Sciences* 304:267–287.

Peters, E., J. Hibbard, P. Slovic, and N. Dieckmann. 2007. Numeracy skill and the communication, comprehension, and use of risk–benefit information. *Health Affairs* 26(3):741–748.

Peto, R., R. Doll, J. D. Buckley, and M. B. Sporn. 1981. Can dietary beta-carotene materially reduce human cancer rates? *Nature* 290:201–208.

Pines, J. M. 2006. Profiles in patient safety: Confirmation bias in emergency medicine. *Academic Emergency Medicine* 13(1):90–94.

Prentice, R. L. 1989. Surrogate endpoints in clinical trials: Definition and operational criteria. *Statistics in Medicine* 8(4):431–440.

Psaty, B. M., D. S. Siscovick, N. S. Weiss, T. D. Koepsell, F. R. Rosendaal, D. Lin, S. R. Heckbert, E. H. Wagner, and C. D. Furberg. 1996. Hypertension and outcomes research: From clinical trials to clinical epidemiology. *American Journal of Hypertension* 9(2):178–183.

Rao, G. 2008. Physician numeracy: Essential skills for practicing evidence-based medicine. *Family Medicine* 40(5):354–358.

Redelmeier, D. A., and E. Shafir. 1995. Medical decision making in situations that offer multiple alternatives. *Journal of the American Medical Association* 273(4):302–305.

Robins, J. M., and S. Greenland. 1992. Identifiability and exchangeability for direct and indirect effects. *Epidemiology* 3(2):143–155.

Robins, J. M., and S. Greenland. 1994. Adjusting for differential rates of prophylaxis therapy for PCP in high- versus low-dose AZT treatment arms in an AIDS randomized trial. *Journal of the American Statistical Association* 89(427):737–749.

Roswarski, T. E., and M. D. Murray. 2006. Supervision of students may protect academic physicians from cognitive bias: A study of decision making and multiple treatment alternatives in medicine. *Medical Decision Making* 26(2):154–161.

Rothman, R. L., R. Housam, H. Weiss, D. Davis, R. Gregory, T. Gebretsadik, A. Shintani, and T. A. Elasy. 2006. Patient understanding of food labels—The role of literacy and numeracy. *American Journal of Preventive Medicine* 31(5):391–398.

Ruberman, W., E. Weinblatt, J. D. Goldberg, C. W. Frank, and S. Shapiro. 1977. Ventricular premature beats and mortality after myocardial infarction. *New England Journal of Medicine* 297(14):750–757.

Ruskin, J. N. 1989. The Cardiac Arrhythmia Suppression Trial (CAST). *New England Journal of Medicine* 321(6):386–388.

Schatzkin, A., and M. Gail. 2002. The promise and peril of surrogate endpoints in cancer research. *Nature Reviews Cancer* 2(1):19–27.

Schneeman, B. 2007. FDA's review of scientific evidence for health claims. *Journal of Nutrition* 137(2):493–494.

Schwartz, L. M., S. Woloshin, and H. G. Welch. 2009. Using a drug facts box to communicate drug benefits and harms: Two randomized trials. *Annals of Internal Medicine* 150(8):516–527.

Selden, C. R., M. Zorn, S. C. Ratzan, and R. M. Parker. 2000. *National Library of Medicine current bibliographies in medicine: Health literacy*. NLM Pub. No. CBM 2000-1. Bethesda, MD: National Institutes of Health.

Shi, Q., and D. J. Sargent. 2009. Meta-analysis for the evaluation of surrogate endpoints in cancer clinical trials. *International Journal of Clinical Oncology* 14(2):102–111.

Shrank, W. H., A. Patrick, P. P. Gleason, C. Canning, C. Walters, A. H. Heaton, S. Jan, M. A. Brookhart, S. Schneeweiss, D. H. Solomon, M. S. Wolf, J. Avorn, and N. K. Choudhry. 2009. An evaluation of the relationship between the implementation of a newly designed prescription drug label at Target pharmacies and health outcomes. *Medical Care* 47(9):1031–1035.

Subcommittee on Science and Technology. 2007. *FDA science and mission at risk*. Washington, DC: Food and Drug Administration.

Taylor, C. L., and V. L. Wilkening. 2008. How the nutrition food label was developed, part 2: The purpose and promise of nutrition claims. *Journal of the American Dietetic Association* 108(4):618–623.

Temple, R. 1999. Are surrogate markers adequate to assess cardiovascular disease drugs? *Journal of the American Medical Association* 282(8):790–795.

Temple, R. J. 2009. *Qualification of Biomarkers as Surrogate Endpoints of Chronic Disease Risk*. Paper read at Institute of Medicine Committee on Qualification of Biomarkers and Surrogate Endpoints in Chronic Disease: Meeting 2, Washington, DC, April 6.

Trumbo, P., and K. Ellwood. 2009. *Developing a Framework for Biomarker Qualification for Chronic Disease*. Paper presented at Institute of Medicine Committee on Qualification of Biomarkers as Surrogate Endpoints for Chronic Disease Risk, Washington, DC, January 12.

Tversky, A. And D. Kahneman. 1974. Judgment under uncertainty: Heuristics and biases. *Science* 185(4157):1124–1131.

Volberding, P. A., S. W. Lagakos, J. M. Grimes, D. S. Stein, H. H. Balfour Jr., R. C. Reichman, J. A. Bartlett, M. S. Hirsch, J. P. Phair, R. T. Mitsuyasu et al. 1994. The duration of zidovudine benefit in persons with asymptomatic HIV infection. Prolonged evaluation of protocol 019 of the AIDS Clinical Trials Group. *Journal of the American Medical Association* 272(6):437–442.

Wagner, J. A. 2002. Overview of biomarkers and surrogate endpoints in drug development. *Disease Markers* 18(2):41–46.

Wagner, J. A. 2008. Strategic approach to fit-for-purpose biomarkers in drug development. *Annual Review of Pharmacology and Toxicology* 48:631–651.

Wagner, J. A., S. A. Williams, and C. J. Webster. 2007. Biomarkers and surrogate end points for fit-for-purpose development and regulatory evaluation of new drugs. *Clinical Pharmacology and Therapeutics* 81(1):104–107.

Wang, T. J., and R. S. Vasan. 2005. Epidemiology of uncontrolled hypertension in the United States. *Circulation* 112(11):1651–1662.

Williams, B. 2005. Recent hypertension trials. *Journal of the American College of Cardiology* 45(6):813–827.

Williams, S. A., D. E. Slavin, J. A. Wagner, and C. J. Webster. 2006. A cost-effectiveness approach to the qualification and acceptance of biomarkers. *Nature Reviews Drug Discovery* 5(11):897–902.

Wittes, J., E. Lakatos, and J. Probstfield. 1989. Surrogate endpoints in clinical trials: Cardiovascular diseases. *Statistics in Medicine* 8(4):415–425.

Wolff, T., and T. Miller. 2007. Evidence for the reaffirmation of the U.S. Preventive Services Task Force recommendation on screening for high blood pressure. *Annals of Internal Medicine* 147(11):787–791.

Appendix C

Committee Biographies

Jane E. Henney, M.D. (*Chair*), is Professor of Medicine and Public Health Sciences at the University of Cincinnati College of Medicine. Previously, she was Senior Vice President and Provost of Health Affairs at the University of Cincinnati. Her experience and expertise lie in managing complex organizations that provide direct health services, regulate science-based products, educate the next generation of health professionals, and conduct biomedical research. She has served in a series of senior health policy leadership positions including Commissioner of the U.S. Food and Drug Administration (FDA) (1999–2001), Deputy Commissioner for Operations (1991–1994), Deputy Director of the National Cancer Institute, Vice Chancellor of Health Programs of Kansas Medical Center, Interim Dean of the University of Kansas School of Medicine, and the first Vice President for Health Sciences at the University of New Mexico. Dr. Henney has served on many boards and committees including the Advisory Committee to the Director for the National Institutes of Health (NIH), the National Advisory Research Resources Council, and the American Cancer Society National Board of Directors. She has served on several Institute of Medicine (IOM) committees including the planning committee for *The IOM Drug Safety Report: Resource Implications* workshop, Committee on Improving Mammography Quality Standards, and the IOM Membership Committee. She has served as the Chair of the IOM Committee on Strategies to Reduce Sodium Intake. Dr. Henney received her undergraduate degree from Manchester College and her medical degree from Indiana University, and she completed her subspecialty training in medical oncology at the M.D. Anderson Hospital and Tumor Institute and the National Cancer Institute. She is an IOM member.

Timothy B. Baker, Ph.D., is a Professor of Medicine in the University of Wisconsin School of Medicine and Public Health. His principal research goals are to increase understanding of the motivational bases of addictive disorders and to develop and evaluate treatments for such disorders. He is also highly interested in developing and using technological and methodological advances to deliver and evaluate treatments for addictive disorders and cancer. Dr. Baker has served as the editor of the *Journal of Abnormal Psychology*, is the Principal Investigator of the University of Wisconsin Transdisciplinary Tobacco Use Research Center award (NCI/NIDA), has a K05 Senior Scientist Award from NCI, and is the recipient of the James McKeen Cattell Award from the Association for Psychological Science.

Rebecca Bascom, M.D., M.P.H., is a professor of medicine at the Milton S. Hershey Medical Center at Penn State University. Her areas of expertise include lung diseases and inhalation toxicology. Dr. Bascom led an analysis team to evaluate the cardiorespiratory health effects on New York City police officers exposed during the 9/11 terrorist attack. She has served on three National Research Council committees, the Committee on the Evaluation of the Department of Defense Comprehensive Clinical Evaluation Protocol, the Committee on Occupational Safety and Health in Research Animal Facilities, and the Committee on Health Effects of Indoor Allergens. Dr. Bascom earned her M.D. from the University of Oregon Health Sciences Center and her M.P.H. in occupational medicine from the Johns Hopkins Bloomberg School of Public Health. She trained in internal medicine, as well as pulmonary and critical care medicine at the Johns Hopkins Hospital.

Shyam Biswal, Ph.D., is Professor at the Department of Environmental Health, Division of Toxicology at the Johns Hopkins University Bloomberg School of Public Health. He also has a joint appointment in the Department of Oncology, Pulmonary and Critical Care Division, Johns Hopkins University School of Medicine. Dr. Biswal was a Research Associate, Division of Pharmacology and Toxicology, at the College of Pharmacy, University of Texas at Austin. Dr. Biswal did his postdoctoral research in the same department, and he has a Ph.D. in biotechnology from the Indian Institute of Technology. Dr. Biswal studies the mechanisms of gene-environment interactions and susceptibility to environmental lung diseases involving tobacco smoke exposure, such as COPD and lung cancer. His group has identified that transcription factor, Nrf2, is a critical modifier of inflammation that determines susceptibility to these diseases based on which novel therapies are being developed. Dr. Biswal has more than 100 publications in this area, and he is the principal investigator on several research grants supported by the National Institutes of Health that are related to COPD and lung cancer.

Daniel Carpenter, Ph.D., is Allie S. Freed Professor of Government and Director of the Center for American Political Studies in the Faculty of Arts and Sciences at Harvard University. He taught previously at Princeton University and the University of Michigan. He joined the Harvard University faculty in 2002.

Dr. Carpenter's primary interest is in the theoretical, historical, and quantitative analysis of American political development, public bureaucracies, and government regulation, particularly regulation of health products. His book, *The Forging of Bureaucratic Autonomy: Reputations, Networks and Policy Innovation in Executive Agencies, 1862-1928* (Princeton University Press, 2001), was awarded the APSA's Gladys Kammerer Prize as well as the Charles Levine Prize of the International Political Science Association. His newly published book on pharmaceutical regulation in the United States is entitled *Reputation and Power: Organizational Image and Pharmaceutical Regulation at the FDA* (Princeton University Press, 2010). He received his doctorate in political science from the University of Chicago in 1996.

Constantine Gatsonis, Ph.D., is Henry Ledyard Goddard University Professor and Chair of the Department of Biostatistics at Brown University. He is a leading authority on the design and analysis of clinical evaluations of screening and diagnostic imaging modalities and has worked extensively on methodological questions in diagnostic medicine and health services and outcomes research. Dr. Gatsonis is the Network Statistician for the American College of Radiology Imaging Network (ACRIN), which conducts multicenter trials of diagnostic imaging in cancer and other diseases. Dr. Gatsonis is the chief statistician of the Digital Mammography Imaging Screening Trial, of ACRIN's arm of the National Lung Screening Trial, and of several other studies of the role of imaging for diagnosis and staging, monitoring, and prediction of response to therapy.

Dr. Gatsonis has served on the IOM Committee on Comparative Effectiveness Research Prioritization, the IOM Immunization Safety Review Committee, the NAS Committee on Identifying the Needs of the Forensic Sciences Community (co-chair), the NAS Committee to Study Engineering Aviation Security Environments, the NAS Committee on Applied and Theoretical Statistics, the Commission on Technology Assessment of the American College of Radiology, the Research Development Committee of the Radiology Society of North America, the HSDG Study Section of the Agency for Health Care Policy Research, review panels of the Center for Devices and Radiological Health of the FDA, and technical expert panels for the Health Care Financing Administration/Center for Medicare and Medicaid Services. He is the founding editor-in-chief of *Health Services and Outcomes Research Methodology*, an associate editor of the *Annals of Applied Statistics, Clinical Trials, Academic Radiology*, and an editor of the *Diagnostic Test Accuracy Reviews* of the Cochrane Collaboration. Dr. Gatsonis was elected fellow of the American Statistical Association. He received his BA in mathematics, *magna cum laude,* from Princeton and his Ph.D. in mathematical statistics from Cornell.

Gary H. Gibbons, M.D., is the director of the Morehouse Cardiovascular Research Institute, a NIH-National Heart, Lung, and Blood Institute-sponsored Research Center of Excellence. He is also an attending cardiologist in the Division of Cardiology at the Morehouse School of Medicine. Dr. Gibbons earned his undergraduate degree from Princeton University and his medical degree from Harvard Medical School. He completed his residency and cardiology fellowship at the Harvard-affiliated Brigham & Women's Hospital in Boston.

Dr. Gibbons has been selected as a Robert Wood Johnson Foundation Minority Faculty Development awardee, a PEW Foundation biomedical scholar, and an established investigator of the American Heart Association (AHA). Dr. Gibbons was a member of the faculty at Stanford University (1990–1996) and Harvard Medical School (1996–1999) before becoming director of the Morehouse Cardiovascular Research Institute in July 1999. He has served on several editorial boards for journals in cardiovascular medicine as well as grant review committees for the NIH, Juvenile Diabetes Foundation, and the AHA.

Dr. Gibbons directs NIH-funded research in the fields of vascular biology and the pathogenesis of vascular diseases. The innovations derived from his research resulted in the receipt of several U.S. patents. His bibliography lists more than 70 reviews and original reports in the fields of vascular biology, gene therapy, hypertension, atherosclerosis, and cardiovascular medicine.

Bonnie L. Halpern-Felsher, Ph.D., is a Professor in the Division of Adolescent Medicine, Department of Pediatrics at the University of California, San Francisco (UCSF). She is also a faculty member at UCSF's Psychology and Medicine Postdoctoral Program, the Center for Health and Community, and the UCSF Helen Diller Family Comprehensive Cancer Center. Dr. Halpern-Felsher is a developmental psychologist whose research has focused on health-related decision making, perceptions of risk and vulnerability, and health communication. She has also conducted research on the relationships among parenting practices, peer relationships, adolescents' self-perceptions, and risky behavior, including tobacco use. She has served as a consultant to a number of community-level adolescent health promotion programs and has been a member on several national campaigns to understand and reduce adolescent risk behavior. Dr. Halpern-Felsher served on the National Research Council Committee on Developing a Strategy to Prevent and Reduce Underage Drinking, the IOM Committee on Reducing Tobacco Use: Strategies, Barriers, and Consequences, and the Committee on Contributions from the Behavioral and Social Sciences in Reducing and Preventing Teen Motor Crashes.

Stephen S. Hecht, Ph.D., is Wallin Professor of Cancer Prevention and American Cancer Society Professor at the Masonic Cancer Center, and Professor in the Department of Laboratory Medicine and Pathology, University of Minnesota. Dr. Hecht serves as head of the Carcinogenesis and Chemoprevention Program of the Masonic Cancer Center. He is also a member of the Medicinal Chemistry graduate program. The focus of the Hecht laboratory is mechanisms and prevention of tobacco-induced cancer. The Hecht laboratory studies mechanisms by which carcinogens are metabolically activated and detoxified in humans, and uses this knowledge to develop practical strategies for cancer prevention, including the validation of tobacco carcinogen and toxicant biomarkers, with a particular focus on nitrosamines, aldehydes, and polycyclic aromatic hydrocarbons. Studies in laboratory animals are used to understand metabolic pathways. Then methods are developed to quantify metabolism of these carcinogens in humans, typically by employing GC-MS, LC-MS, or related methods to analyze carcinogen metabolites in urine, or carcinogen DNA or protein adducts in tissue or blood. These methods are applied in molecular epidemiology studies designed to determine factors that influence susceptibility to cancer development in people who use or are exposed to tobacco products, or who are exposed to carcinogens via other routes.

Peter K. Honig, M.D., M.P.H., is currently Head, Global Regulatory Affairs at AstraZeneca. Dr. Honig received his baccalaureate, medical, and public health degrees from Columbia University in New York. He has postgraduate training and is board-certified in internal medicine and clinical pharmacology and has authored numerous peer-reviewed publications and book chapters. He has held senior leadership positions at the U.S. Food and Drug administration and Merck Research Laboratories. He is and has been the PhRMA representative to the International Conference on Harmonization (ICH) Steering Committee since 2002 and the current co-chair of the ICH Global Cooperation Group (GCG) whose mission it is to promote regulatory harmonization in non-ICH countries and regions. Dr. Honig is also an Associate Editor of *Nature Clinical Pharmacology and Therapeutics*.

Richard J. O'Connor, Ph.D., is Associate Member in the Department of Health Behavior, Division of Cancer Prevention and Population Sciences, Roswell Park Cancer Institute. Dr. O'Connor's research focuses on the interaction between tobacco products and consumers, from how cigarettes are designed and how those designs affect smokers' perceptions and use of the product, to how best to inform policy makers crafting tobacco product regulations. Ongoing work includes developing and applying filter-based methods for assessing cigarette smoke exposure, characterizing physical properties and design features of international tobacco products, assessing smokers' interest in alternative nicotine delivery systems, and smokers' reactions to novel tobacco products. He is a co-investigator on the International Tobacco Control Policy Evaluation Project (ITC Project), which is an international collaboration of tobacco control researchers seeking to evaluate the psychosocial and behavioral effects of national-level tobacco control policies throughout the world. He is also a co-investigator on an NCI-funded Program Project, directing research to evaluate the impact of tobacco product regulations on product design and performance, as well as smoking behavior. He is Principal Investigator on an NCI grant examining approaches to assessing current smokers' interest in using alternative nicotine sources, such as smokeless tobacco and nicotine replacement.

Joel L. Schwartz, D.M.D., D.M.Sc., is Professor and Director of Oral Maxillofacial Pathology in the Colleges of Dentistry and Medicine at the University of Illinois at Chicago. Dr. Schwartz's research interests include screening and prevention of tobacco- and environment-induced head and neck cancers and associated cancers. Dr. Schwartz develops various oral cancer models to understand and quantify the earliest genetic and molecular events that change a normal cell to a cancer cell following exposure to a virus or chemical carcinogen. First, in laboratory models Dr. Schwartz studies viral and chemical oral carcinogenesis using normal, transformed premalignant, and malignant cells. Second, in animals such as hamster, rat, and mouse, he validates laboratory findings and further assesses genetic and molecular progression as he changes a normal cell to a cancer cell. His approach also involves noninvasive screening of this process using oral cytology samples and RNA microarray. This method is unique because he harvests from the identical animal samples throughout the process of oral carcinogenesis. An identical approach is translated to clinical human populations to validate findings, to study early prevention, or to monitor various therapies to improve the quality of life for the oral cancer patient. Methods include human cells and various animal models (hamster, rat, mouse) to prevent carcinogenesis and tumorigenesis following administration of carotenoids/tocopherols/polyphenols, or specific genetic regulation using peptide/viral vector vehicles. Dr. Schwartz also has previous experience as a senior pathology core expert to enhance epidemiology or basic chemistry approaches to study tobacco product activities in an NIH-funded cancer center.

Donna-Bea Tillman, Ph.D., M.P.A., joined Microsoft in 2010 after 16 years at the U.S. Food and Drug Administration. She held numerous positions within the FDA's Center for Devices and Radiological Health, culminating in her 2004 appointment to the position of Director of the Office of Device Evaluation, where she oversaw the medical device premarket review program. During her tenure at the FDA, she played a pivotal role in the development of guidance documents, standards, and policy frameworks for medical device software and health information technology. At Microsoft she is the Director of Policy and Regulations in the Health Solutions Group. Dr. Tillman received her B.S.E. in engineering from Tulane University, her Ph.D. in biomedical engineering from the Johns Hopkins University, and her Master's in Public Administration from the American University.

Alastair J. J. Wood, M.D., FACP, was Professor of Medicine and Pharmacology, Assistant Vice Chancellor, and Associate Dean at Vanderbilt Medical School before being appointed Emeritus Professor of Medicine and Emeritus Professor of Pharmacology in 2006. His current academic appointments are Professor of Medicine and Professor of Pharmacology at Weill Cornell Medical College, New York. He is a Partner at Symphony Capital LLC, a New York–based private equity company. Dr. Wood is a member of the National Academies' Institute of Medicine; the American Association of Physicians (AAP); the American Society for Clinical Investigation (ASCI), Honorary Fellow; American Gynecological and Obstetrical Society (AGOS); and Fellow of the American College of Physicians. Dr. Wood served on the *New England Journal of Medicine* editorial board and was the *NEJM* Drug Therapy Editor for many years. He authored the chapter in Harrison's *Principles of Internal Medicine* on Adverse Drug Reactions from the 9th through the 15th edition. He was the chairman of the FDA's Nonprescription Drugs Advisory Committee until 2006. He previously served as a member of the FDA's Cardiovascular and Renal Advisory Committee, and the Nonprescription Drugs Advisory Committee. His research interests have been focused on understanding the mechanisms for interindividual variability in drug response and toxicity. His research has resulted in more than 300 publications.

Anna H. Wu, Ph.D., received her undergraduate degree in physiology at the University of California, Berkeley and her doctoral degree in public health (Epidemiology) from the University of California, Los Angeles in 1983. She became Assistant Professor at University of Southern California in 1984, Associate Professor in 1994, and Full Professor with tenure in 2002. Early in her career, she conducted a series of lung cancer studies to determine the role of indoor air pollution from passive smoking, cooking and heating fuels/fumes, and other factors in explaining the high rates of lung cancer in Chinese women when few were active smokers. She continues to have a strong interest in lung cancer research, particularly to better understand hormone-related effects on lung diseases. Dr. Wu's current research activities are focused in two main areas. One area is devoted to studying the etiology of breast and ovarian cancers. A second area of Dr. Wu's current research is cancers of the gastrointestinal tract, including adenocarcinomas of the colon, stomach, and esophagus. These studies are aimed at identifying environmental and genetic determinants of these cancers. Although most of these studies are aimed at understanding causes of cancers, Dr. Wu is now expanding her work to also identify lifestyle and genetic factors that may influence treatment response and cancer outcome.

Appendix D

Meeting Agendas

Thursday, February 3, 2011

Keck Center of the National Academies
500 Fifth Street, NW
Washington, DC 20001

1:00–1:15 pm	Committee Introductions and Chair's Opening Statement *Jane Henney* *Committee Chair*
1:10–1:30 pm	Charge to the Committee *Lawrence Deyton, M.D., M.S.P.H.* *Director* *Center for Tobacco Products* *U.S. Food and Drug Administration (FDA)*
1:30–2:00 pm	Discussion About the Charge *Committee and FDA Representatives*
2:00–3:00 pm	Public Comment
3:00 pm	Adjourn

Monday, May 9, 2011

Embassy Suites
900 Tenth Street, NW
Washington, DC 20001

9:45–10:00 am Welcome and Introductions
Jane Henney
Committee Chair

10:00–11:30 am Panel Discussion I: Tobacco Manufacturers
Moderator: Peter Honig

10:00–11:00 am: Presentations by industry representatives
11:00–11:30 am: Questions from the Committee

- Introduction
  Standards for pre-clinical studies
  Mike Ogden
  Senior Director of Regulatory Oversight
  R.J. Reynolds Tobacco Company

- Standards for studies on in vitro models of disease
  Chris Proctor
  Chief Scientific Officer
  British American Tobacco

- Standards for clinical studies and biomarkers
  Mohamadi Sarkar
  Senior Principal Research Scientist
  Altria Client Services

- Population communication and risk perception
  Lars Erik Rutqvist
  Senior Vice President for Scientific Affairs
  Swedish Match

- Summary
  J. Daniel Heck
  Principal Scientist
  Lorillard Tobacco Company

11:30 am–12:00 pm	*Andrew Salmon* *Senior Toxicologist and Chief* *Air Toxicology and Risk Assessment Section* *Office of Environmental Health Hazard Assessment* *California Environmental Protection Agency*
12:00–1:00 pm	Lunch
1:00–1:30 pm	*David Jacobson-Kram* *Associate Director of Pharmacology and Toxicology* *Office of New Drugs* *Food and Drug Administration*
1:30–2:00 pm	*Peter G. Shields* *Professor of Medicine and Oncology* *Deputy Director, Lombardi Comprehensive Cancer Center* *Georgetown University* *Mirjana Djordjevic* *Health Scientist Administrator* *Tobacco Control Research Branch* *National Cancer Institute*
2:00–2:30 pm	*John Baron* *Professor of Medicine* *Dartmouth Medical School*
2:30–2:45 pm	Break
2:45–3:15 pm	*Brenda Edwards* *Associate Director* *Surveillance Research Program* *National Cancer Institute*
3:15–3:45 pm	*Ruth S. Day* *Director* *Medical Cognition Laboratory* *Duke University*
3:45–4:15 pm	*David Mendez* *Associate Professor* *Department of Health Management and Policy* *University of Michigan School of Public Health*

4:15–4:30 pm Break

4:30–6:00 pm Panel Discussion II: public health representatives
Moderator: Dan Carpenter

4:30–5:30 pm: Presentations by public health representatives
5:30–6:00 pm: Questions from the Committee

- *David Abrams*
  Executive Director
  The Schroeder Institute for Tobacco Research and Policy Studies
  Professor, Department of Health, Behavior and Society
  The Johns Hopkins Bloomberg School of Public Health

- *Tom Glynn*
  Director, Cancer Science and Trends
  Director, International Cancer Control
  American Cancer Society

- *Mark Greenwold*
  Director, Regulatory Affairs
  Campaign for Tobacco-Free Kids

- *Rose Marie Robertson*
  Chief Science Officer
  American Heart Association

- *Mitch Zeller*
  Vice President for Policy and Strategic Communications
  Pinney Associates

6:00 pm Adjourn